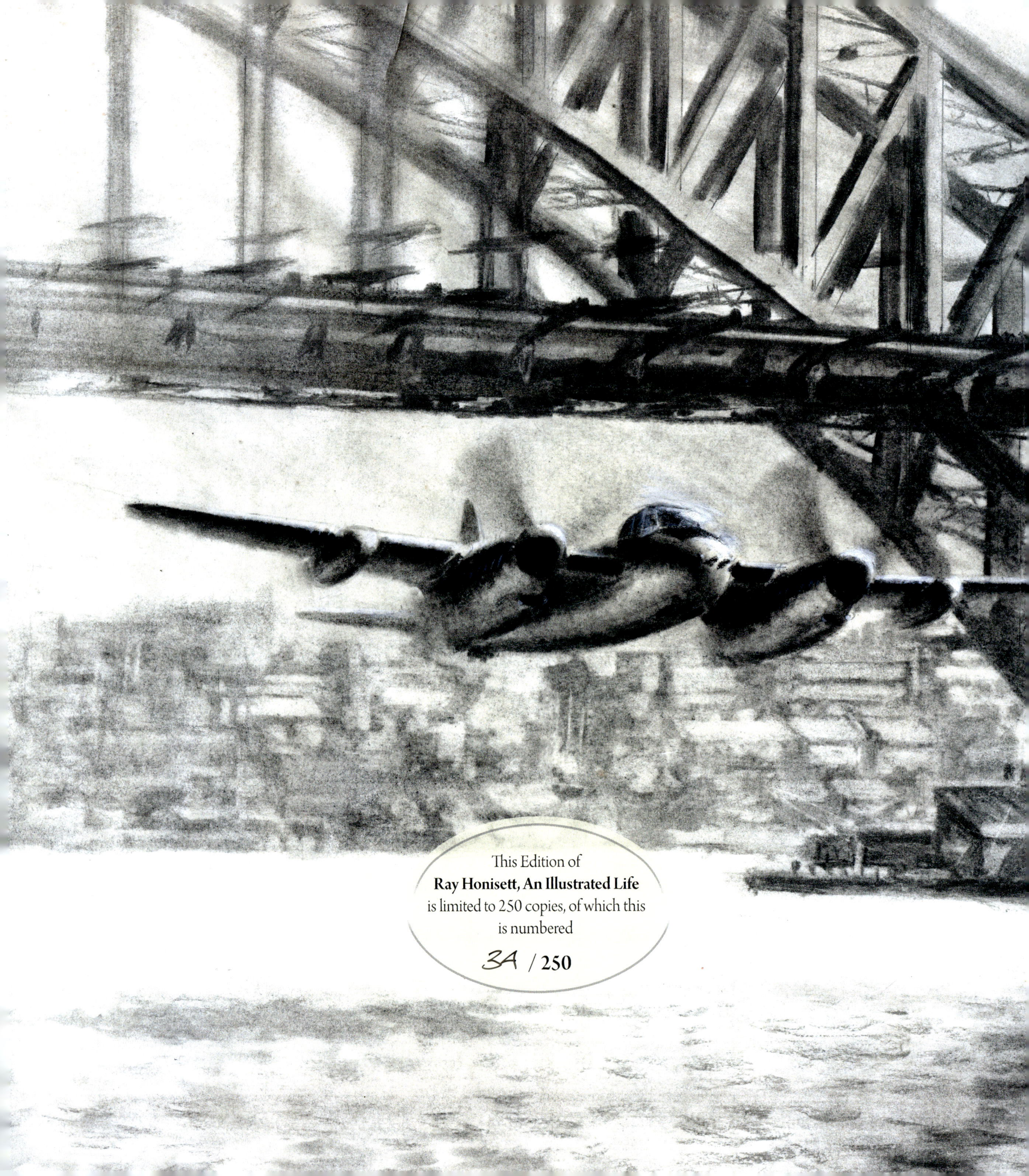

This Edition of
Ray Honisett, An Illustrated Life
is limited to 250 copies, of which this
is numbered

3A / 250

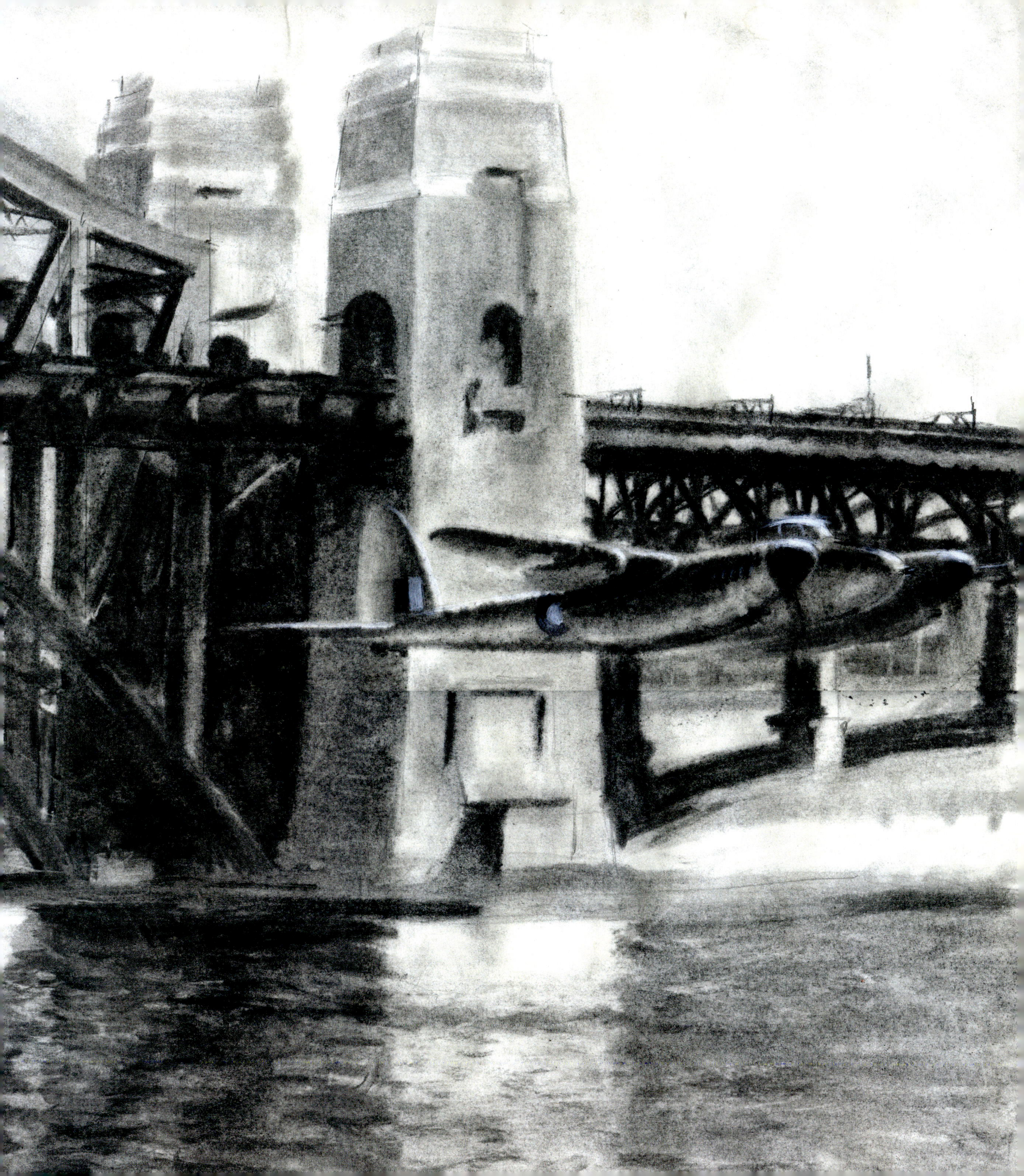

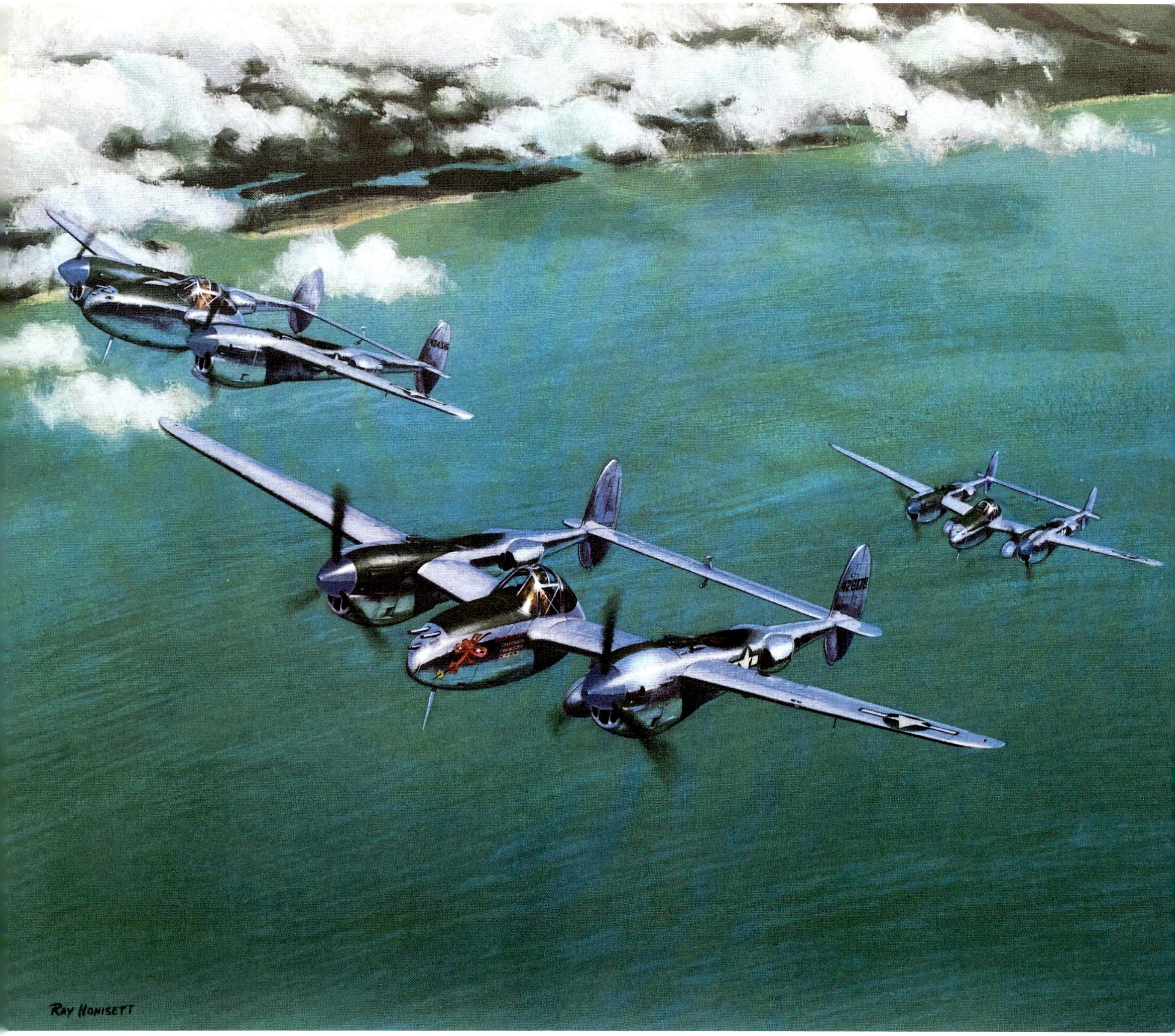
RAY HONISETT

RAY HONISETT

AN ILLUSTRATED LIFE

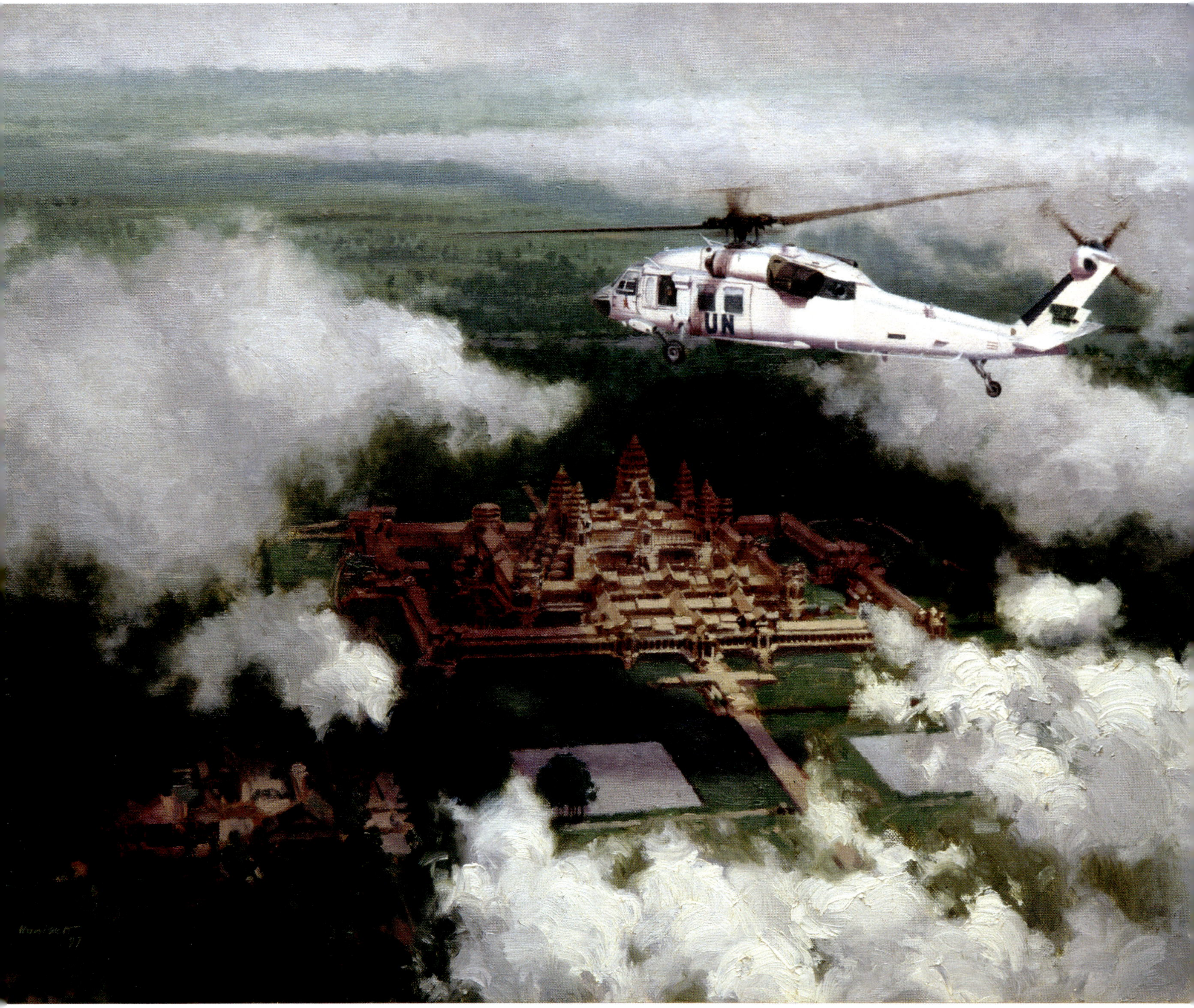
UN

RAY HONISETT

AN ILLUSTRATED LIFE

GAVIN FRY

E+A

EDMUND & ALEXANDER

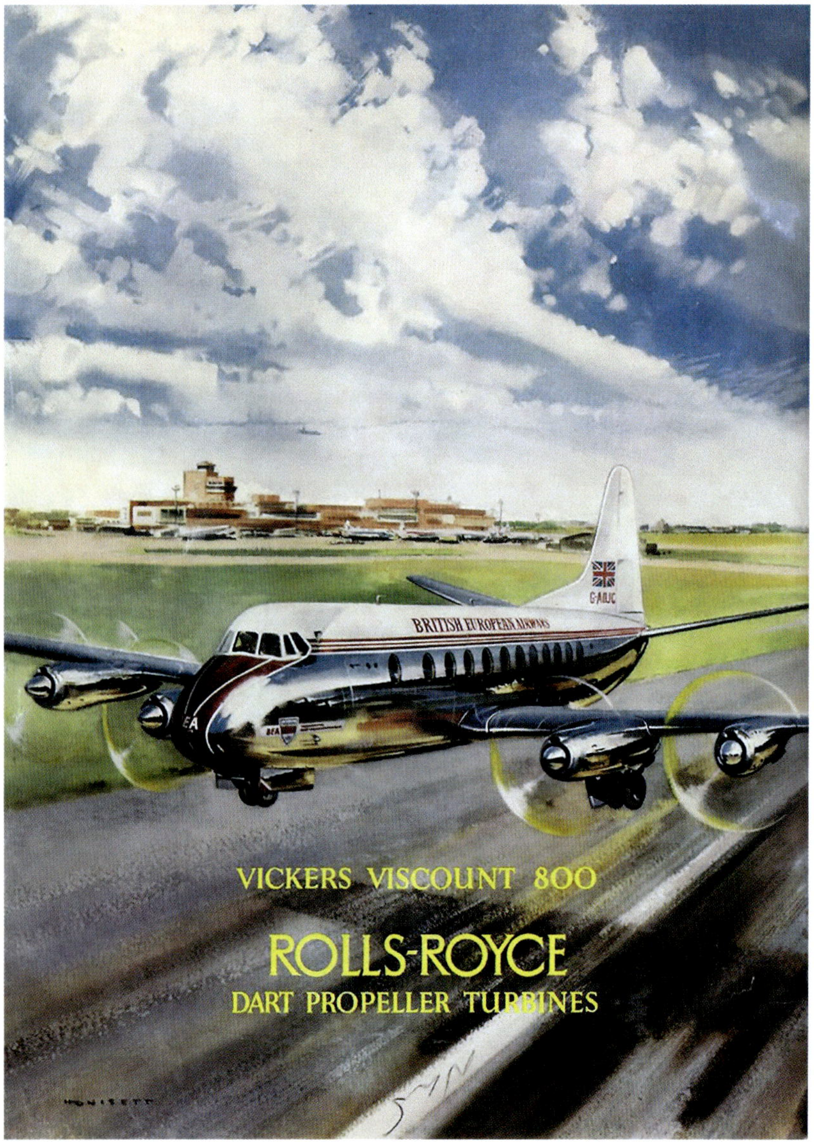

Ray Honisett, illustration for Rolls Royce engines, circa 1956

Page 4
Ray Honisett, ***Thundering Lightnings****, (1966)*
Aircraft Illustrated Pacific Zone 1939-1945, Vol 1, No7

Lockheed P-38L Lightnings of 7th Fighter Sqdn., 49th Fighter Group, 5th Air Force, USAAF, climb over Lingayen Bay, Western Luzon, Philippine Islands, early 1945. Foreground aircraft is flown by Captain R.M. De Haven.

Page 6
Ray Honisett, ***Sikorsky Blackhawk over Angkor Wat, Cambodia*** *(1997)*
Oil on canvas
Formerly Sikorsky Aircraft Company collection

Endpapers
Ray Honisett, ***Mosquitoes under Sydney harbour Bridge****,*
(date unknown.), Charcoal and watercolour on paper

As part of the Victory over Japan celebrations of 15 August 1945 three Mosquito FB40s of 94 Squadron RAAF were to fly under the Sydney Harbour Bridge — two made the daring manoeuver while the third opted out at the last minute.

Opposite, Ray Honisett, ***Camels and Albatrosses****, c.2000*
Mixed media on card
Private collection, Melbourne

ISBN 978-0-9875351-9-1
Designed and published by
Edmund & Alexander Books 2020
Printed by Courtney Colour, Dandenong South

CONTENTS

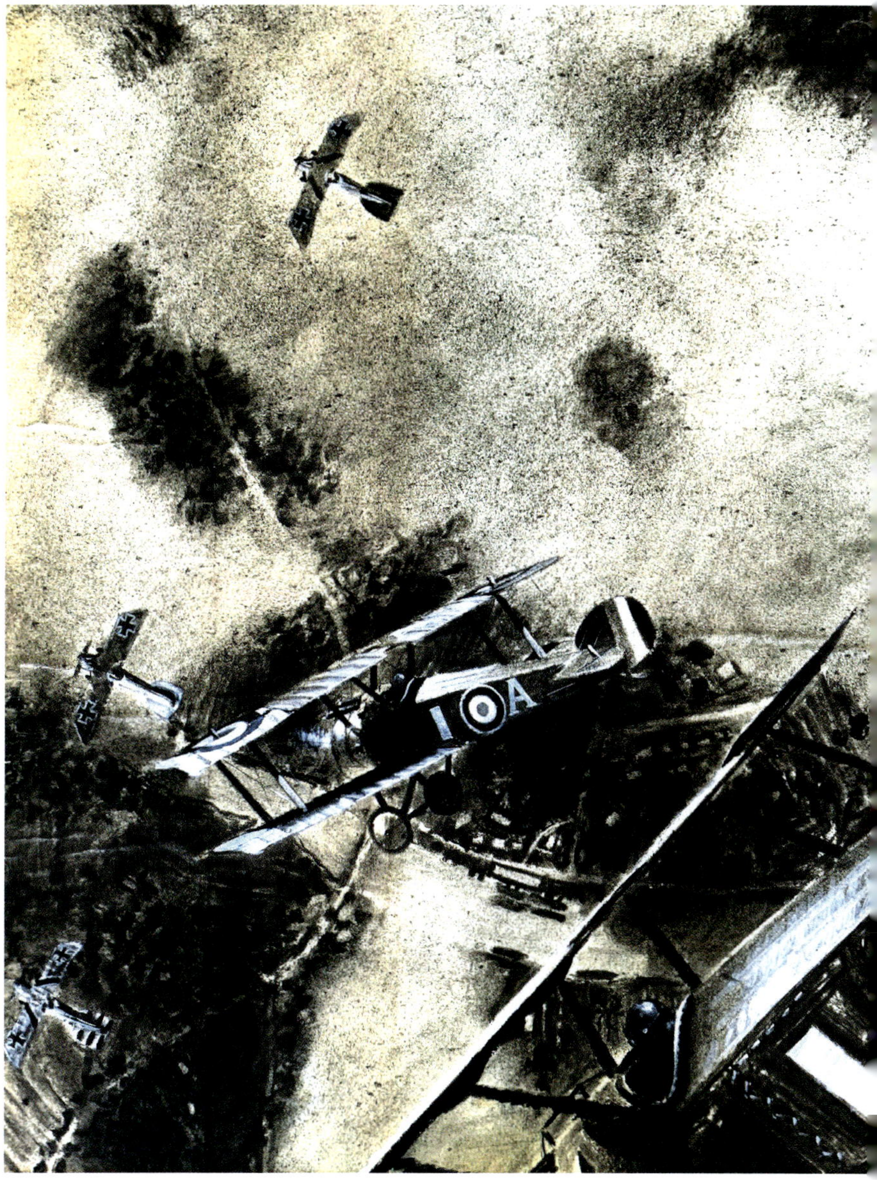

PREFACE

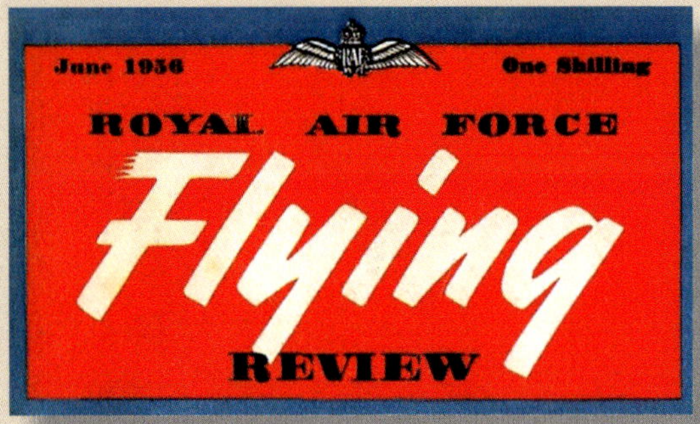

Illustrations for the RAF Flying Review, very much in the idiom of the 'Boys' Own Adventures' that had been Ray Honisett's passionate interest in his boyhood days in wartime Perth.

Below, Avro Anson and Messerschmitt 109s Royal Air Force Flying Review, March 1957

Opposite, Lancaster Bailout Royal Air Force Flying Review, January 1957

This book has had a remarkably long gestation, having commenced as an idea some 25 years ago. In May of 1996 Ray Honisett attended a party in Melbourne to celebrate my 50th birthday. It was a memorable event, especially as Ray had presented me with a beautiful painting — the windy beach scene reproduced at page 153. In a quiet moment on that day we discussed the possibility of producing a book of his work, as I had started on a regular program of writing and publishing, building on the work started at the Australian War Memorial in 1982. We talked about it on various occasions when I returned 'down south', but the project seem to be constantly sidetracked by one pressing demand or another. What I didn't know, was that Ray had gone home and commenced writing a memoir of his life, written with great humour, honesty and insight. It transpired that he kept this work a closely guarded secret and the manuscript was not discovered until after his death in 2019. It is dated as having been commenced in 1996 and, at some 42,000 words in length, the memoir tells in detail of his time growing up in Perth, his move to Melbourne and on to his working life in London. But the narrative stops abruptly in 1955, with a suddenness that suggests that there was either another cache of missing papers, or that more likely, he simply put it aside meaning to continue, but never got around to it. I transcribed the entire document and found it charming, at times disturbing, and an altogether fascinating document. It is however not entirely appropriate for publication in a book of this type and, in consultation with Ray's family, the decision was made to keep it aside for the present. I have produced a short precis of the work for the introduction to this publication and have quoted directly from it, to give a voice to the artist's ideas and personality. Some of Ray's published writing, generously made available by Australia Post and the Australian Society of Aviation Artists has also been included. These pieces add both colour and an authentic voice when studying the work of Australia's pre-eminent artist in his field. In order to make clear the authorship of those pieces, they are presented here against a light blue background. Pieces by other writers have also been presented against a toned background. Finally, while I am aware of the convention of many 'scholarly' works that always refers to the subject by their surname only, in this case there is an element of informality and personal connection that means on occasion the artist is referred to by his first name.

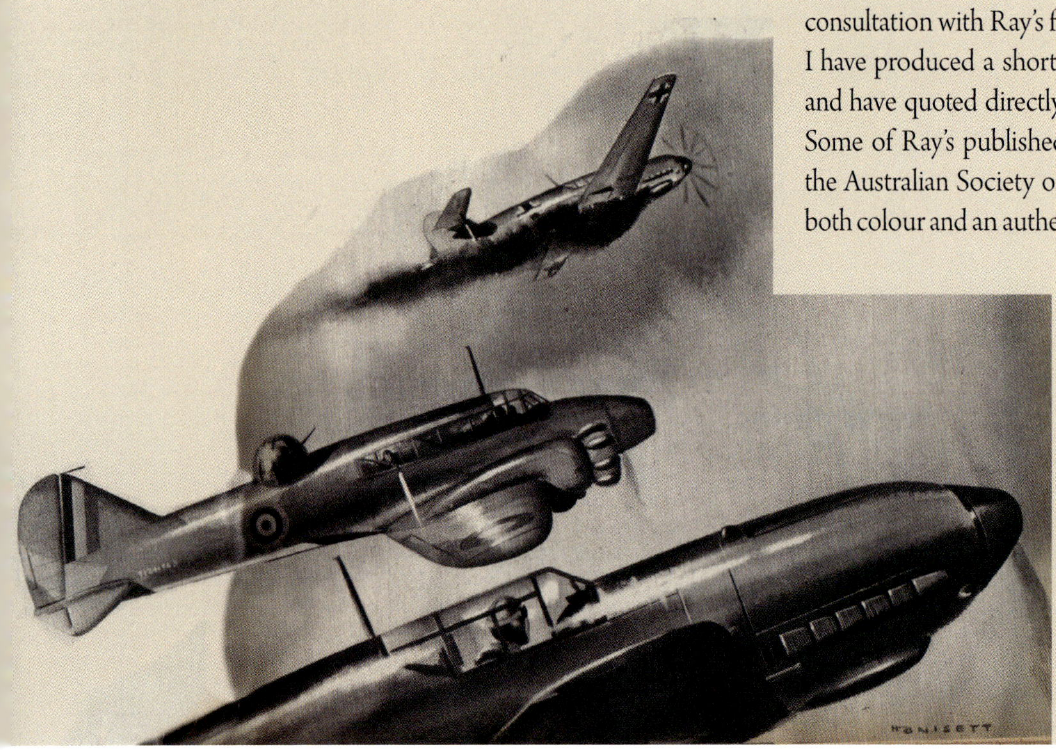

Gavin Fry
2020

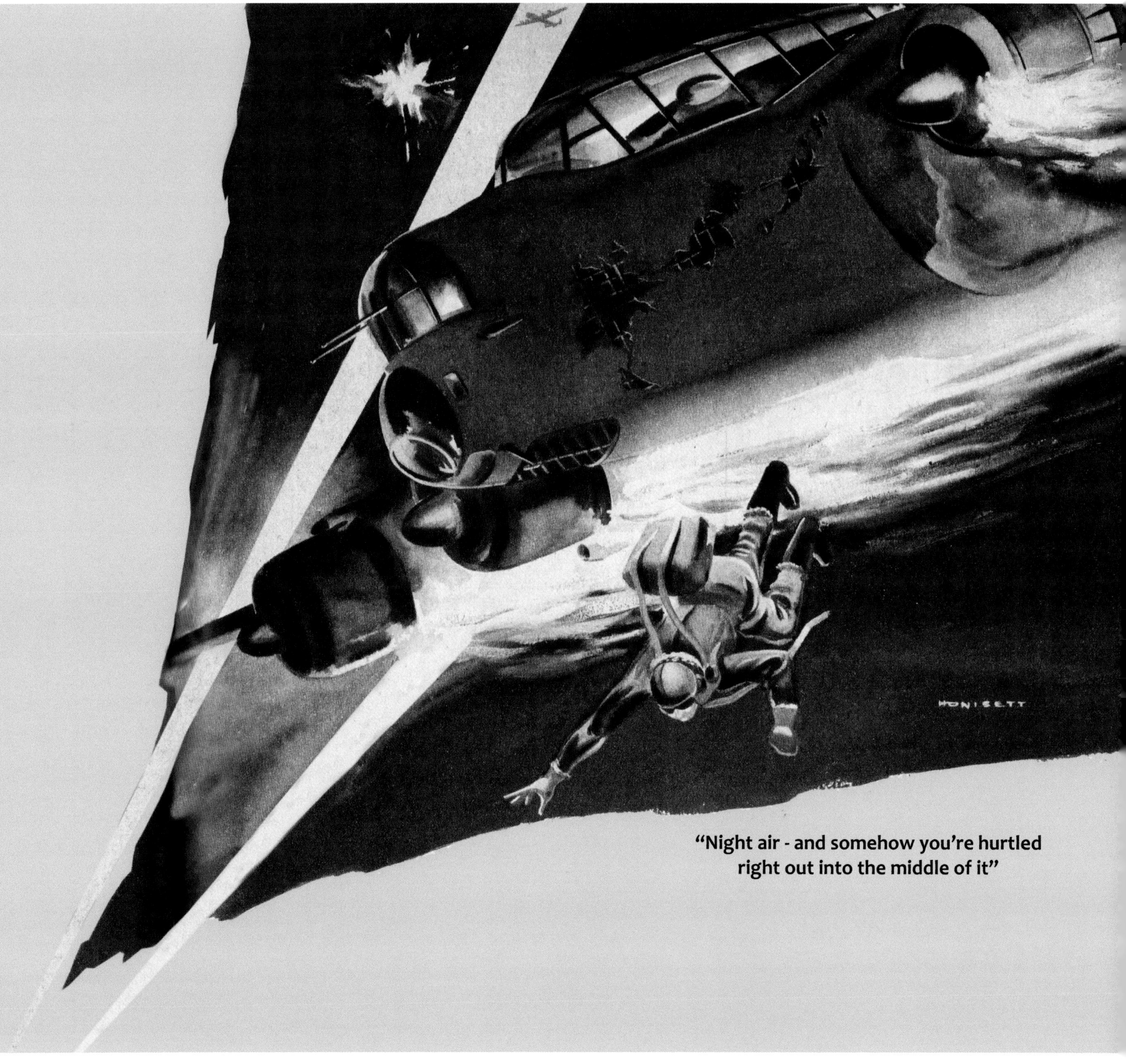

"Night air - and somehow you're hurtled right out into the middle of it"

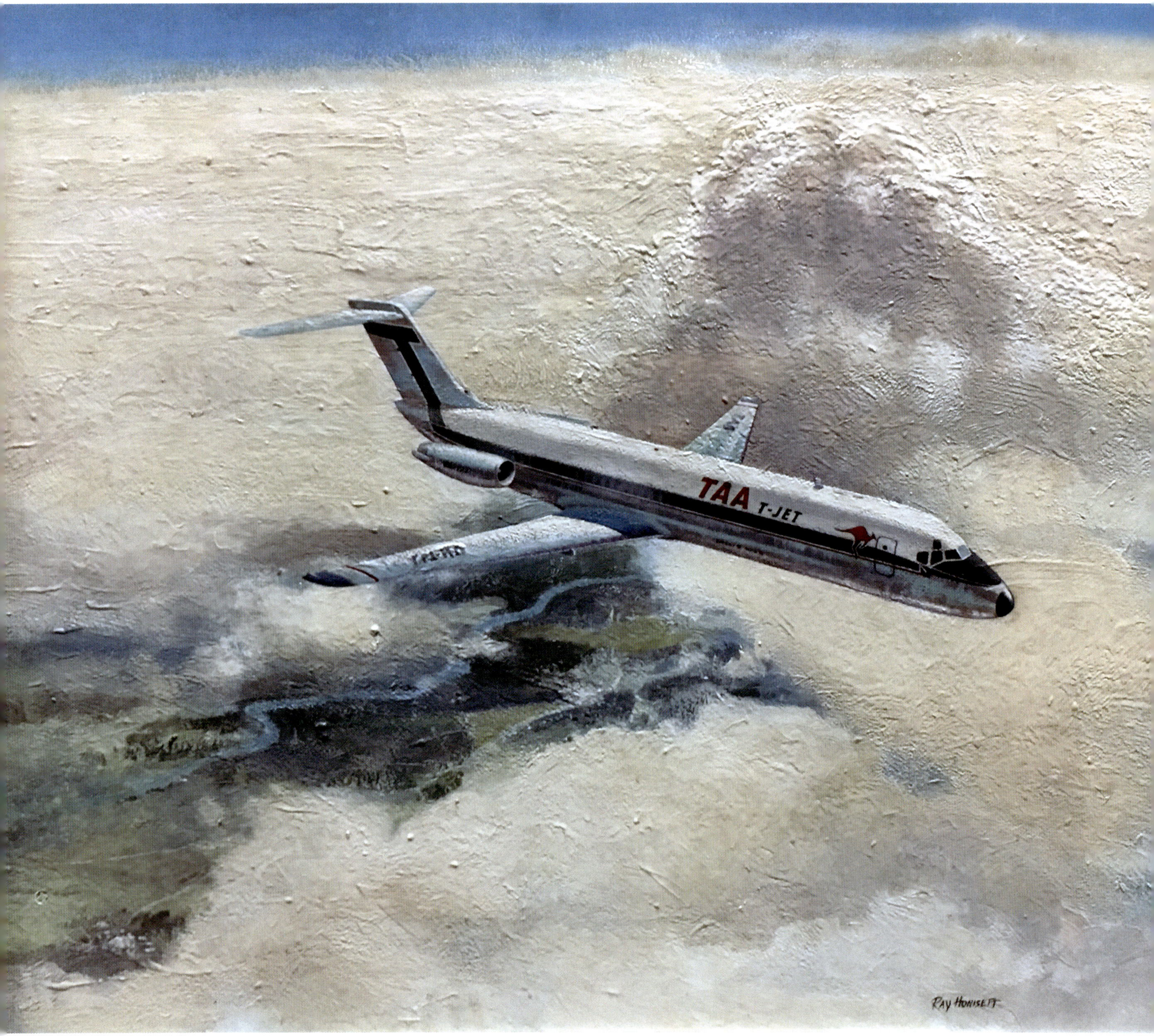
TAA T-JET
RAY HONISETT

INTRODUCTION

Ray Honisett's work has been seen by millions of Australians and international visitors, filling as it does a prominent place in the nation's most visited and best-loved public institution, the Australian War Memorial in Canberra. His stamps have been used and admired by countless people across Australia and the Pacific for more than 25 years and, from the 1950s onwards, his art graced the covers of the world's most prestigious aircraft industry journals. His name is however missing from nearly every account of art in Australia and even those in the art business are unfamiliar with his work. The reason for that is simple enough, for however skilled and professional he might have been, he operated in what is now seen as a lesser art form, the world of illustration and commercial art.

Ray's rise to prominence in this specialised niche of the art world coincided with major changes in the training and education of artists in the years following world war two. In the first half of the 20th century, it was not uncommon for so called 'fine artists' to also produce work for advertising and the publishing industry. Norman Lindsay might have been the best-known artist of his generation, but he is remembered as much for his wonderful illustrated book *The Magic Pudding* as he is for his robust nudes and nostalgic maritime art. Harold Herbert, the best selling watercolourist of his time, also maintained a busy commercial practice, producing illustrations for the newspapers and magazines of the day. With no government support for the arts and precious few teaching positions, artists had to make a living where and as they could, with such high flyers as Sidney Nolan and Albert Tucker getting their start in the busy art department of a hat manufacturer under the commercial illustrator Vernon Jones.

In the 1950s and 1960s, there was a rapid expansion in tertiary education in Australia, necessitated by the large cohort of 'baby-boomers' moving through the system and also the heavy intake of new migrants displaced by the war in Europe. Where in pre-war years there were only a handful of art schools in the country, there were now literally dozens springing up in all the major cities and regional towns. As a new curriculum was developed and the teaching became more professional, students were diverted into either the fine art or the commercial stream. The latter group would study illustration, graphic design, lettering and the technical aspects of the printing industry. That led to a new class system, with the painters, printmakers and sculptors being groomed for a place in the exhibiting world as 'real' artists, while the commercial practitioners were steered toward the advertising agencies and publishing houses. Technology was also changing and many of the tasks undertaken by commercial artists lessened in demand as photography came to dominate the pictorial space, and new products such as Letraset and photo-typesetting were in the hands of technicians rather than trained artists.

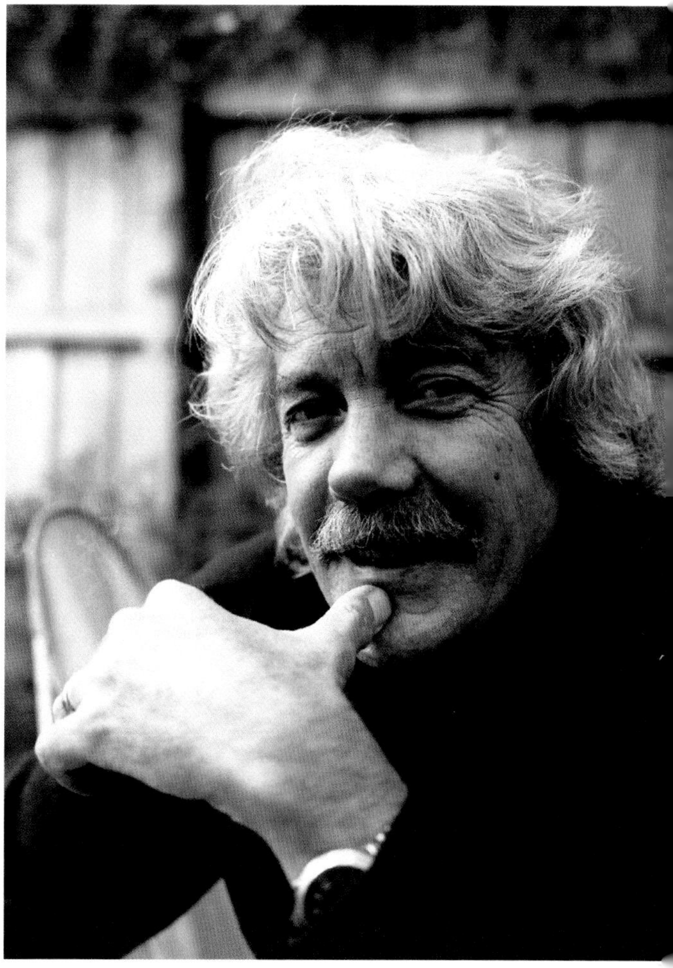

Ray Honisett, 1984, photograph by the author.

***TAA DC-9**, 1967*
Acrylic on card,
Cover art for Trans-Air in flight magazine, February -March1967
TAA Museum collection, Essendon

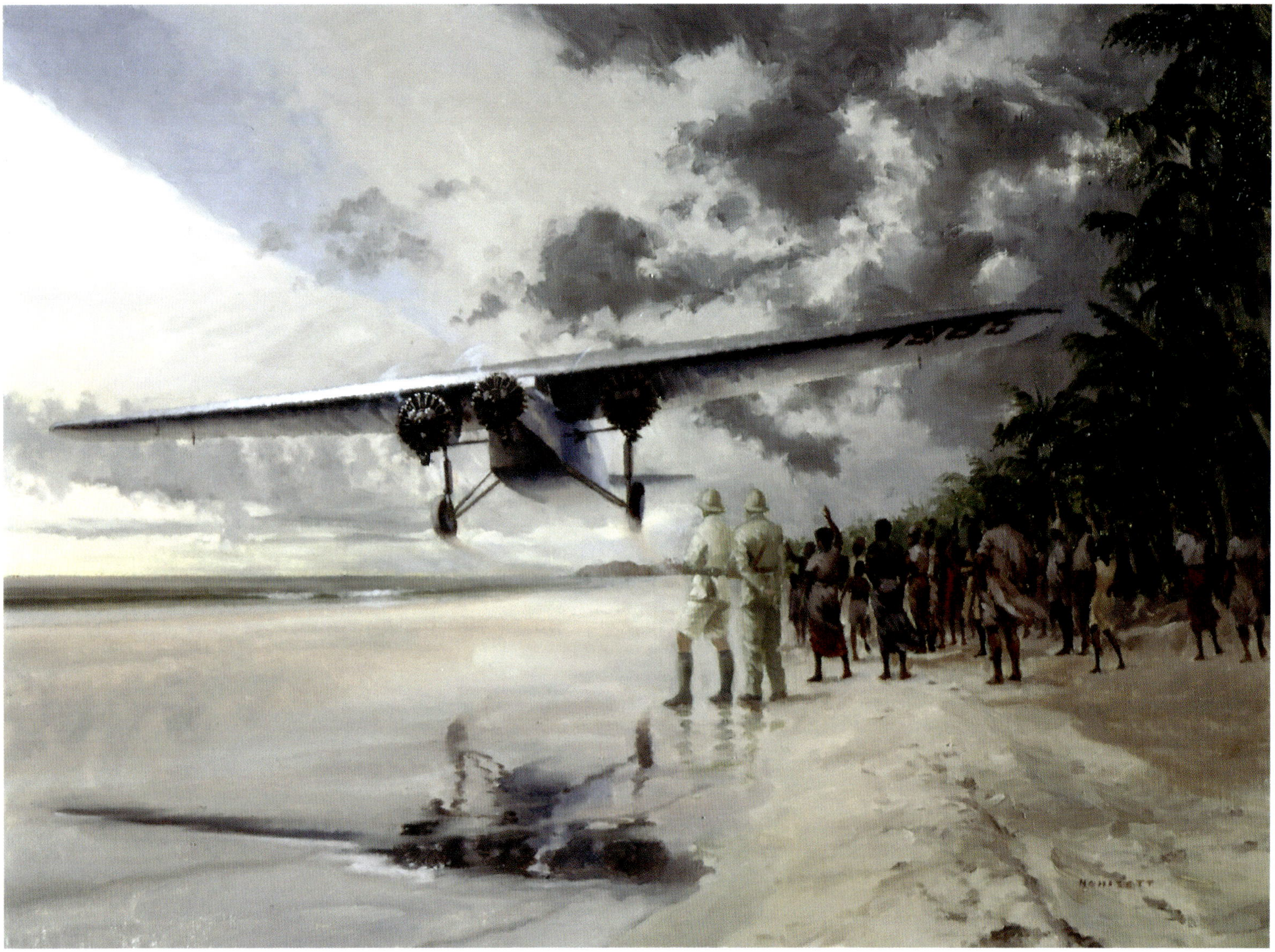

The Last Leg,
Oil on canvas, 75.5 x 98.5 cms
Powerhouse Museum Collection
Gift of McDonnell Douglas Limited, 1988
'Southern Cross' taking off from Suva on 8 June 1928 on the first trans-Pacific flight from America to Australia, flown by Charles Kingsford Smith, Charles Ulm, Harry Lyon and James Warner.

What constitutes the art of illustration is a curious and ill-defined idea, for in a great many cases there is no apparent difference between a work created as 'art' and a near identical one considered 'mere illustration'. Certainly Ray Honisett's skills as a painter were the equal and more of those of many of the painters of his generation. It would seem that the real difference, and it is a very fine point indeed, is that the artist chooses his or her own subject, whereas the illustrator is responding to an external brief and will comply with a range of set conditions, or the work is not acceptable and considered 'incorrect'. In the end it is really neither here nor there where a subject comes from — for who indeed would criticise Raphael for a painting of the Madonna and Child, made to a strict list of conditions laid down by his papal patrons?

Many artists have included motorcars in their paintings, for example, and it doesn't really matter if it is recognisable as a particular make or model — it is simply 'a car'. But for the illustrator, if it is to be a 1958 Holden FC Special sedan, it must be instantly recognisable as such, and correct in every detail, so it is not seen as a 1956 FE model, or with the wrong trim or accessories. Similarly if commissioned to paint a wartime incident, as Ray Honisett was many occasions, which included, for example, a Bristol Beaufighter MkX, it has to be absolutely correct and not to be confused with a MkVI, however similar they might look. Ray took to heart the words of his assigned mentor at the Australian War Memorial, Sir William Dargie, who impressed upon him that such works needed to be ... "good art and good history".

For the art purist, the illustrator is a slave to the desires and wishes of those who have commissioned him, as it is indeed he who pays the piper calling the tune. But as Ray Honisett would argue, and many similar artists would attest, the real secret is to not only satisfy the brief, but to also produce a fine work of art at the same time. To put that difference in its simplest terms, when, on more than one occasion, Sidney Nolan had Ned Kelly's trigger the wrong around on his gun, it was seen as a quirky invention — had an illustrator been guilty of the same error they would have been laughed out of court. There are of course moments when the obsessions of the illustrator can be taken too far — the 'rivet-counters' who take accuracy and fidelity to a point where any creativity and 'art' is lost in the pursuit of detail perfection and accuracy. The purpose of this book is therefore to present the art and life of Ray Honisett in the same manner and context as any study of a first rate artist who has made a significant contribution to the cultural life of the nation. Like all artists, his career was filled with highs and lows, the battle for acceptance and the ability to work as a professional, recognised by his peers and acknowledged in the public domain.

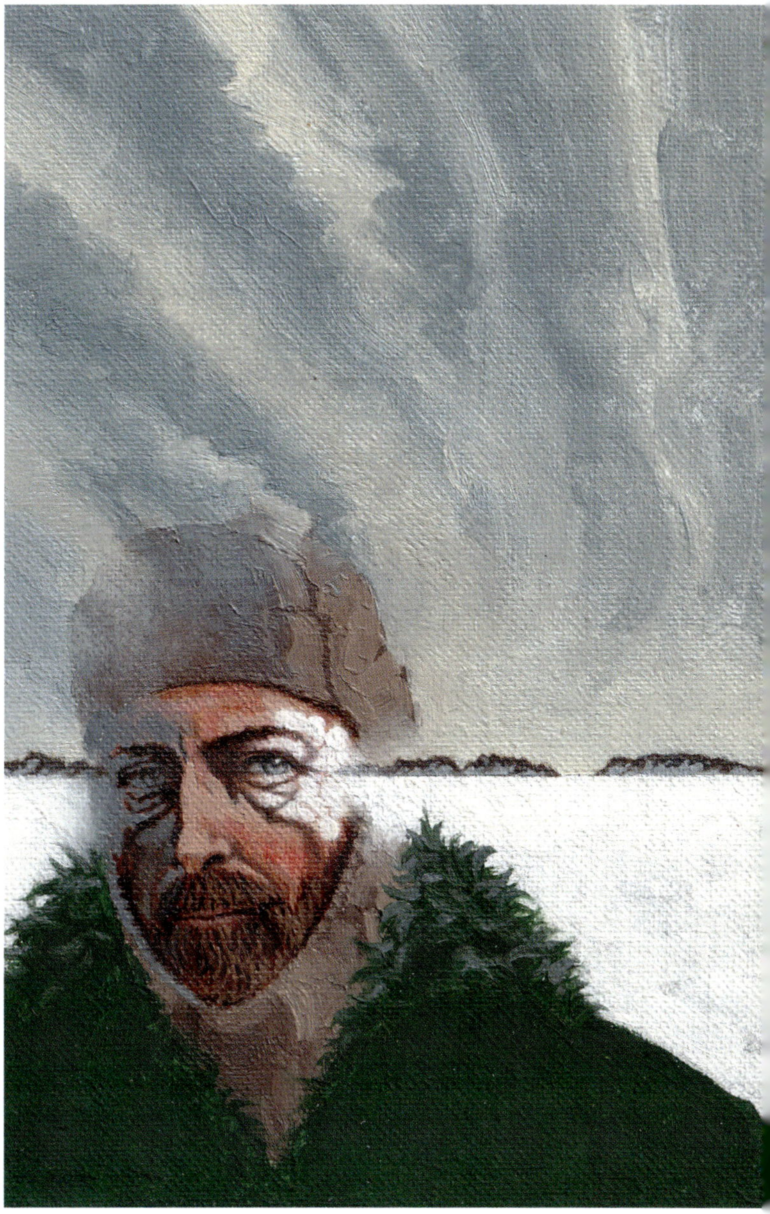

Ray Honisett, ***Sir Douglas Mawson,***
design for 50th anniversary commemorative stamp, 1982
Acrylic on card, 12 x 8 cm
Australia Post Archives, Melbourne

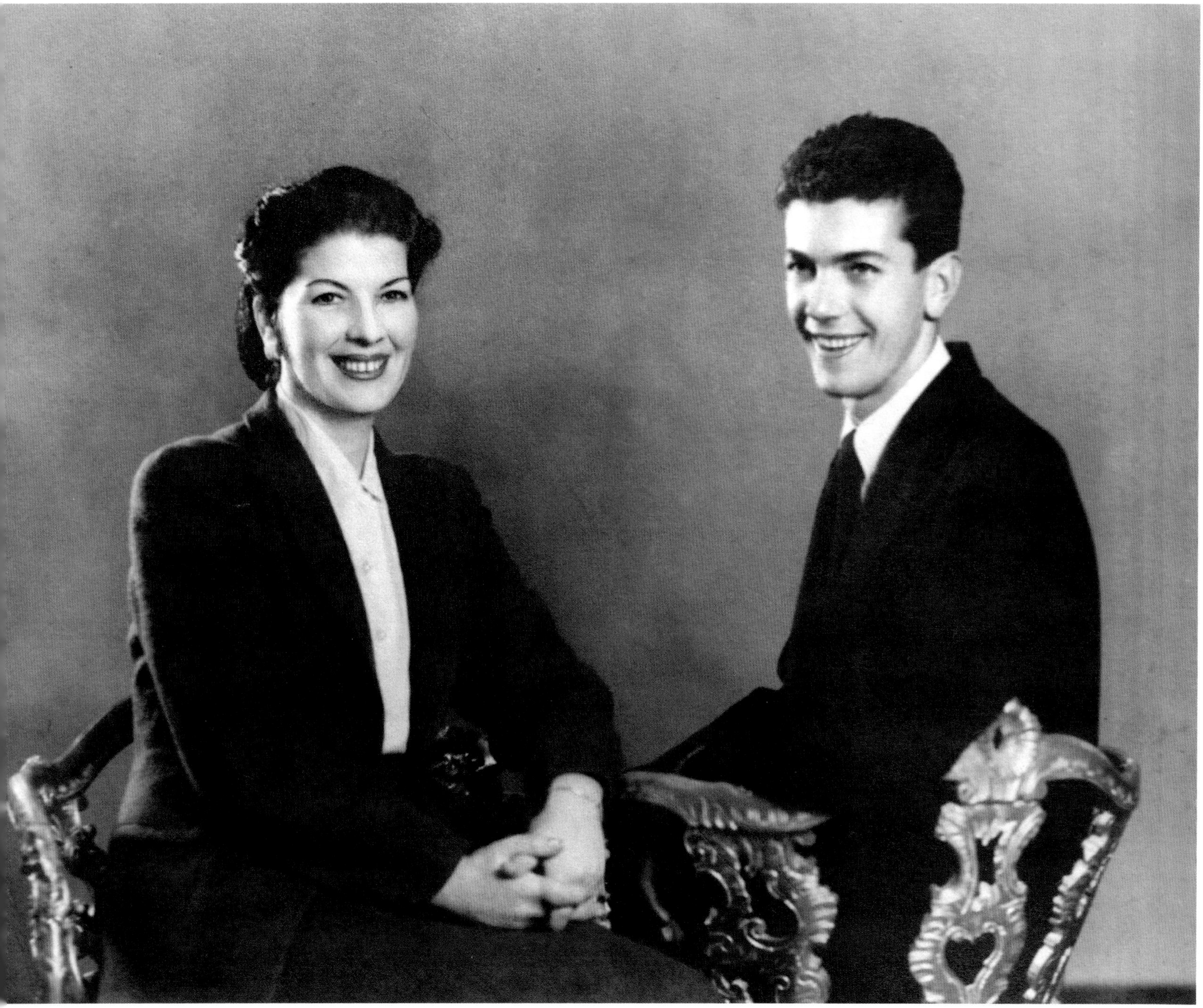

Mother and son, Carmen and Ray Honisett, Circa1947

THE BOY IN PERTH

In penning a memoir of his early life, Ray Honisett gives us the simple clear facts of his birth. 'I was born at 10pm on 11th April, 1931 in Perth, Western Australia'. His full name, Ramon Francis Honisett, gives the first clue to the fundamentals of his conception — a mother of Spanish heritage and a father steeped in the British tradition. The Honisetts had come to Australia from the south of England, with Ray's Grandfather Frank Honisett born in the village of Battle, near Hastings. His side of the family had connections to Sir Christopher Wren and the Wren name was perpetuated down the generations, including with Ray's father Frank Wren Honisett. Ray's Mother was born to a Spanish father, John Arenas and a mother of Scottish descent, one of the Argyle clan, resident in Perth since the 19th century. It was a love match that would lead to problems when John Arenas died young in the great influenza epidemic of 1919-20. Despite her father having been a sophisticated and educated gentleman, the newly widowed mother and daughter found life difficult when forced to return to the strictly protestant Argyle household. Carmen Arenas had led a sheltered childhood and was educated in a Catholic convent after the death of her father. Suddenly released into the wider world, she quickly succumbed to the charms of the young Frank Honisett. The two were married when Carmen was barely 18 years old, with Ray being born not long after.

The young Honisett family settled in Commonwealth Avenue, North Perth, not far from the centre of the city. Like all of Australia, and indeed most of the world, the great depression made life a struggle for all but the wealthiest of people. Frank Honisett did have a job, working as a salesman and delivery driver for a tyre company, using a motorcycle and sidecar to make deliveries around the city. In 1937 he suffered a serious accident when a car struck his outfit, crushing his leg. The injury was so serious the limb was amputated below the knee. The accident would have a profound effect on Frank's life, and with it the happiness of his family. The accident, and a later loss of an eye, led to serious depression and family life was intensely stressful for Carmen and her only child. Frank Honisett used his compensation payout from the accident to purchase a general store and cafe in the Western Australian wheat belt town of Dowerin, 160 kilometres north east of the city. While the business was successful, the work was all consuming and, after a year of hard slog, the family sold up and returned to Perth. With war imminent, Frank applied to join the Australian Army and, despite his obvious disability, was accepted and posted to the Pay Corps.

On their return to the state capital the family rented a home in South Perth, reached either by a long causeway, or by ferry direct from the north side of the river. Ray attended the local primary school, enjoying the experience and getting to know his cohort of local friends. He later recalled that his times in South Perth were the best of his young life, with the freedom to play and explore the river, the beach at Como and the surrounding areas, still succumbing to urbanisation.

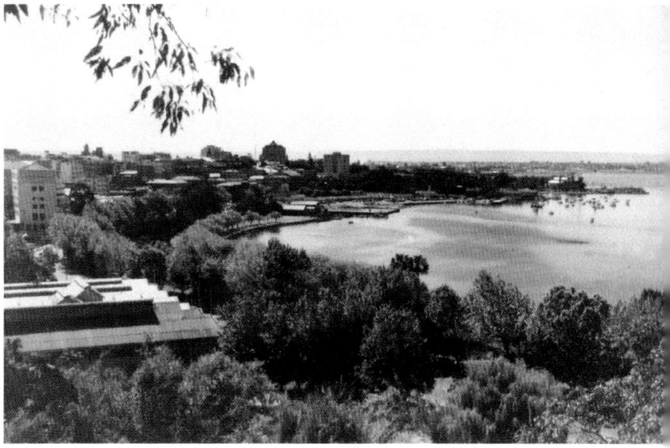

Perth, Western Australia, 1940 — Just before WW2 Western Australia's population was 466,000 out of a national tally of 7 million, with Perth accounting for more than half of the state's residents.

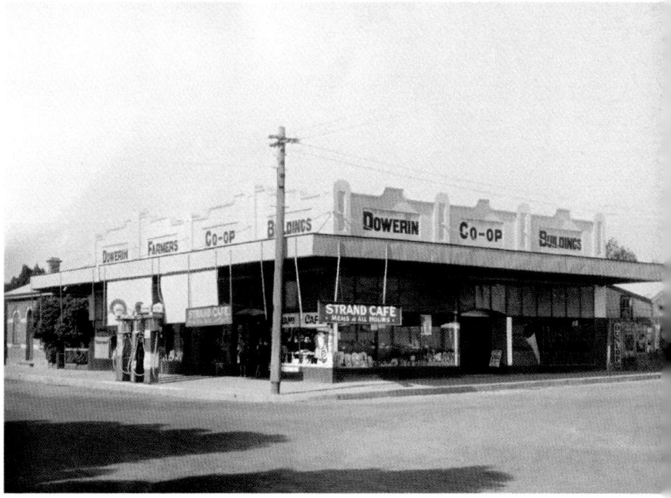

The Strand Cafe, Dowerin, home to the Honisett family in 1937-38.
Photograph courtesy Dowerin District Museum

If Christian Brothers College, Perth, excelled in anything it was in the realm of discipline. They wrote the book. It was appropriately titled "A Tight School is a Snappy School". The snap referred to the sound of 2cm thick, stitched leather straps coming down on young open hands. CBC was a school for boys only; ostensibly they didn't bruise as easily as girls but it wasn't for want of trying. There were few amongst us who did not regularly feel the strap for some transgression of expected academic performance, behaviour or the dress code (we wore uniforms of course). Really, the only enjoyable aspect of my year there was the trip on the steam ferry across the 1½ kilometre width of Perth Water.

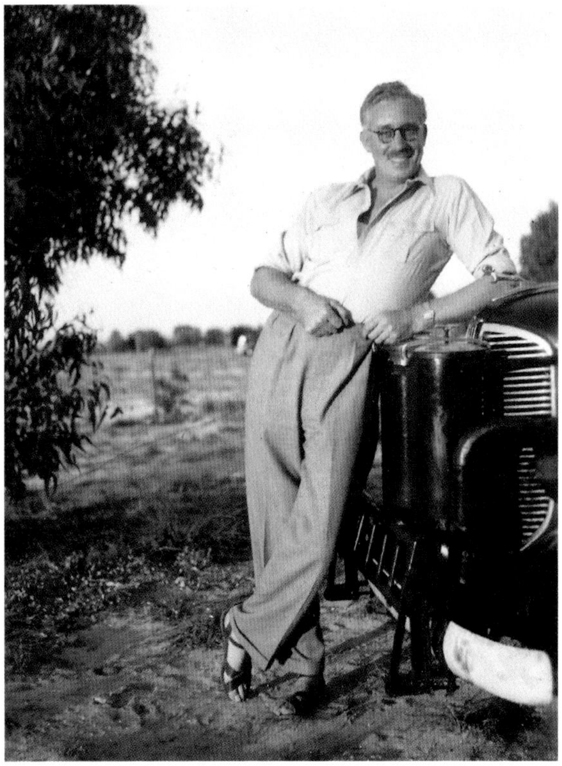

Rollo Watts, with his Dodge sedan, fitted with a gas producer attached to the front bumper. His influence was crucial in gaining Ray his first job after leaving school at the age of 15.

Because of her education and strict Catholic upbringing, Carmen Honisett had ambitions for her only son and, despite the hardships involved, she insisted that when the question of secondary schooling arose, that he would attend the Christian Brothers College in Perth. Even though this meant a daily commute on the ferry, the expense of uniforms and tuition fees, she felt it was important that he be given that start in life. But however earnest the intentions and his desire to please his mother, he found the college absolutely the antithesis of all he desired as an education.

After a miserable year with the Christian Brothers, Ray was eventually able to persuade his mother that he would be much better off attending school on their own side of the river. Not to be deterred, Carmen was determined her boy would receive a serious education and, ignoring the religious change, enrolled him at the new Wesley College, a short walk from their Waverley Street home. The study of religion was less prominent at Wesley, a school where sport was the main focus of attention. For a strapping young man a good head taller than his peers, the young Ray Honisett found a lot to like about the school.

While things had improved on the education front, matters at home had deteriorated markedly, especially after Frank Honisett was posted to the Army Base at Northam, just on 100 kilometres north east of the city, ironically on the road to Dowerin. Their relationship reached a crisis point and the couple separated, with Carmen and Ray crossing the river to live with her mother and her new husband in Leederville, just north of Perth. While it was a difficult time for all concerned, there was one positive outcome, with Ray now able to attend Perth Boys High School, near the centre of the city.

Ray's talent for drawing had been apparent at an early age and he spent much of his hard earned pocket money on aircraft magazines as a source of inspiration for his drawing. With Australia already at, war there was a heightened interest in the ships and aircraft that started to arrive in Perth. Some were on their way to the east, while others used it as a base for operations in the Indian Ocean. Ray was particularly taken with the American Catalina flying boats based near the city and excitedly watched them take off and land on the broad reaches of the Swan River. He became particularly enamoured with the work of the English war artist and illustrator Frank Wootton, even making the effort to write to the artist, care of the magazines that displayed his work. He was delighted when the famous artist took the time to send an encouraging reply and, from then on, he became determined to make a career as an illustrator and painter of aircraft. While many a young lad has dreamed of making pictures, the young Ray Honisett made it an obsession he would not easily give up. While Ray was attending Perth High School, Carmen struck up a relationship with a gentleman named Rollo Watts, a businessman who worked as the Western Australian representative of the Aladdin Lamp Company, which sold kerosene lamps to the far flung farmers, graziers and miners who were still years away from access to electric power and lighting.

With his mother distracted by a new relationship, Ray made a bold but subversive move that would define his career and future life. Realising that he could never advance in his chosen career if he went through the daily grind at school, he quietly slipped out of class and began his own self-directed art education. Because he was just over the age of 15 years, the school assumed he had left to get a job and, as he diligently went off each morning in his uniform, his mother assumed he was off to Perth Boys High as usual. In fact he would go into the city each day, drawing the people, buildings and vehicles that took his eye around the town. If the weather turned bad he would go to the public library and study every art book he good find, all the time keeping to the expected hours of a boy at school. The ruse worked right until the end of the year when the plan was undone by the non-appearance of his expected school report. His mother was understandably shocked, disappointed and angry by what she saw as a betrayal of trust of a high order. But luckily Rollo Watts was able to see that this was not a mere passing fad, but a deeply held belief in what his future might hold. Watts had good connections in the city and spoke on Ray's behalf to a friend at Art Photo Engravers, one of the major operators in the print support industry in Western Australia.

The head of the art department at APE was sufficiently impressed with the young lad's work that he organised for him to undertake some test pieces, which involved producing copies of two professional advertising illustrations. Ray gave it is best and, to his delight, the work was of a sufficient quality that he was offered a job in the art studio. The position was essentially an apprenticeship, but it was a foot in the door and, for a 15 year old with no professional training, it was an opportunity to be grasped with enthusiasm.

After two years working at Art Photo Engravers, Ray was ready to start moving ahead in the profession. He had learned a great deal in his first job, but he was eager to undertake more illustration work, rather than the more mundane graphic and layout work that was his role at APE. The opportunity came when one of his senior colleagues, Athol Snook, crossed over to the *West Australian*, then the leading news publisher in the state. Six months later Ray was offered a position there and, with the higher pay guaranteed by the Journalists Association, he made the move. The work was demanding in time and hours, but again he relished the chance to learn from the experts around him and have the opportunity to show his capabilities.

Although he was still a teenager, Ray had four solid years of work experience under his belt and life was good on all fronts. His mother's divorce from Frank Honisett was eventually settled and she married Rollo Watts. Ray's own social life was running apace and he was leading the life he'd hoped for when he first sneaked out of Perth Boys High School to make his own way in the world. For all the excitement of his work and the advances he was able to make in his profession, Perth was still a small place on the national scene and he began to think about the possibilities that might open up in the eastern states, 3,500 kilometres away on the other side of the continent. As the national economy gradually grew out from under its wartime restrictions, new opportunities would present themselves and Ray was ready take the next step forward.

Going to work every day was sheer bliss. I executed my set chores happily and was open-mouthed at the products of each of the staff. I'd never known a professional artist before, let alone seen any of their work in original form. Some were graphic designers, others were illustrators (my applied direction), others specialised in lettering. All were heroes. All, without exception, were kind and helpful too. I watched how they went about commencing each job and I kept an eye on their progress during the day and would practise small programmes suggested by the more senior members and head artist.

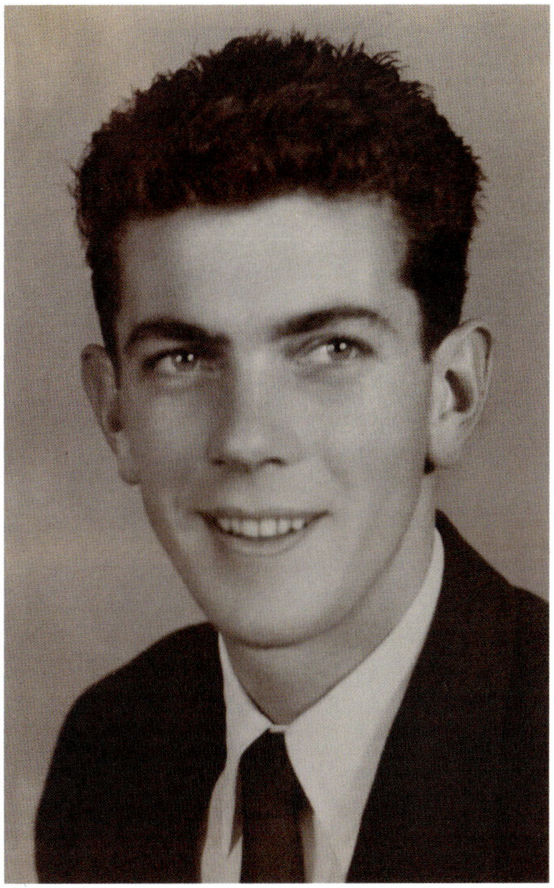

A young Ray Honisett ready to enter the working world, c.1949

EAST TO MELBOURNE

Athol Snook had arranged that I stay with his friend Bill Martin at Sassafras, in the Dandenong Ranges, East of Melbourne for the first couple of weeks. Bill was a brilliant figure illustrator, in demand by the biggest national advertisers. His success was attributable almost as much to speed of execution as his skill and originality. Speed had always been a trademark of the better artists I have known and have known of. It seems to signify a freshness and spontaneity in their work, but for my own product it often made for a glib and superficial quality that I detest.
Bill's work was just damn good, whether it was sophisticated fashion drawings for the wool board or a series of whimsical heads, like an up-beat Norman Rockwell, for Cadburys.

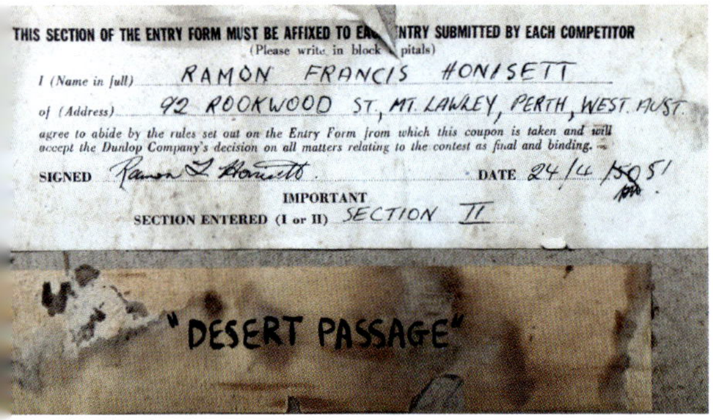
THIS SECTION OF THE ENTRY FORM MUST BE AFFIXED TO EA ENTRY SUBMITTED BY EACH COMPETITOR
(Please write in block pitals)
I (Name in full) RAMON FRANCIS HONISETT
of (Address) 92 ROOKWOOD ST, MT. LAWLEY, PERTH, WEST. AUST.
agree to abide by the rules set out on the Entry Form from which this coupon is taken and will accept the Dunlop Company's decision on all matters relating to the contest as final and binding.
SIGNED Ramon F. Honisett. DATE 24/4 1951
IMPORTANT
SECTION ENTERED (I or II) SECTION II

Opposite
*Ray Honisett, **Desert Passage**, 1951*
Watercolour and gouache on card
Private collection, Melbourne

Ray Honisett's decision to make the move east was again prompted by the actions of a senior colleague. When his friend from APE, Denny Ericson, took the plunge and headed to Melbourne, Ray felt his time had come. Ericson quickly landed work with the Myer Emporium, Melbourne's largest retail store and wrote back enthusiastically of the job prospects in the nation's second largest city. Ray made plans to make the trip east, but before he went he made one bold statement in line with his personal ambitions as an illustrator. In 1950 the Dunlop Rubber Company announced a major art prize and the competition was promoted nationally. Substantial prize money was on offer, along with great publicity and prestige for the winner. Many of the top names in Australian art were entered and it was a measure of the young artist's confidence in his promise and ability that he put his work forward. His work was a gouache of a DC4 being refuelled at night, on the tarmac at Perth Airport. It was in dark blues and purples, with yellow lights throwing a faint glow that picked up the form of the men and machines. He was more than pleased when the company itself purchased the painting for its aviation division, a satisfying endorsement that some in high places were taking notice of his work. It was not surprising that Sidney Nolan took out the major prize, given he was just beginning his rapid rise to prominence, not just in Australia but in London as well. Ray entered again in 1951, this time choosing to paint a striking illustration of a Lockheed Constellation airliner taking off from a desert airfield, somewhere in the Middle East. The work was very much in the Wootton idiom, but it remains a remarkably sophisticated presentation for a young man just out of his teens.

Ray made his farewells to family and friends, heading east by train not long before his twentieth birthday. The four and a half day journey gave him plenty of time to think on his future and to reflect too on choices made to that point. Certainly his unilateral decision to leave school early had proven to be a good one, and his application in the workplace had shown his ability for hard work while also nurturing his creative spirit. When reaching Melbourne his friend Athol Snook had again paved the way, organising accommodation for him while he settled in to a challenging new life.

After a few weeks finding his feet, the young artist was ready to push ahead with his career. He applied for, and won, a position with a rapidly expanding industrial firm headquartered in the inner city suburb of Richmond.

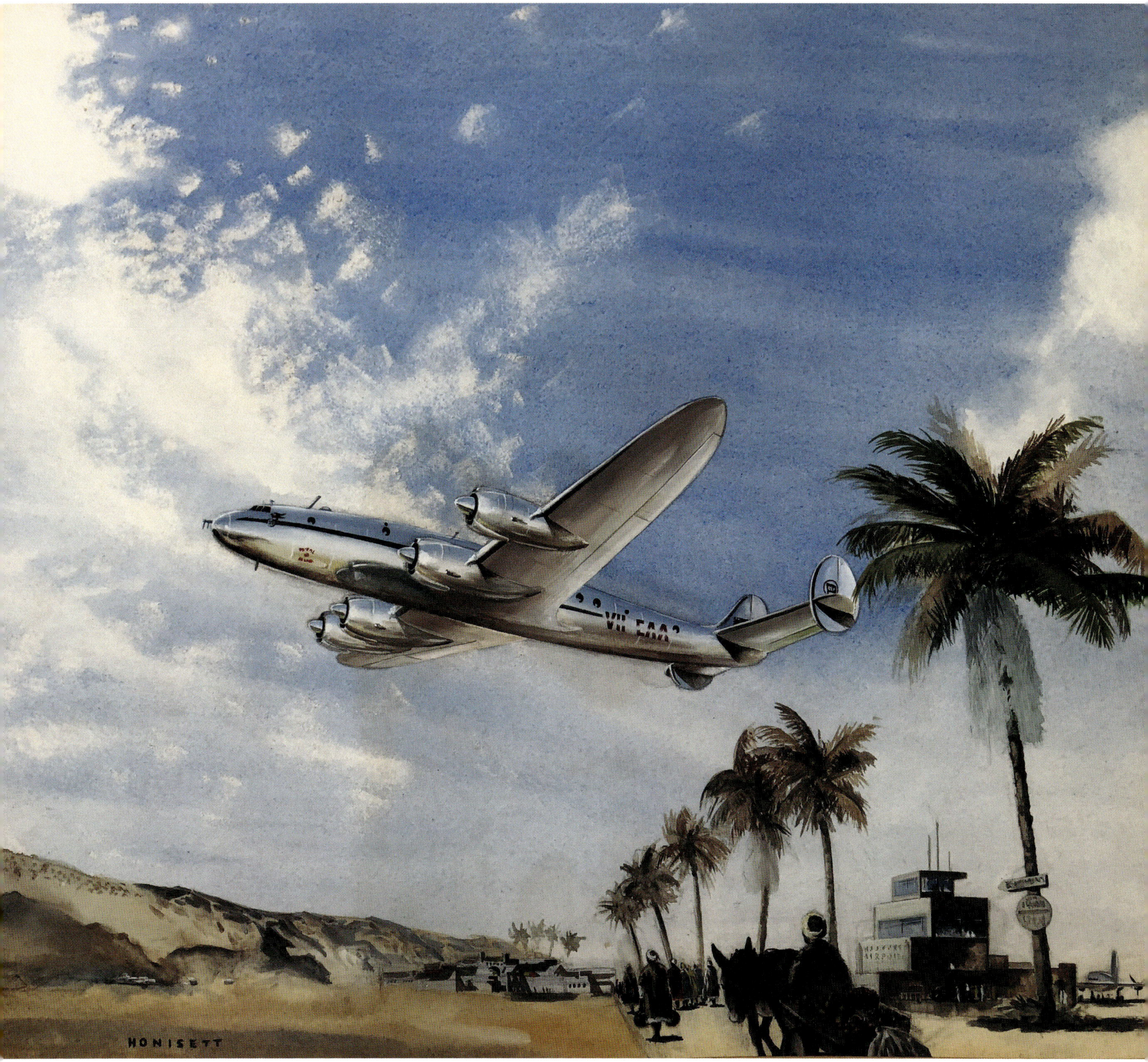
HONISETT

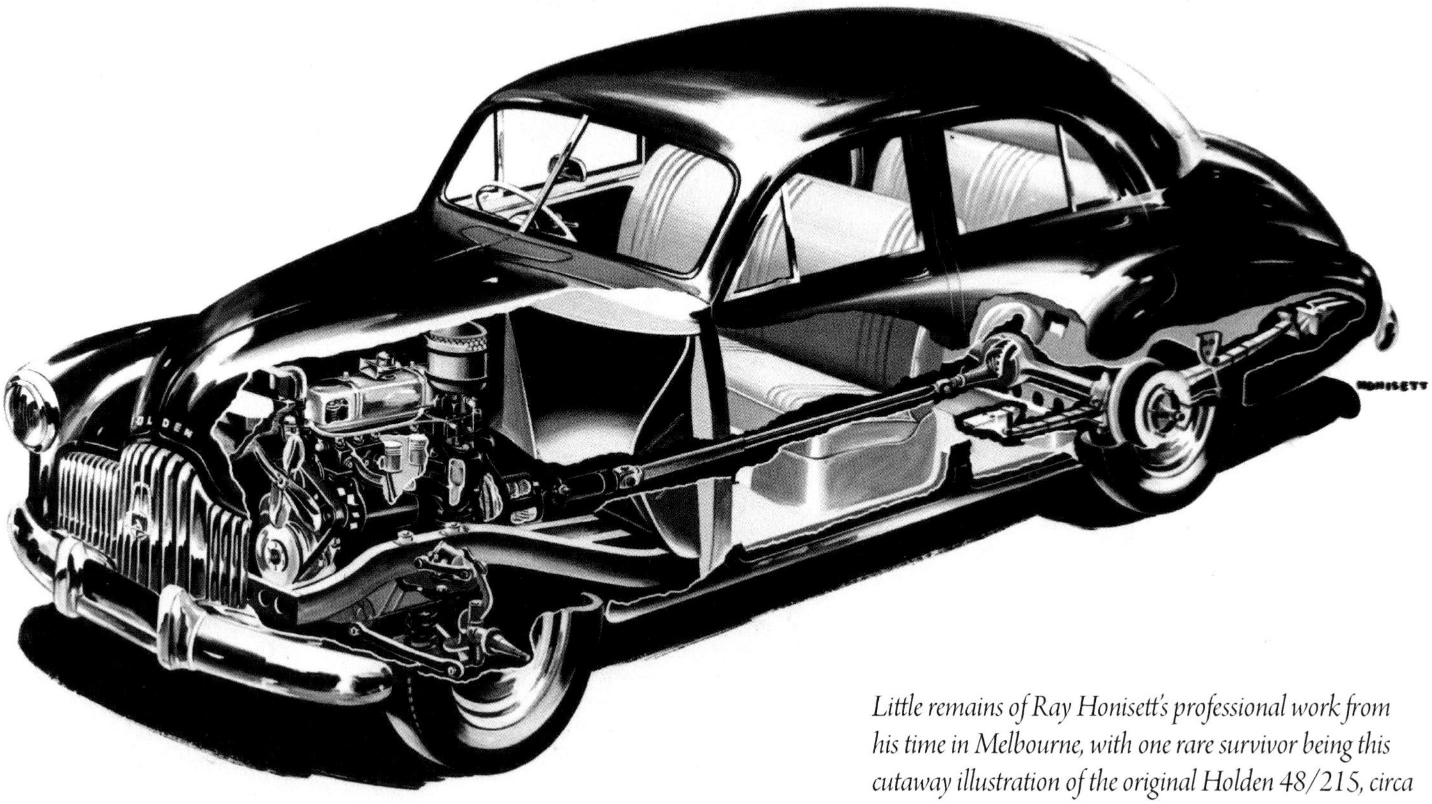

Little remains of Ray Honisett's professional work from his time in Melbourne, with one rare survivor being this cutaway illustration of the original Holden 48/215, circa 1951

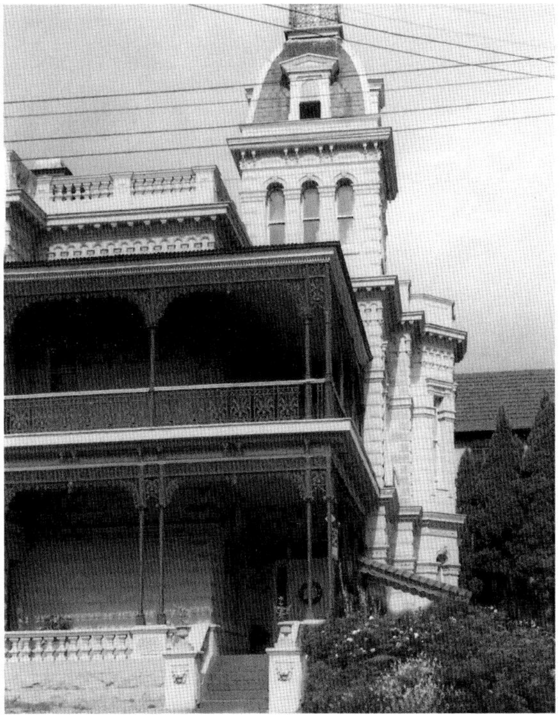

Moulded Products was a manufacturer of domestic and industrial products in the new technology of plastics. In the boom years of the early 1950s the company was expanding rapidly, laying the groundwork for its new entity as the Nylex Corporation, soon to become a household name in Australia. Again it was a good learning experience for a young artist and he threw himself into the working life, as well as the social life that thrived alongside it. Ray found boarding house accommodation in nearby South Yarra, just across the river. His life took another major step when a young woman he had known in Perth, Lois Wheeler, made the journey across the Nullabor and the two struck up a relationship.

LONDON CALLING

Life in Melbourne developed along predictable lines, with Ray's professional work becoming better known in the city and, in parallel, his relationship with Lois developing to the point where the pair decided to marry and, despite Ray's relative youth, to make a life together. Ray admitted later that part of the motivation for the union was that both saw advantage in the marriage as a way of speeding up their desire to travel to the United Kingdom to work. They moved in together, settling into rooms in the old Victorian mansion *Goodrest,* in Toorak Road West, South Yarra. (left)

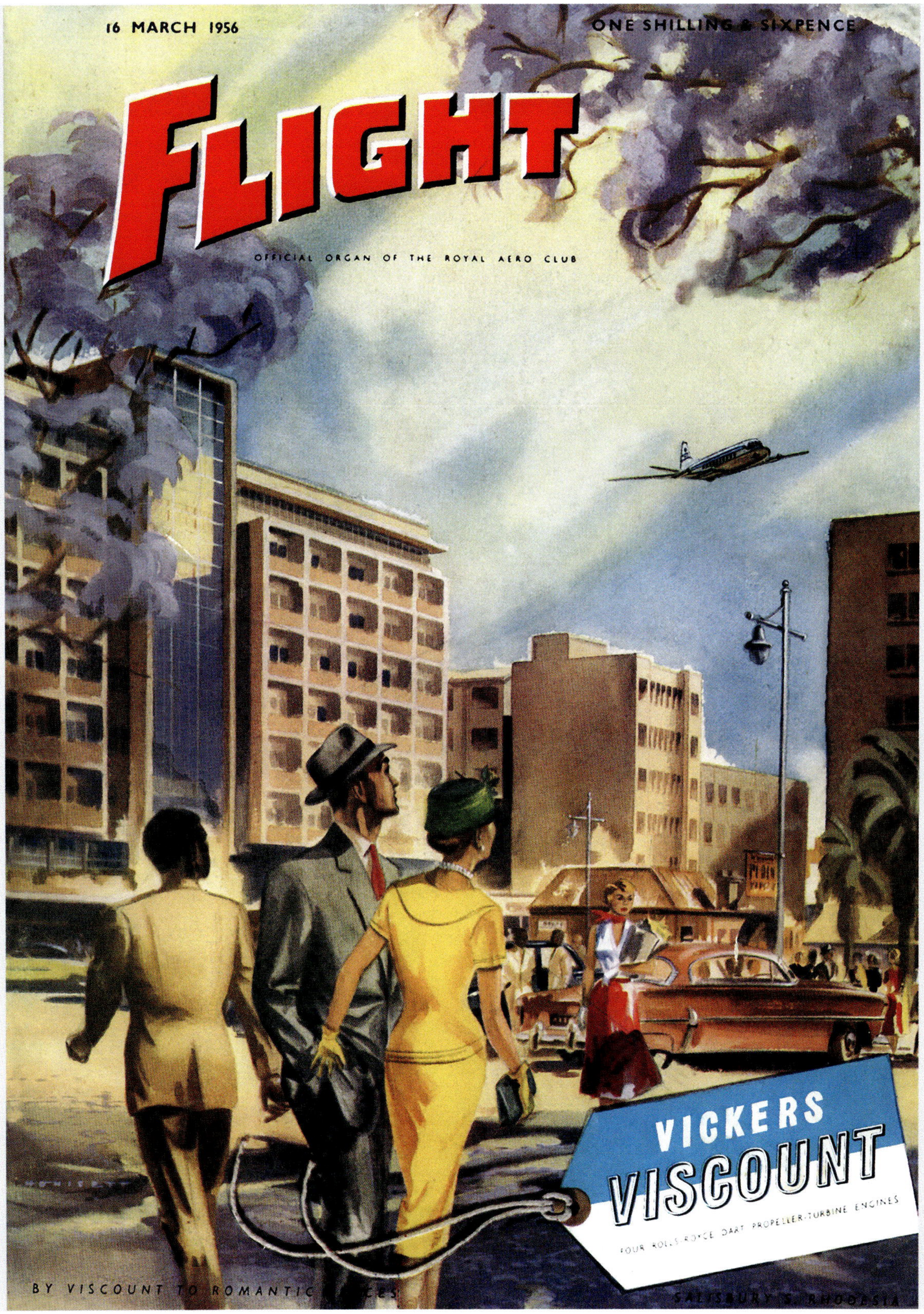

It may have been the family feel of the company or the many nice people working there, but either way, the whole place was most congenial and the studio at Moulded Products produced a consistently high standard of successful and effective work. The studio also housed industrial design staff who were responsible for both original product design and the adaption to local requirements of many American items produced under licence. The creative mix seemed to add a stimulus to both areas.

While my duties mainly centred on graphic design, I soon learned to maintain an illustrative balance by designing a fair amount of illustration into my work. As we were all, for the most part, responsible for the finished art of our design work I satisfied the needs of both Moulded Products and myself.

Ray Honisett, Cover illustration for ***Flight****,*
16 March 1956
It was a testament to Ray's tenacity, ability and hard work that he was able to realise his boyhood ambition, to produce the cover artwork for the world's most important aviation journal, within nine months of arriving in London. For a 24 year old from remote Western Australia it was a remarkable achievement.

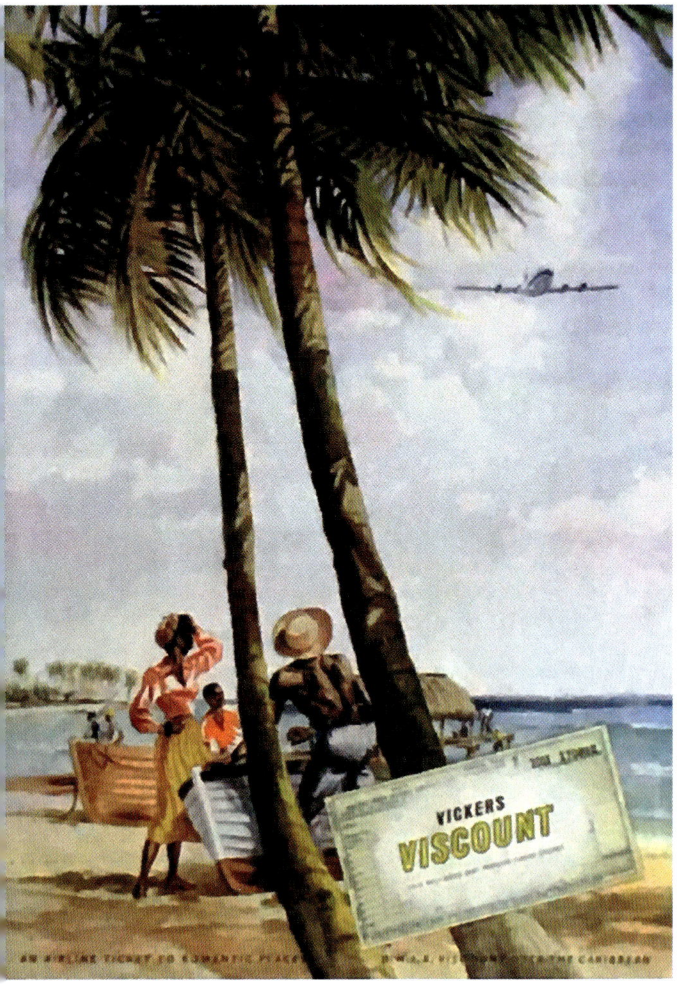

Stopping off at exotic island locations soon paid dividends, for within a year of his arrival in London Ray would be producing brightly coloured advertising material promoting overseas air travel.
*Advertisement for Vickers Viscount aircraft, used in **Flight** and **Aircraft** magazines.*

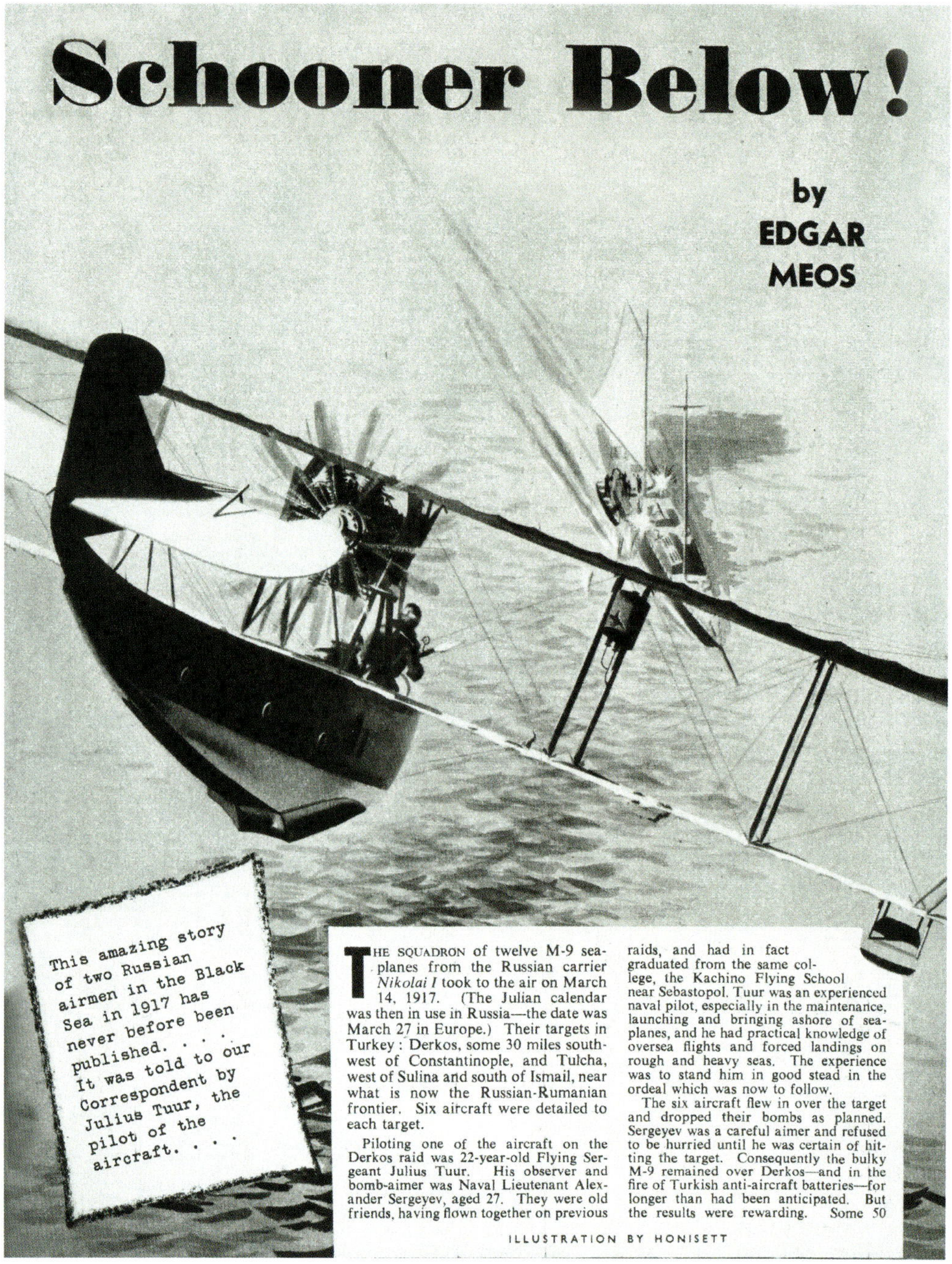

Schooner Below!

by
EDGAR
MEOS

This amazing story of two Russian airmen in the Black Sea in 1917 has never before been published. . . . It was told to our Correspondent by Julius Tuur, the pilot of the aircraft. . . .

THE SQUADRON of twelve M-9 seaplanes from the Russian carrier *Nikolai I* took to the air on March 14, 1917. (The Julian calendar was then in use in Russia—the date was March 27 in Europe.) Their targets in Turkey: Derkos, some 30 miles southwest of Constantinople, and Tulcha, west of Sulina and south of Ismail, near what is now the Russian-Rumanian frontier. Six aircraft were detailed to each target.

Piloting one of the aircraft on the Derkos raid was 22-year-old Flying Sergeant Julius Tuur. His observer and bomb-aimer was Naval Lieutenant Alexander Sergeyev, aged 27. They were old friends, having flown together on previous raids, and had in fact graduated from the same college, the Kachino Flying School near Sebastopol. Tuur was an experienced naval pilot, especially in the maintenance, launching and bringing ashore of seaplanes, and he had practical knowledge of oversea flights and forced landings on rough and heavy seas. The experience was to stand him in good stead in the ordeal which was now to follow.

The six aircraft flew in over the target and dropped their bombs as planned. Sergeyev was a careful aimer and refused to be hurried until he was certain of hitting the target. Consequently the bulky M-9 remained over Derkos—and in the fire of Turkish anti-aircraft batteries—for longer than had been anticipated. But the results were rewarding. Some 50

ILLUSTRATION BY HONISETT

*Illustration for the **RAF Flying Review***
January 1957

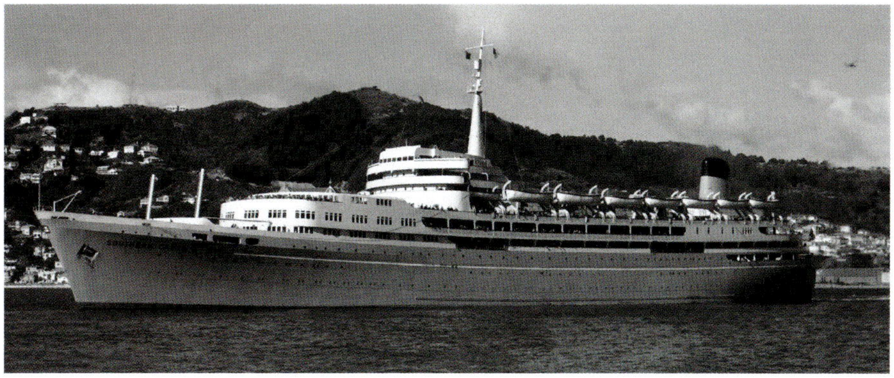

The lure to work overseas became ever stronger and they made the commitment in 1955, booking passage on board the brand new liner *Southern Cross*, then making its maiden west-bound voyage around the world. They returned to Perth to make their family farewells, joining the ship in Fremantle for the voyage to Southampton.

The journey was an adventure in itself and the couple made a cohort of interesting friends, including the notorious East-end gangster Billy Hill and his partner Gypsy, who were being forced back to London after unsuccessfully attempting to relocate his operations to Sydney. After stopovers in Capetown and the Canary Islands, the young couple reached London ready to start the next adventure.

Ray Honisett had a substantial portfolio of work ready to show potential employers and, perhaps more importantly, a recommendation from the world's leading practitioner in his chosen field, Frank Wootton. If anything Ray's work was perhaps too much in the style of the English master and he realised he would have to work hard if he was to find his own artistic voice, rather than forever being cast as a Wootton clone. One thing his mentor was able to do was to put in a word for him with London's leading commercial artist's agency, the Astral Arts Group. His work impressed the firm and they put him on their books straight away. Astral was different from most art studios in that it acted as the agent for individual artists, rather than as an employer taking in work for production in-house. Astral's role was to match the artists with the clients and take a percentage for all the successful introductions made. Ray soon settled into the Astral way of working and found he was getting commissions for exactly the sort of work he wanted, including his most cherished ambition, the production of cover art for *The Aeroplane* and *Flight*, arguably the world's two most influential and prestigious aircraft industry publications. Within 12 months of landing into London, Ray had seen his first cover work, an advertisement for the new Vickers Viscount. Many similar commissions would follow, including illustrations for the *RAF Review* and advertising images for a range of aircraft and automobile manufacturers. He continued to work through the Astral Arts Group, enjoying each professional challenge that came his way. Unfortunately his personal life had not met with the same success and he and Lois separated while in London. Ray returned home in 1962 ready to recommence his life and work on home soil.

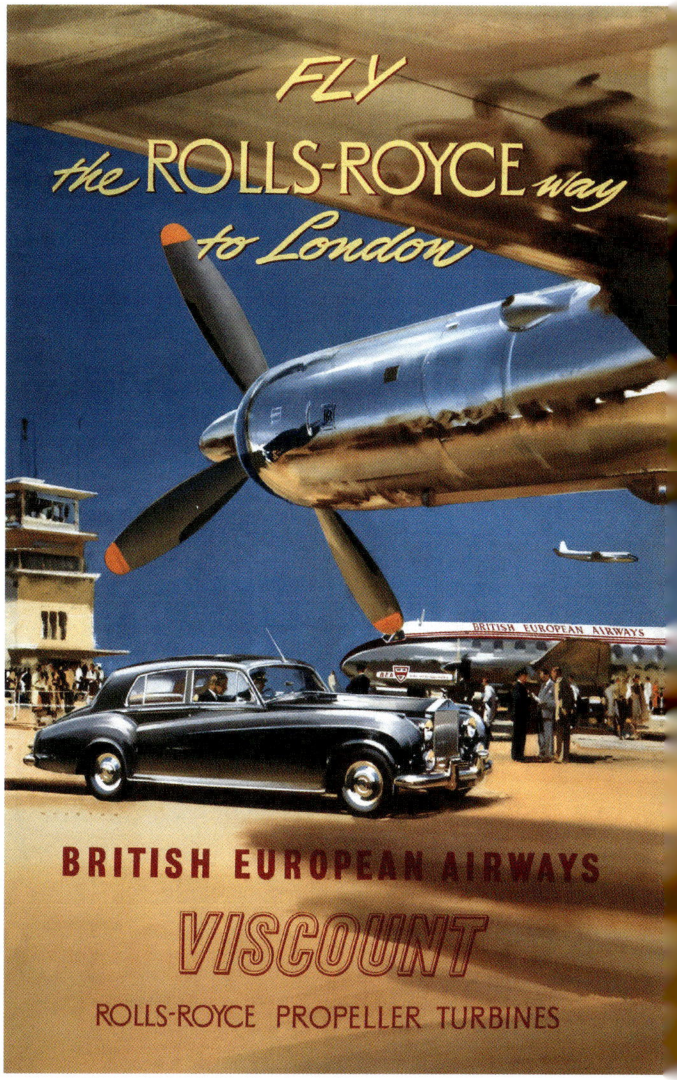

Ray Honisett's mentor Frank Wootton had transitioned smoothly from Official War Artist to advertising illustrator, quickly winning contracts for all the major aircraft and motor manufacturers. Rolls Royce, which had a foot in both camps, was an obvious fit for Wootton's elegant style. For a young artist starting out in the field of aircraft illustration it was not difficult to fall under the spell of Wootton's exemplary style.

Above, left, The Shaw Savill liner Southern Cross leaving Wellington, New Zealand on its maiden voyage to Australia and on to the UK. The ship was the ultimate modern liner, offering one class accommodation and no provision for freight. All the machinery was placed at the stern of the ship to maximise space and minimise noise and vibration.

With that decision everything changed for the better. More challenging assignments being offered, increased income and, most satisfying of all, a commission from Vickers Armstrong to produce a series of full colour illustrations for advertisements promoting the Viscount. These would show the different airlines that had the aircraft already in service, or had them on order. At last, the front-cover of The Aeroplane and Flight.

The ironic aspect of this little story is that, for ten years, that cherished ambition had imagined manifestations of conventional aviation art — quite probably sleek, shiny aircraft traversing a dramatically clouded sky, or some similar idea. The reality was that, to fulfill the brief, the illustrations were each a ground level scene, typical of the particular country, with a group of large foreground figures, suitably attired, looking skyward at a very small over-flying Viscount.

Have you, too, noticed that even when the Gods deign to grant a wish to a mortal they seem loath to do so entirely as requested? They simply cannot resist inserting their unexpected and mischievous twist.

Ray Honisett

Newsletter of the Australian Society of Aviation Artists, September 2009

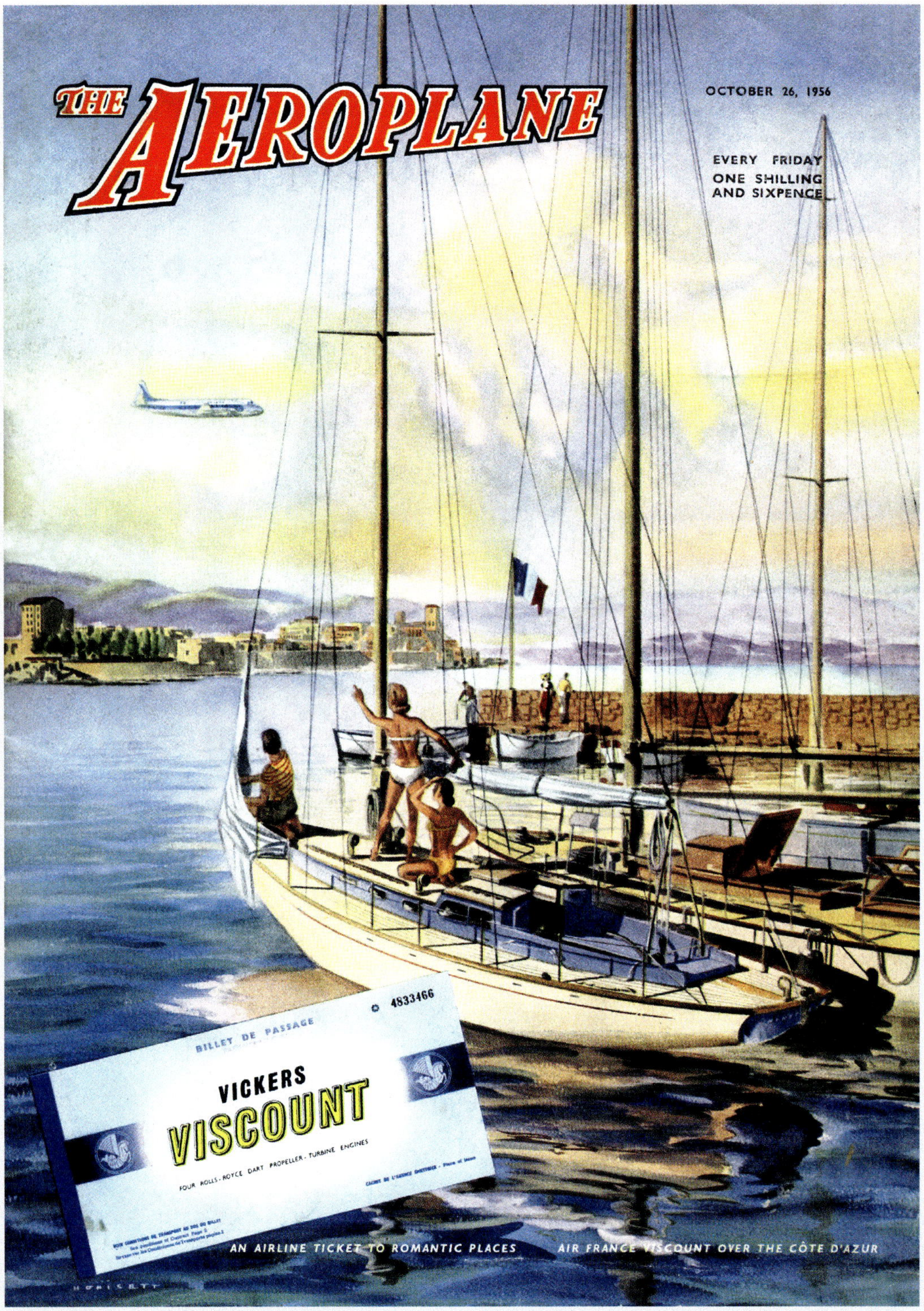

THE ART OF ILLUSTRATION

AIRCRAFT ILLUSTRATED

Ray Honisett arrived back in Perth on a Qantas flight from the United Kingdom on 19 April 1962. Once he had caught up properly with the family, he headed east to Melbourne where he set about reestablishing himself in the local art world. From a professional point of view it was a difficult moment to land back into the land of his birth. In the first two years of the new decade Australia was in the middle of a financial boom, with wool prices climbing and the general economy expanding to accommodate a rapidly growing population, driven by immigration and the maturing of the baby boomer generation born in the years immediately after the war. The decision to make fundamental changes to import policies saw rampant inflation as Australians rushed to take advantage of much lower prices on imported goods, especially consumer items. Federal Treasurer Harold Holt had pressed hard on the economic brakes, pushing up taxes and interest rates while curtailing lending in what became known as the 'credit squeeze'. Given the small scale of the aircraft industry in Australia, Ray had hoped he might gain work with the automotive and consumer goods manufacturers, but this was precisely the market niche most affected by the credit squeeze.

Ray Honisett's name first appears in 1966 with a publication that played to his strengths, a folio of twelve reproductions of paintings he had made of military aircraft in the Pacific Theatre of World War Two. The folio was published by the Melbourne Book Company, with an introduction by then Minister for Air, The Hon. Peter Howson MHR. While not destined to be a big seller, the folio was important in that some retailers, notably Melbourne's Technical Bookshop, made the individual prints available separately, thereby expanding the market, especially to the young. That particular specialist store was significant in being immediately across the street from the Royal Melbourne Institute of Technology where Ray would gain a teaching appointment in May of 1968. Initially a three day per week job, the part-time position allowed him plenty of time to pursue his freelance illustration work in parallel with his lecturing commitment. The publication did receive some valuable publicity, with a two page colour spread in *The Australasian Post*, a popular magazine aimed a wide audience, particularly among men.

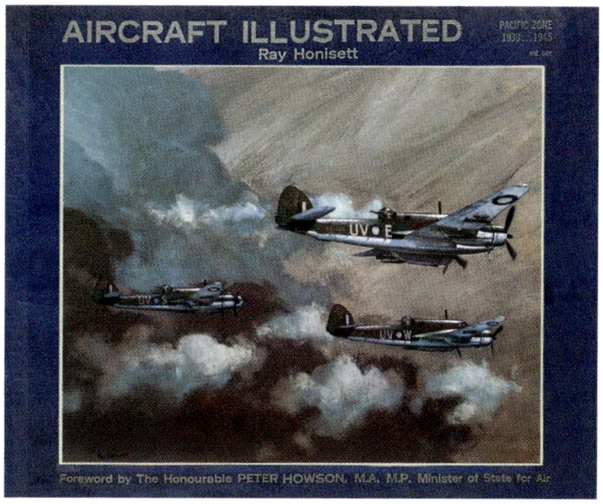

***Aircraft Illustrated**, published in 1966 by the Melbourne Book Company and printed locally by Frank Daniels Pty. Ltd. in North Melbourne. While only a slim volume, the printing was of an extremely high quality for the time, with the matte paper giving the colours a richness that has stood the test of time. All the images in this section have been reproduced from an original copy of the book.*

*Opposite, **The Aeroplane**, 26 October 1956*
*When Ray Honisett landed back into Melbourne, ready to restart his career on home soil, he brought with him a solid portfolio of professional work. Most important perhaps were his covers for **Flight** and **The Aeroplane**, the world's two most prestigious aircraft industry journals.*

FOREWORD TO AIRCRAFT ILLUSTRATED

Melbourne (Aust.) artist, Mr. Ray Honisett, has embarked on a most interesting and ambitious undertaking with his paintings, "Aircraft Illustrated". I understand that his intention is to produce paintings of all aircraft which flew during World War 2. This first volume of 12 illustrations comprises aircraft which were flown in the Pacific theatre by the Air Forces of Australia, the United States and Japan. Subsequent volumes will also include aircraft which flew in all other theatres of the war.

It is obvious to me that Mr. Honisett has put much thought into the preparation of this project. To ensure technical accuracy he has been given access to official photographs and documents at the Australian War Memorial and the Department of Air. He has also discussed the markings of squadrons, and even individual aircraft, with the men who flew them in operations. In addition, and of no less importance, Mr. Honisett has gone to considerable effort to recapture the atmosphere of the terrain from which the aircraft operated. This is particularly evident in the painting of Spitfire Vc's as they take off from Livingston strip, Northern Territory, and the Catalina as it battles its way through a storm over the Timor Sea.

It is believed that this is the first attempt by any artist in the world to paint the complete series of aircraft of World War 2. This is an undertaking of some magnitude and obviously will take many years to complete. Posterity will benefit, as these aircraft are gradually going out of existence. By their very nature, aircraft deteriorate rapidly and are quickly replaced by their more sophisticated successors. They are not easily preserved and the only permanent record must be in paintings such as these. Not only, therefore, has the artist provided an important series of paintings for collectors, but he has also provided an historical record which will be invaluable to generations to come.

As a World War 2 airman myself, I look forward with pleasure to future volumes of Mr. Honisett's work.

The Hon. Peter Howson MA MP, Minister of State for Air
1966

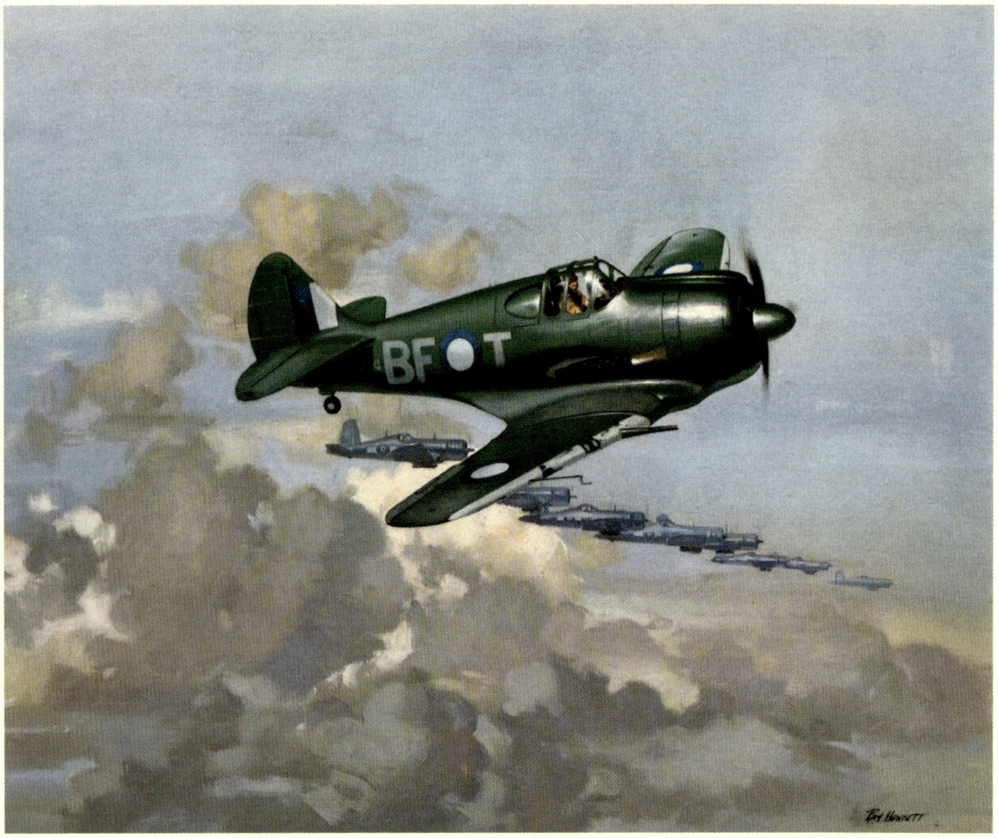

*Ray Honisett, **Boomerang over Bougainville**, (1966)*
Aircraft Illustrated No 1

Operating out of Torokina, Bougainville Island, during January 1945, a C.A. 19 Boomerang of No.5 Squadron RAAF leads in Corsairs of No.21 Squadron RNZAF.

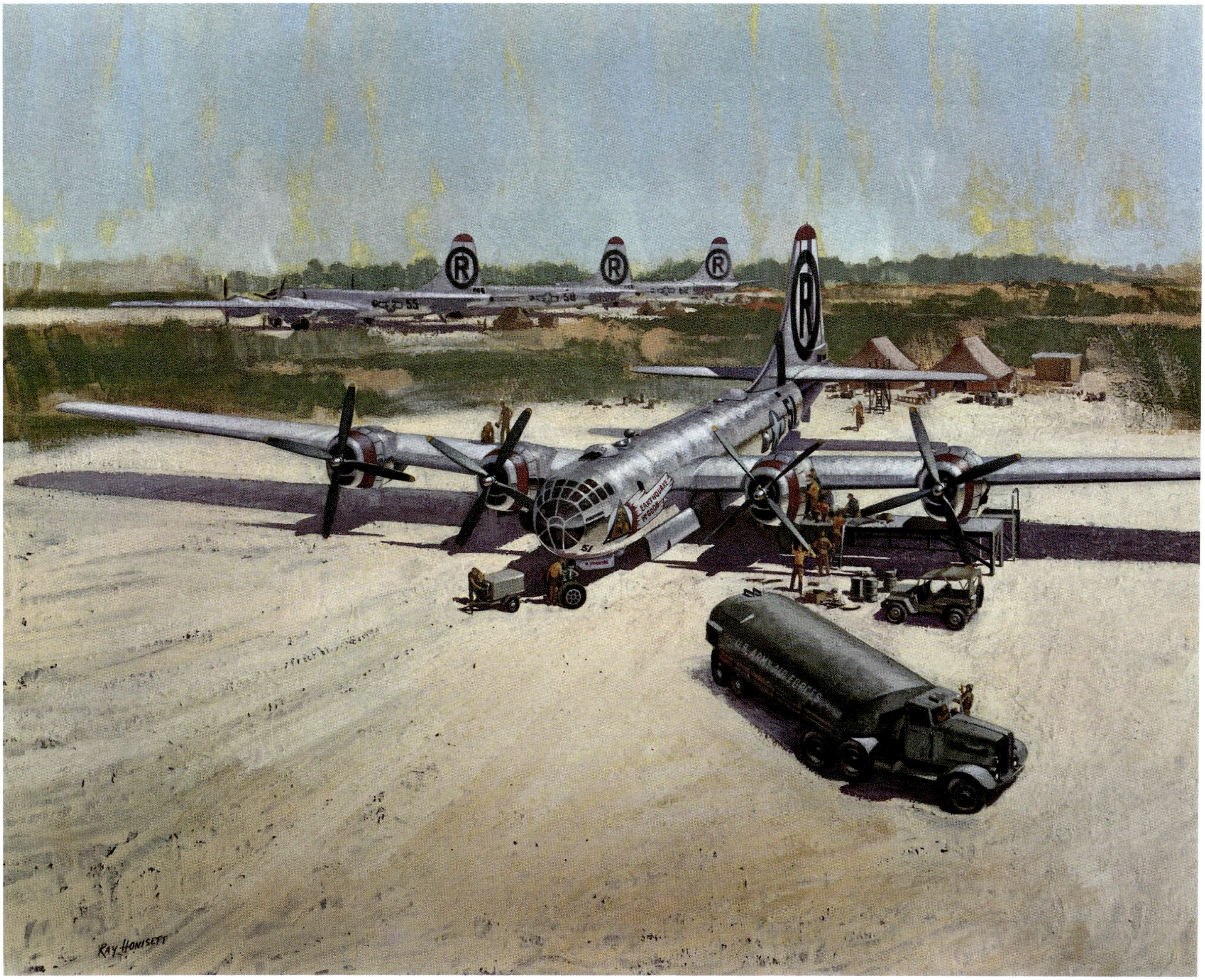

Ray Honisett, ***The Calm Before****, (1966)*
Aircraft Illustrated Number 9

'Earthquake McGoon', Boeing B-29 Superfortress of the 6th Bombardment Group, 313th Bombardment Wing basks lazily in the sun on North Field, Tinian while ground-crew ready her for the next mission.

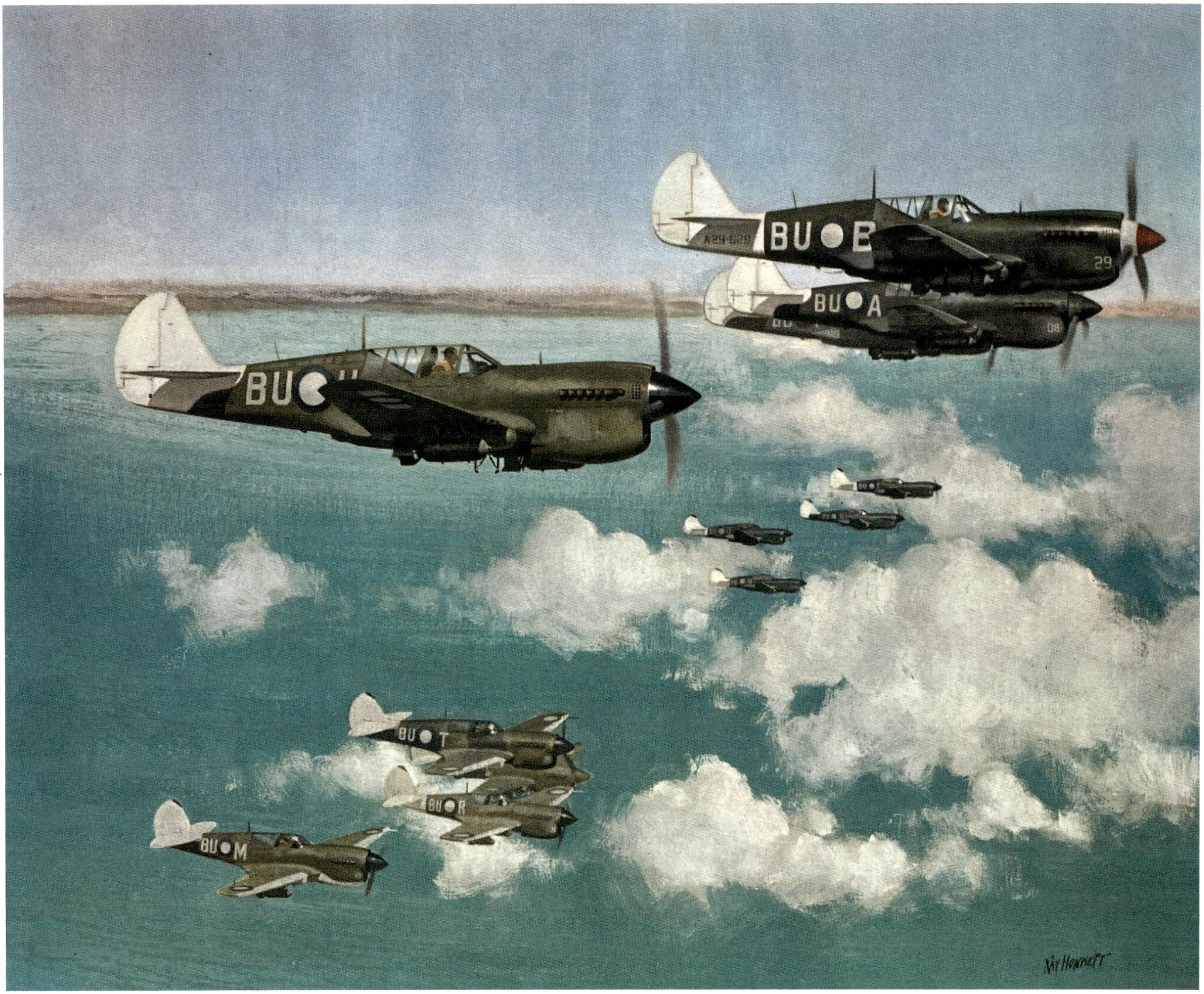

Ray Honisett, ***A Flock of Kittyhawks*** *(1966)*
Aircraft Illustrated No 3
Displaying a variety of colour schemes and marking details, Kittyhawk IVs (P.40 Ns) of No.80 Sqdn. RAAF form up off the coast of Moratai Is., late in 1944. BU-B, the Wing Commander's aircraft, is the well known Cleopatra III.

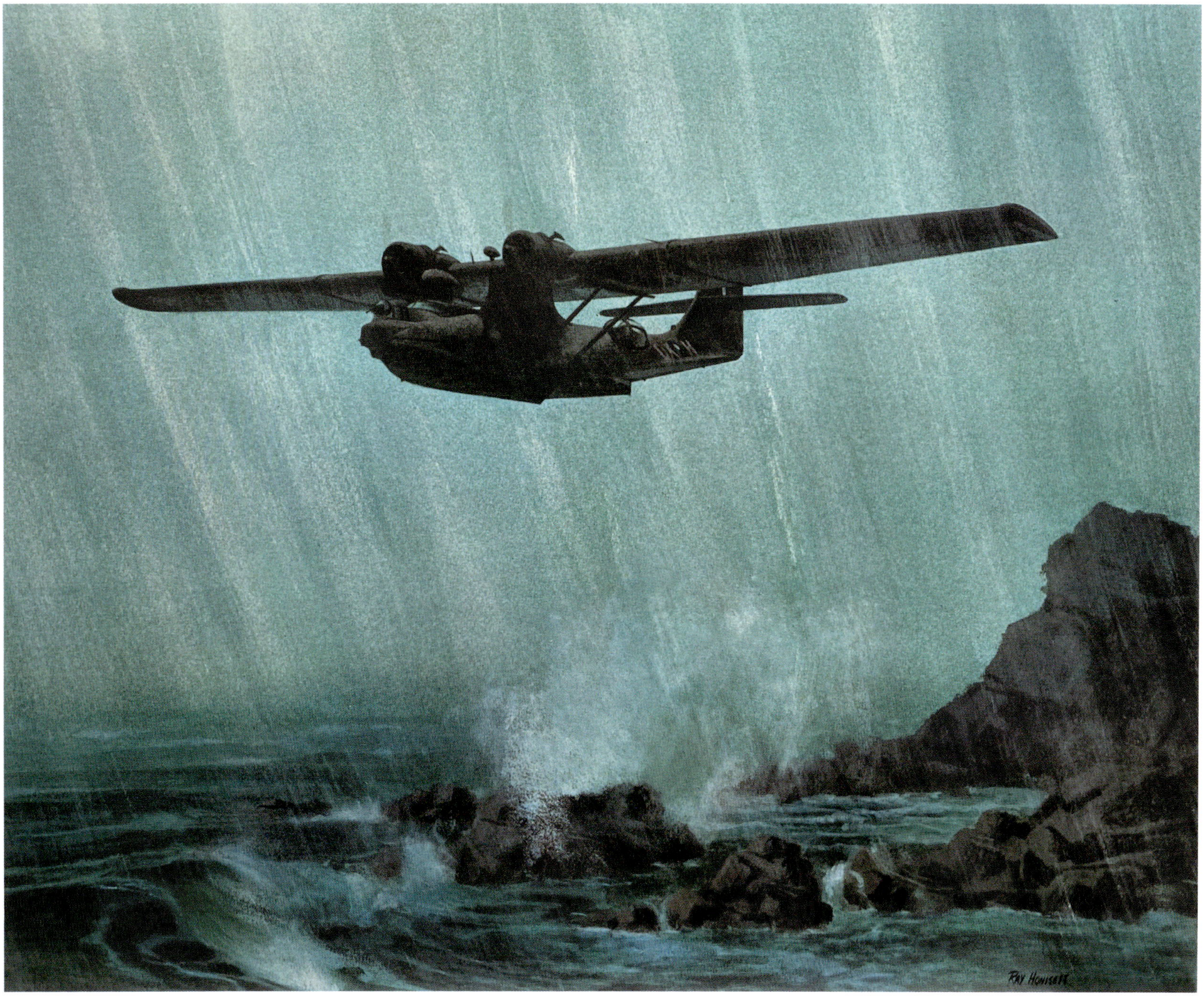

*Ray Honisett, **The Long patrol**, (1966)*
Aircraft Illustrated No 5
A Darwin based Catalina of No.43 Squadron RAAF heads out over the Timor Sea, 1945

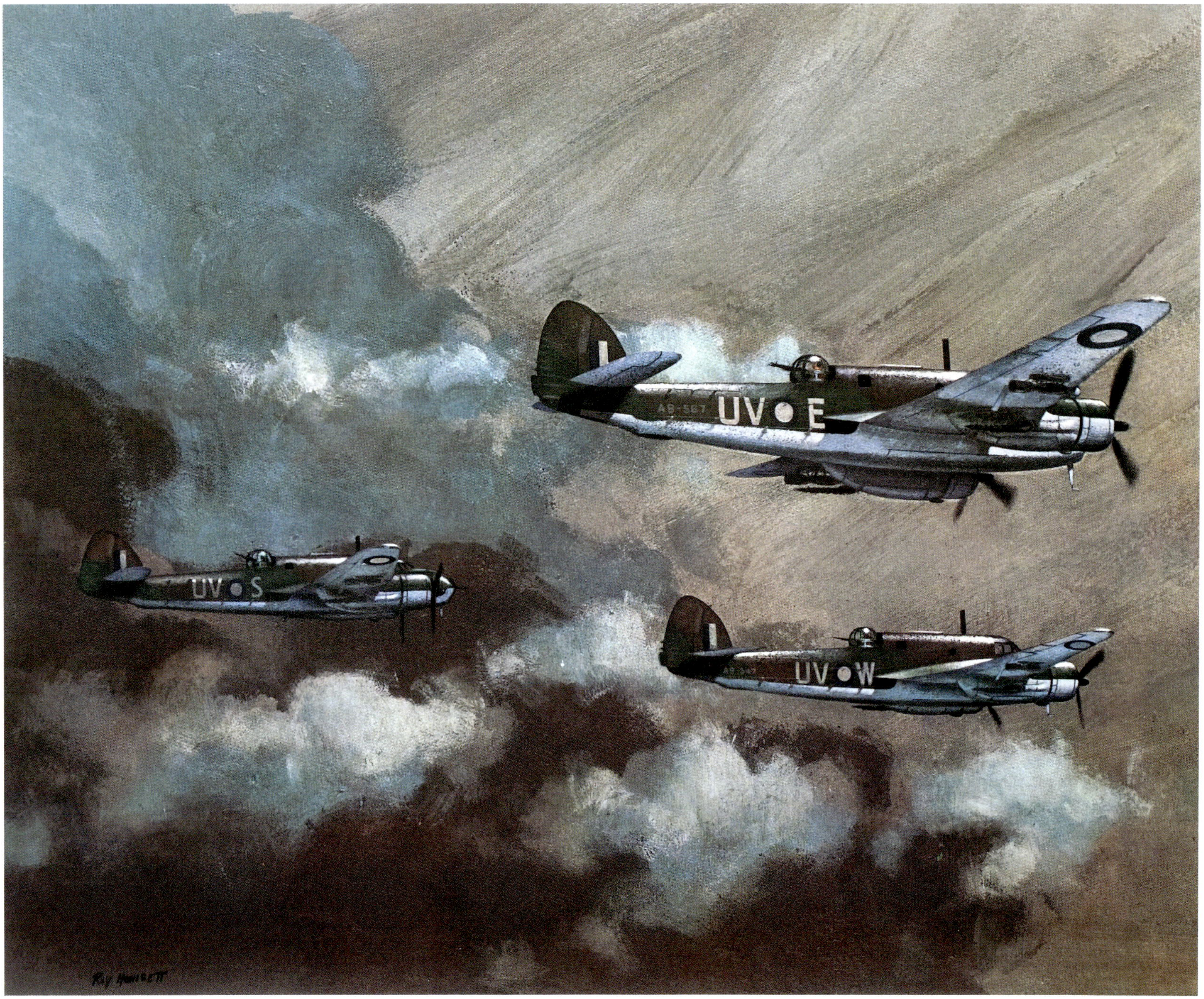

Ray Honisett, ***Beaufort Show****, (1966)*
Aircraft Illustrated No 8

Australian-built Bristol Beaufort VIIIs of No8 Squadron RAAF, on the 90 mile run between their base at Tadji, Northern New Guinea and Wewak, where they gave close support to the Australian 6th Division, late 1944

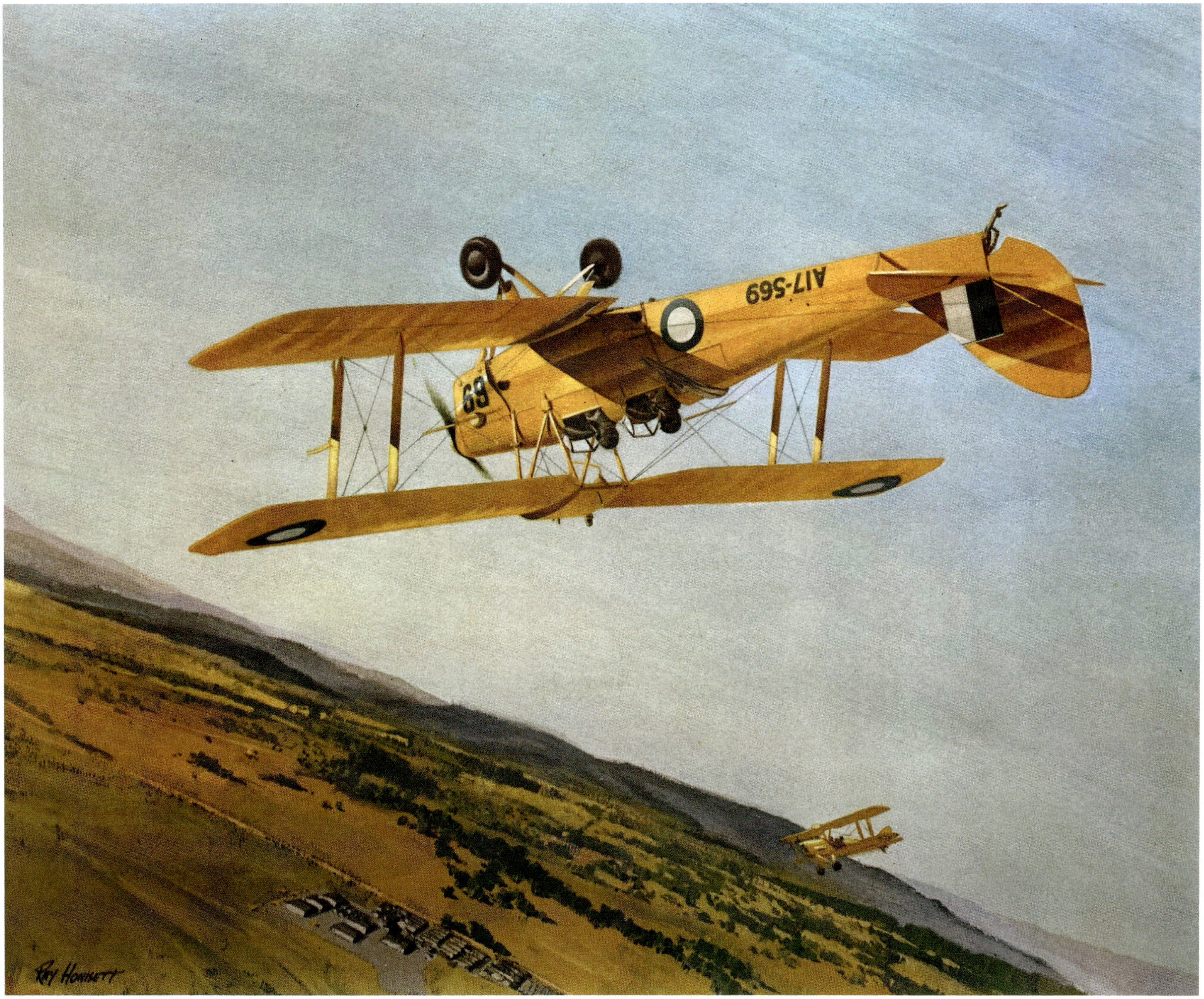

Ray Honisett, ***Tiger Country****, (1966)*
Aircraft Illustrated No 4

De Havilland Tiger Moths over a typical RAAF wartime training field — Benalla, Victoria.

THE AUSTRALIAN WAR MEMORIAL, CANBERRA

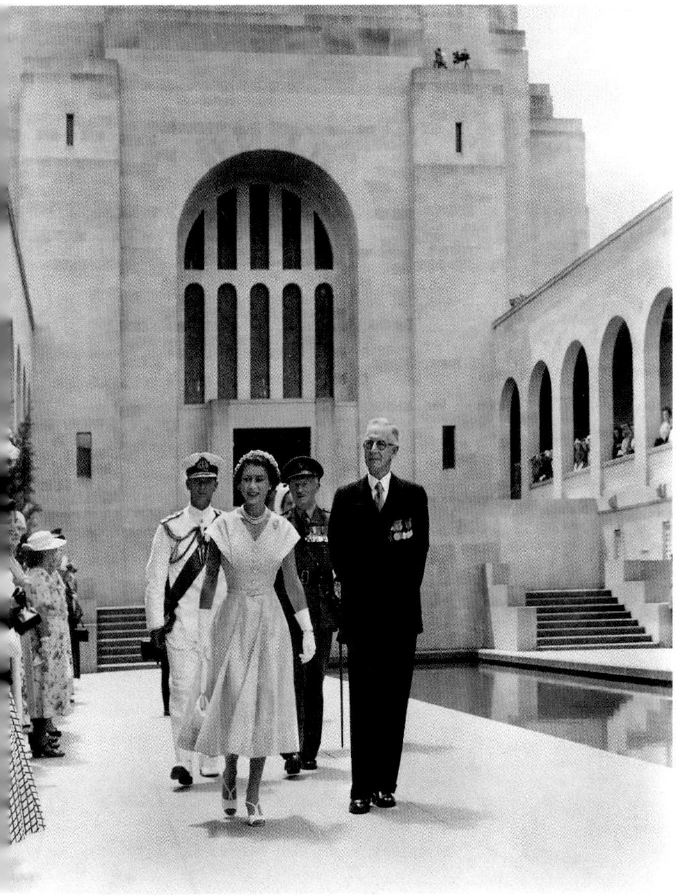

Charles Bean, founder of the Australian War Memorial, with Queen Elizabeth II in 1954. The AWM was, and remains, perhaps the single most important cultural institution in the nation, combining as it does a museum of world standing within Australia's National Memorial to those who served in armed conflict.

If the appearance of his work on the covers of *Flight International* and *Aircraft*, the two most prestigious aeronautical journals published in the United Kingdom, marked a point of achievement in Ray Honisett's early career, the commissioning of a work by the Australian War Memorial was the next level, and point of departure, in his career as an artist. Instead of being at the beck and call of advertising agents and magazine publishers, the Memorial commission gave him recognition as an artist in his own right, producing work that was an end in itself, not merely an attractive way to sell ideas and products. It was not all plain sailing, especially when dealing with the chain of command and layers of bureaucracy inherent in an old and conservative public institution, but nonetheless an achievement of which he was justly proud. It would also build his reputation and lead to many new opportunities, some of which he could never have dreamed of as a young boy sneaking out of class to try his hand as 'an artist'.

The exact timing and method of his coming to the attention of the Australian War Memorial is not documented, but can be deduced from comments made in correspondence between the Memorial's then-director, Bill Lancaster, and the artist. Wishing to capitalise on the interest shown in his 1966 publication of twelve paintings of World War Two aircraft, Ray sent a copy of *Aircraft Illustrated* direct to Lancaster and, somewhat later, the actual paintings were placed on loan and displayed in the Memorial galleries in Canberra. The fact that the publication had received the endorsement of the Minister of State for Air, The Hon. Peter Howson, would certainly not have gone unnoticed

It should be remembered that the Australian War Memorial was created to commemorate the service of Australians in the Great War of 1914-18. When it opened in November 1941, a full two years after the outbreak of World War Two, the displays were of the previous war, not the one in progress. It would not be until well after the war had ended that the displays began to reflect the WW2 conflict. While there was a large collection of work by the 43 official war artists, there were few large paintings of the type that dominated the WW1 galleries. In the late 1950s and early 60s a number of commissions were placed with artists, with Ivor Hele the dominant figure on the list, undertaking three large paintings and six portraits between 1956 and 1958. While nobody doubted Hele's ability and compelling style, there was a danger of the exhibitions becoming a one-man band. This was perhaps understandable, as Hele and the Memorial's Art Advisor, Sir William Dargie, had won the Archibald Prize 13 times between them in the 16 years from 1941 to 1957. In the establishment art world there were no two bigger names and their works dominated not just the Memorial, but also Parliament House and many other public institutions throughout the country.

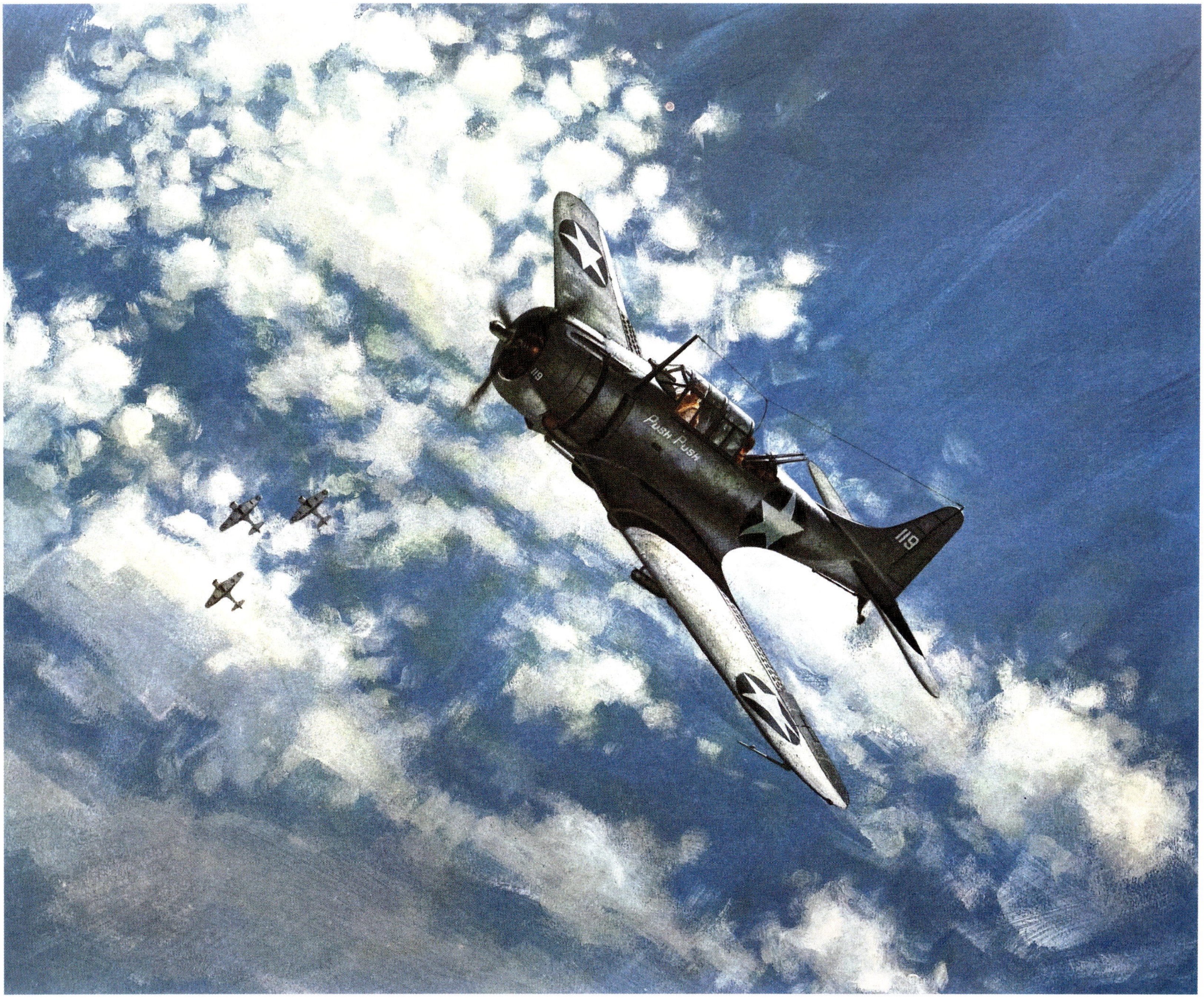

Ray Honisett, ***Beachcomber****, (1966)*
Aircraft Illustrated No 10

'Push Push', a Douglas SBD-4 Dauntless of Marine Squadron VMSB-144 slices through the morning sky searching for targets in support of the landing at Torokina, Bougainville Island, November 1943. After winning fame with the Navy Squadrons in the Battle of Midway, the Dauntless went on, with the F4U Corsair, to form the backbone of US Marine Corps aviation throughout the war in the Pacific.

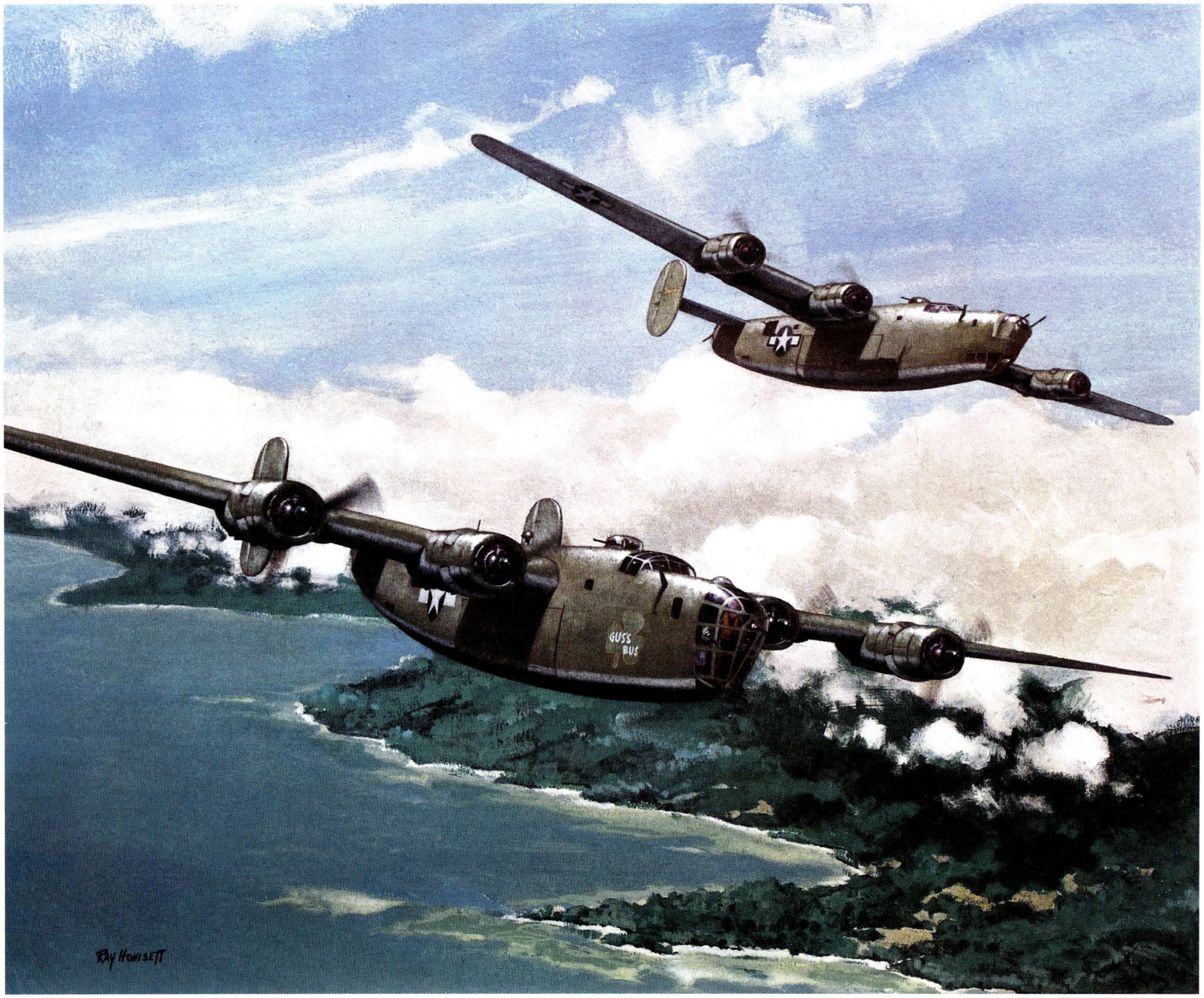

*Ray Honisett, **Long Range Libs.**, (1966)*
Aircraft Illustrated No 12

'Gus's Bus', a Consolidated B.24D Liberator (42-40504) of the 530th Bomb Squadron, 380th Bomb Group, 5th Air Force, USAAF, boring down on the west coast of Dutch New Guinea near Fakfak, late 1943 in company with a B.24D1 (42-40471) of the same squadron - one of the early Air Depot modified B.24s fitted with a rear turret in the nose. These aircraft were based at Fenton, Northern Australia.

*Opposite above, Ray Honisett, **Tokyo Defender** (1966)*
Aircraft Illustrated No 11
A colourfully marked Kawasaki Ki.61 Hein (Swallow), code named 'Tony', of the 244th. Fighter Regiment JAAF, based at Chofu near Tokyo, passes over the airfield perimeter prior to landing, 1945

*Opposite below, Ray Honisett, **Morning Mission** (1966)*
Aircraft Illustrated No 2
Bristol Beaufighters of No.30 Squadron RAAF prepare to take off from Aitape, Northern New Guinea, 1944.

The Memorial planned to build two extra wings on the original building to accommodate the exhibitions of WW2 and the subsequent conflict in Korea, with Vietnam being added to the list even before the new galleries were completed in 1971.

To manage this lack of balance, and to ensure all three services were properly represented, the AWM Planning Committee set out to produce a list of subjects for paintings to be commissioned from a range of artists. Everything done at the Memorial was designed to respect the numbers involved in the various services and the campaigns in which they fought. As most of the Official War Artists had been appointed to work with the Army, there was a need to redress the balance and the RAAF was given the bulk of the subjects on the list – 8 for the RAAF, 6 for the Army and 4 for the Royal Australian Navy. Unsurprisingly, despite their attempts to widen the scope of the collection, the first two commissions went to Dargie and Hele, with the former being allocated the first RAAF subject, the Kittyhawks at Milne Bay in the New Guinea campaign. An early commission went to the senior RAAF Official War Artist, Harold Freedman, with two other subjects going to Alan Moore. Two more picture commissions were allocated to former war artists Geoffrey Mainwaring and Vernon Jones. Dennis Adams, a proven performer in maritime and aviation subjects, was to undertake the operations of RAAF Sunderlands in operations against submarines.

One year on from the original decision to commission the list of 18 subjects, three more had been added, one each for Korea, Malaya and Vietnam. Most of the works had been allocated, with just three subjects remaining, RAAF Beaufighters of 455 Squadron attacking Enemy Shipping, RAAF Lincolns in Malaya and a Naval subject for the Korean War. Charles Bush had declined the Beaufighter commission, and suggested either Vernon Jones or Frank Norton, but both were already committed on other projects. Bill Lancaster was expressing the need for some fresh faces on the list as the youngest of the WW2 Official War Artists, Douglas Watson, had been born in 1920, with the others stretching as far back as the late 19th century. It was then that Ray Honisett's name was put forward, on the basis of his excellent folio of aircraft subjects that were exactly what the Director had in mind.

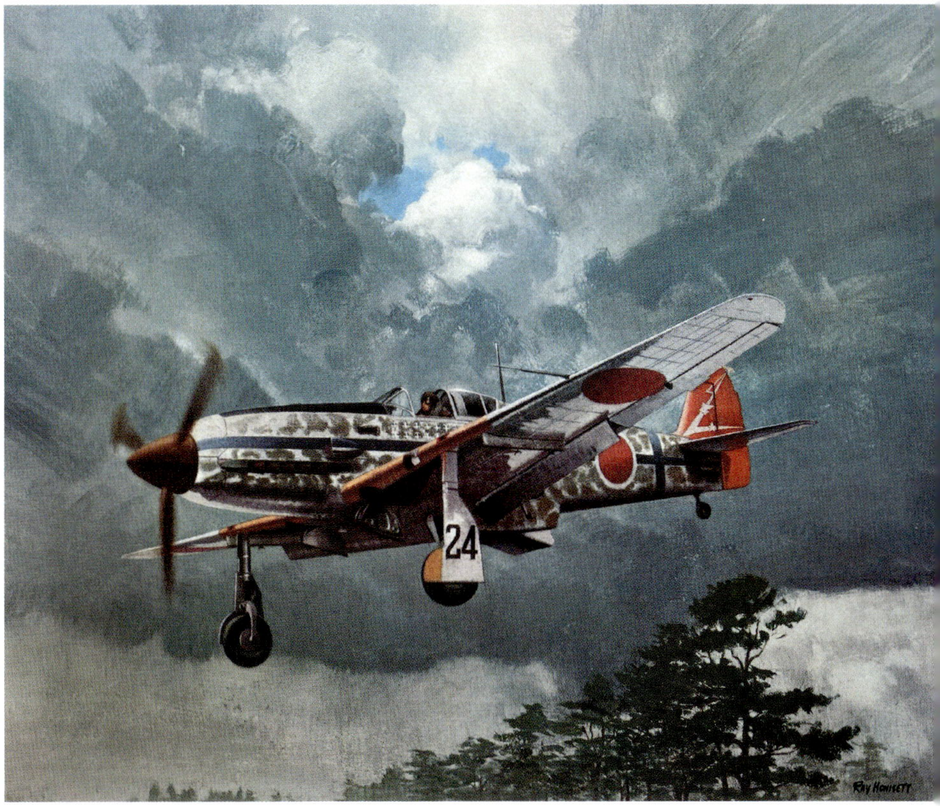

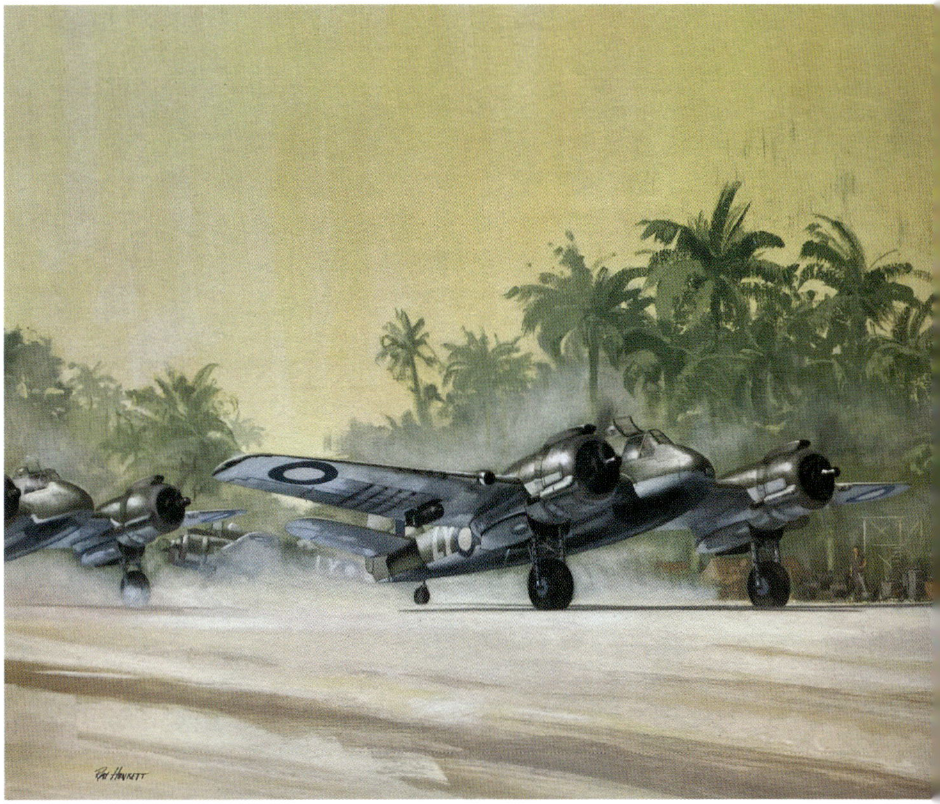

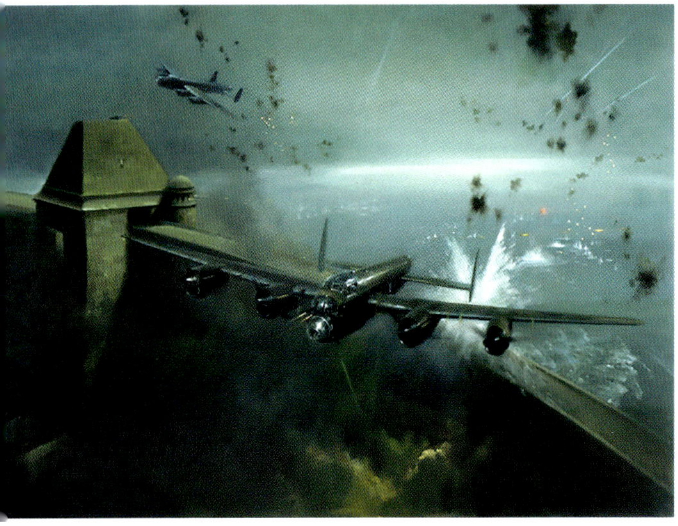

Frank Wootton, **617 Squadron raid on the Mohne Dam,**
16 May 1943, (1980)
Oil on canvas, 91.5 x 122 cm
Australian War Memorial ART40896

The Memorial's acquisition of Frank Wootton's rendition of the famous 'Dambusters' raid was another personal milestone for Ray Honisett. He had idolised Wootton since he was a young lad in Perth during the war, even writing to the artist and sending him copies of his pictures for him to critique — and the elder statesman of aviation art was good enough to respond with encouragement and good humour. Four decades on from those early steps, the two would at last hang side by side in the nation's most visited and loved cultural institution.

Although the original paintings for *Aircraft Illustrated* were relatively small, they had a bold presence and feeling for action that few of the older artists could emulate. This was not surprising, as there were few specialists in the aviation field, even those who had worked with the RAAF. Ray Honisett on the other had made it his forté and had rubbed shoulders with the very best while working with the Astral Arts Group in London.

In February of 1968 Ray was offered the choice of either the Malayan subject or the Beaufighters, and he was quick to seize upon the latter. He was asked to produce an appropriate sketch for consideration by the Art Committee, for which he would be paid $150. The artist negotiated a fee of $2,000 for the final work, which was happily received by the Committee. It was noted in the minutes from the meeting that this was the 'minimum ever paid for a 9' x 5' picture' (Most received $2400, while the 'stars', Hele and Dargie, received $4200 for that size of work). The artist was certainly not complaining, for this was by far the largest commission, and payment, he had achieved in 20 years of professional work. At a personal level the substantial fee was particularly important, for five years earlier he had married a young woman he had met in London, Elaine Simone Clayton, and the couple had a young family to support. Far more important than any fee to be paid, the commission marked a crucial moment for the Memorial, being the first awarded to an artist who had not served as an Official War Artist. The Committee had noted the 'excellent' sketch and, with the approval of Sir William Dargie, the contract was put in place.

The Beaufighters painting was completed by mid 1969, and was formally accepted by the Art Committee. Such was the enthusiasm for his work, Ray Honisett was immediately offered the remaining Air Force subject, RAAF Lincolns in the Malayan Emergency. In offering the commission, Lancaster suggested that the artist might care to vary the size and orientation of the work, moving away from the standard 'wide screen' 9' x 5' format that was beginning to limit display opportunities in the galleries. On 22 July 1969, after having personally inspected the Beaufighter painting for the first time (its formal acceptance had been originally undertaken in Melbourne by Sir William Dargie), Lancaster wrote that ... 'You have produced a picture which we will be proud to hang and the staff at the Memorial, who can be very critical, feel the same way as I do about it'. [AWM 895/2/100] The Director went even further in a letter to Sir William Dargie, written the same day, that 'It would be hard to match it with any painting we have of an aircraft attack'. [AWM 895/2/84, 22.7.1969] The final judgement came in the minutes of the Memorial's Executive and Finance Committee on 16 December 1969. "The Beaufighter painting when received was very highly praised and considered by many to be the finest major work in the Memorial's paintings of RAAF subjects". [AWM 895/2/100] Ray Honisett had passed his first major test with flying colours, beginning an important relationship that would last for the next fifteen years.

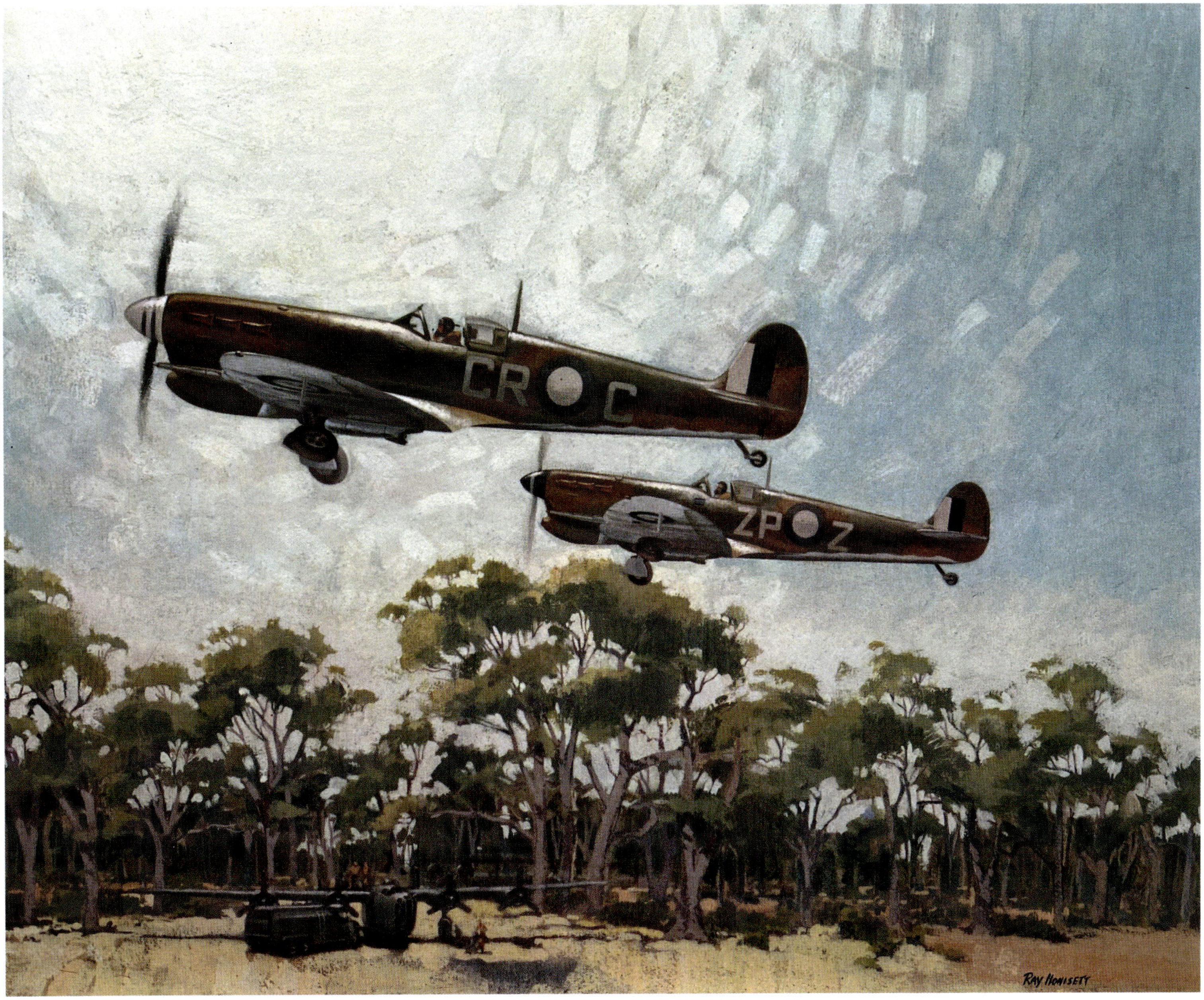

Ray Honisett, ***Illustrious Defenders****, (1966)*
Aircraft Illustrated No.6
Spitfire Vcs flown by Wing Commander C.R. Caldwell, DFC & Bar (Australia's top ace of WWII) and Squadron Leader K.E. James (CO 457 Squadron) lead No457 Sqd. RAAF from Livingston strip to rendezvous with the other squadrons of No.1 Wing (54 and 452) in defence of Darwin, June 1943

THE BEAUFIGHTERS COMMISSION

BY RAY HONISETT

After almost 60 years of earning a living from art it is possible to recall several highlights, but few are more vivid than my first commission for the Australian War Memorial. When the offer was made, and hastily accepted, my thoughts were all of the prospect of my work being in the same collection as so many eminent painters who had been Official War Artists – Streeton, Lambert, McInnes, Dargie, Dobell, Hele, Browning, etc. The year was 1969; the AWM was expanding and filling historical gaps in its painting collection with Ivor Hele, George Browning, Frank Norton, Dennis Adams and others, all executing commissions at much the same time. And, I soon discovered, to my dismay, with canvases at much the same size; around 9 feet by 5 feet. My God, until that time I had never filled an area one-third the size of those dimensions.

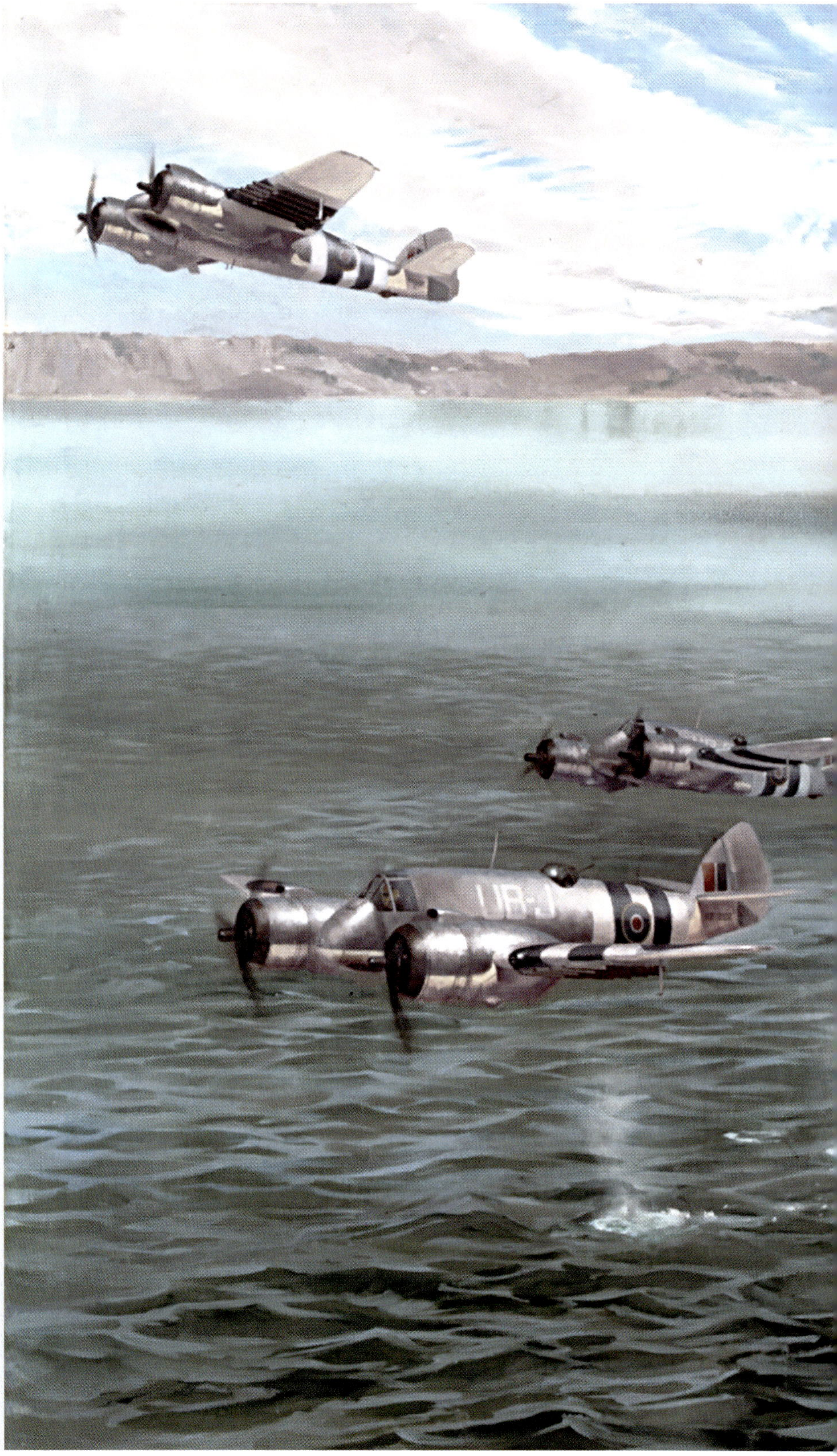

No. 455 Squadron Beaufighters attacking shipping off Norway, 1944 *(1969)*
Oil on canvas, 186.5 cm x 309.5 cm
Australian War Memorial ART 27629

Part of an anti-shipping strike wing of Coastal Command, No. 455 Squadron, Royal Australian Air Force, became highly skilled in this role in sweeps along the French, Belgium and Dutch coasts. In October 1944 the squadron was assigned full-time to strikes against shipping in Norwegian waters. At that stage the Germans were transferring troops and equipment from the occupied countries to help counter the grave situations which had developed elsewhere in Europe. The objective of the anti-shipping squadrons was to halt such traffic. In the Norwegian fiords flying was particularly hazardous, as ships were able to anchor almost at the foot of towering cliffs. Attack was possible from one direction only, enabling a concentration of anti-aircraft fire upon the attackers. In the open sea several Beaufighter aircraft could divide the attention of the ship's gunners, and attack with rockets and cannon simultaneously.

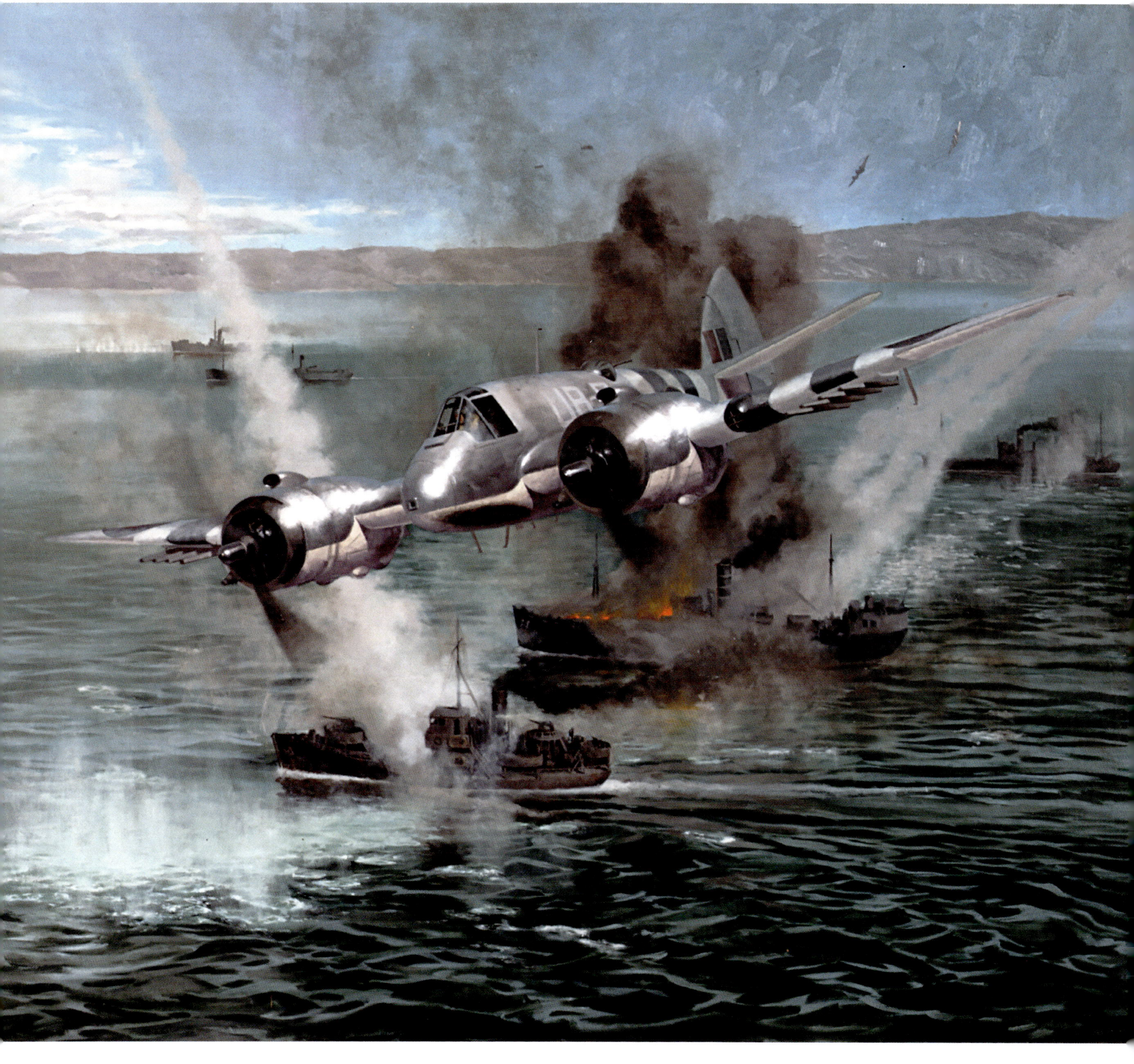

The last RAAF subject on the 'List of 21' had been the hardest to allocate so, after the success of the Beaufighters, Ray was the obvious choice for the Malaya commission as well. The Council and staff had been extremely impressed with the Beaufighters commission and there were no qualms in awarding him the Malayan Emergency subject. The topic itself sounded somewhat bland - *RAAF Lincolns in operations in Malaya*, but as is often the case, it was much harder to pin down a way of telling the story with power and engagement for the viewer. After some lengthy discussion, the Memorial called on RAAF historian Keith Isaacs to assist. He came up with an overall concept, producing a sketch to explain his idea. Ray was able to take the concept and, through the device of turning the composition to the vertical with a tilted horizon, give it real action and drama. Such was the desire for accuracy that the Memorial's director offered to have the artist taken to Malaya to experience the jungle first hand, but for logistical reasons he settled for a compromise arrangement where RAAF photographers took a large number of colour slides as reference. Ray was particularly touched when the Memorial gave a special preview of the finished work to HRH The Duke of Edinburgh when he visited to open the new wing in 1971.

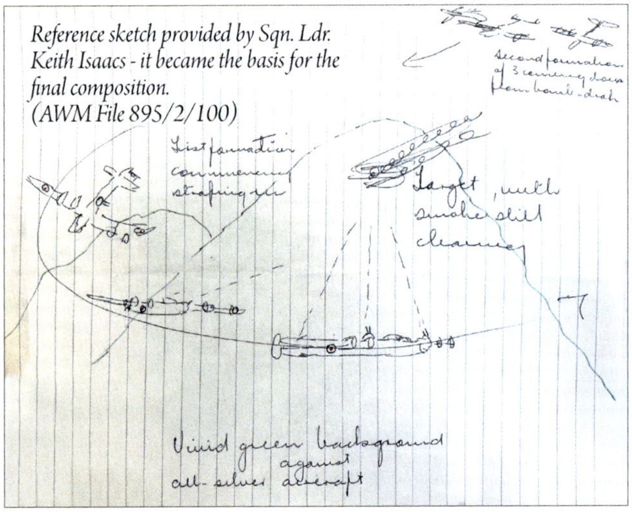

Reference sketch provided by Sqn. Ldr. Keith Isaacs - it became the basis for the final composition. (AWM File 895/2/100)

Lincolns of No. 1 Squadron operating over Malaya during the Emergency, 1957 *(1971)*
Oil on canvas 228 x 182.8 cm (framed)
Australian War Memorial ART27684

The briefing at the Australian War Memorial, Canberra, given by the Director and the Assistant Director was at once comprehensive, supportive and considerate; a welcome change after so many years dealing with the superficiality and often unnecessary pressures so much part of galleries and the advertising industry. Bill Lancaster and Bill Sweeting gave me two days of their undivided time, furnishing me with riches of pertinent reference beyond my wildest dreams of avarice. The painting, in Oils, was to be of 455 Squadron Beaufighters on a North Sea shipping strike in late 1944. Access to the Squadron Record Book enabled us to select a particular raid for depiction, the aircraft involved with their correct codes and serial numbers, the names of their crews, weather on target, heights, convoy make-up and a mountain of black and white strike and camera-gun photographs.

Of course the primary objective was to record a moment in the specific raid that embodied the essence of the almost daily activity carried out by 455 Squadron. We all know how time-consuming, though interesting, research for a painting can be so it is little wonder I remember the support of the AWM staff; the explicit and comprehensive reference, copies of which filled a suitcase and the encouragement. But that was the way the War Memorial operated then, and with all subsequent commissions over the next ten years or so. Nothing was too much trouble to ensure authenticity, accuracy and a contribution to a successful result.

Sir William Dargie, then Chairman of the AWM Gallery Committee, was nominated my mentor, *"available for advice and consultation"*. The calls I made to him have all slipped my memory, except one, during which he seemed to encapsulate the heart of the Australian War Memorial by saying, *"It has to be good art and good history"*.

Probably the most crucial piece of research proved to be contacting one of the pilots who had taken part in the chosen action. The Memorial's records only contained his pre-enlistment address, but I quickly tracked him down because he had called his New South Wales country property by the same name as the base in Scotland from where 455 Squadron had operated. After some initial reticence he seemed to warm to what I was searching for and opened up with considerable candour. His flat unvarnished descriptions, devoid of humorous exaggeration or implied heroics were, nonetheless, extremely revealing and his facial expressions, when he touched nerves in his memory, often revealed as much as his words.

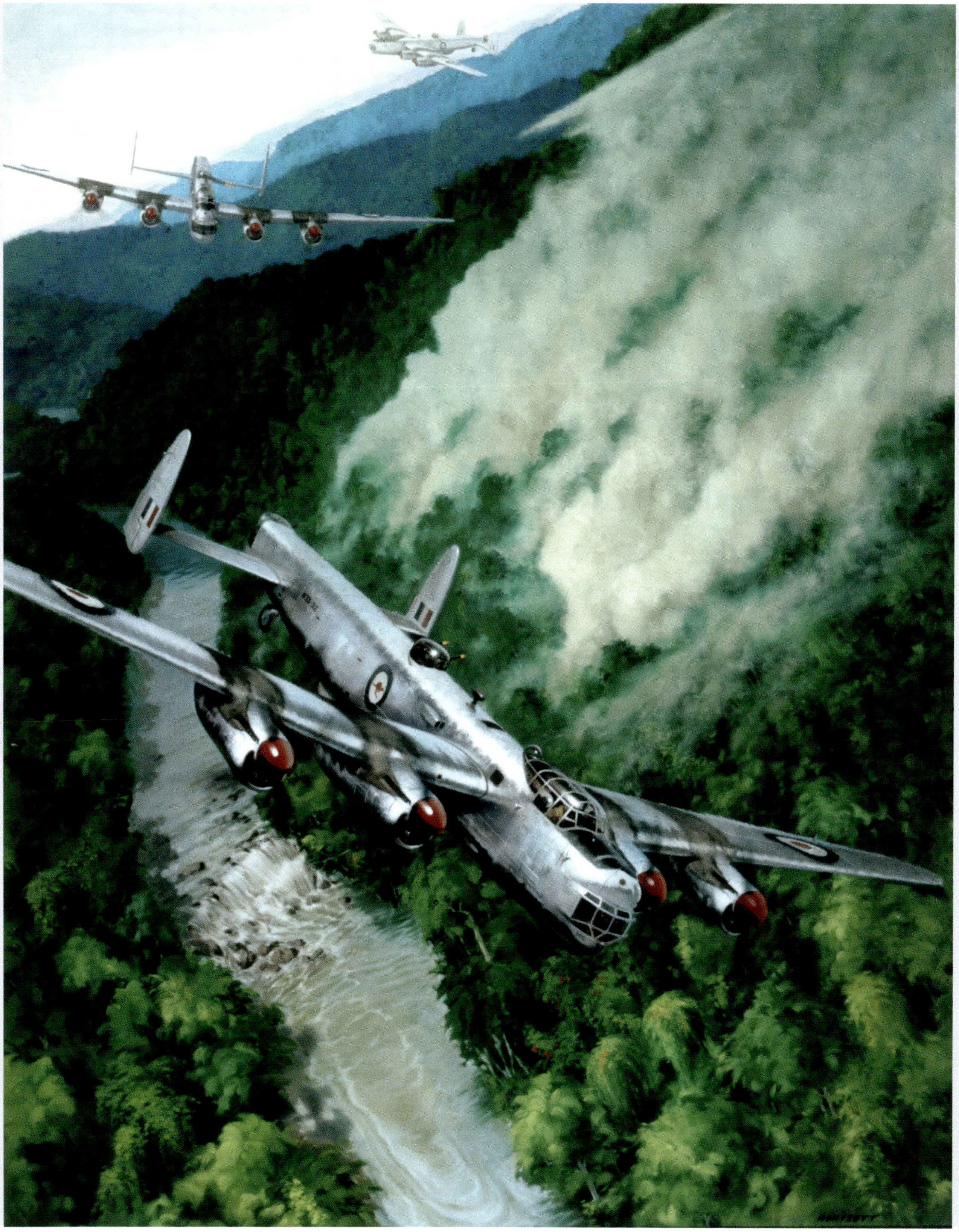

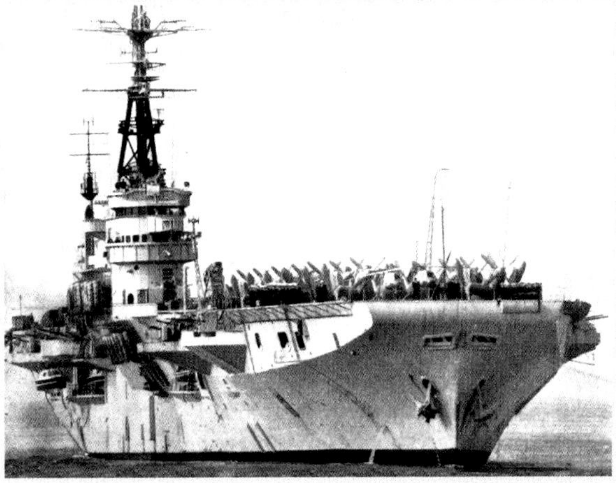

As you will see, the basis of my painting is the shot of Vengeance ,(the same class as Sydney) with the identical hull shape and general similarities up to the deck line. Island superstructure, masts, deck equipment, etc., is according to enclosed general arrangement plan/elevation and those photographs which I could positively verify as having been taken at the time of her Korean tours. All details were double checked from a number of reference sources and personal observations of men who served on Sydney during that period. Similar care was taken with the reference of her visible escorts, HMAS Tobruk and HMCS Sioux.

... the Vengeance photograph ... seemed to embody a feeling of size and solidity which I wanted to convey. To me the essence of the exercise was the fact that Sydney operated with great efficiency and effect in Korea despite extremely bad weather conditions for most of the time. In order to say this, my painting should show, I feel, a massive vessel being thrown about by the greater strength of the sea and the weather; through it all the determination of the crew and flying personnel to mount operational sorties and do their job. These objectives I believe I have achieved. A more conventional angle, showing length rather than concentrated bulk, would not have emphasised another consideration connected with carrier operations; that of the pilots having to take off from and later find and land upon a confined area moving in a very large sea.

Ray Honisett to Noel Flanagan, AWM Director, 14 April 1975
AWM 895/2/120

Thanking him for his time, I came away with an emotional insight and an abundance of detail impossible to glean from cold, printed records, no matter how comprehensive.

The practice of talking and collaborating with central participants became an essential part of my research procedure for all subsequent AWM commissions, most necessary for me as I did not have the background experience of having been a war artist. Fortunately in the sixties and seventies many of those men remained alive with their powers of recall still clear and accurate.

After digesting the surfeit of printed and oral information I submitted a 24" wide, proportionate colour rough in acrylics, its acceptance leaving me with no alternative but to come to grips with what was then an intimidating 45 square feet of virginal white canvas. In my favour was a knowledge of the colours and a feel of the area for, when working in London as a freelance illustrator some years earlier, I had executed painting commissions along the North-West coast of Europe for holiday travel posters.

Now it must be admitted that I am not a fan of articles describing how and why an artist did this or that when painting a picture. They often sound to me like justification or afterthought. In this instance, suffice to say that, as the images evolved and developed, and to my surprise, I seemed able to maintain control, more or less, of the overall painting, the experience became in increasingly enjoyable. Not so for my family, who were forced to endure my working in the loungeroom as my studio was, at the time, too small to be able to stand back far enough to take in a picture of this size. Even in the larger room I often resorted to reversing binoculars to check on overall progress.

In those days my full-time lecturing commitment at RMIT's Faculty of Art absorbed most of the week, consequently the Beaufighter canvas took around a year to complete. After appending my signature, I remember thinking aloud, "Well, the painting is not what I had thought and hoped it would be, but it is as good as I am able to achieve at this time".

The painting was duly delivered to the AWM's Melbourne Receiving Centre at the Exhibition Building, while I tortured myself with imaginings of the criticism and decision that would be coming my way from Canberra. What a relief when two or three days later I received a telephone call from Sir William, enthusiastically congratulating me and ascribing to my painting his earlier words – "It's good art and good history". Oh, boy!

Ray Honisett
Newsletter of the Australian Society of Aviation Artists, March 2009

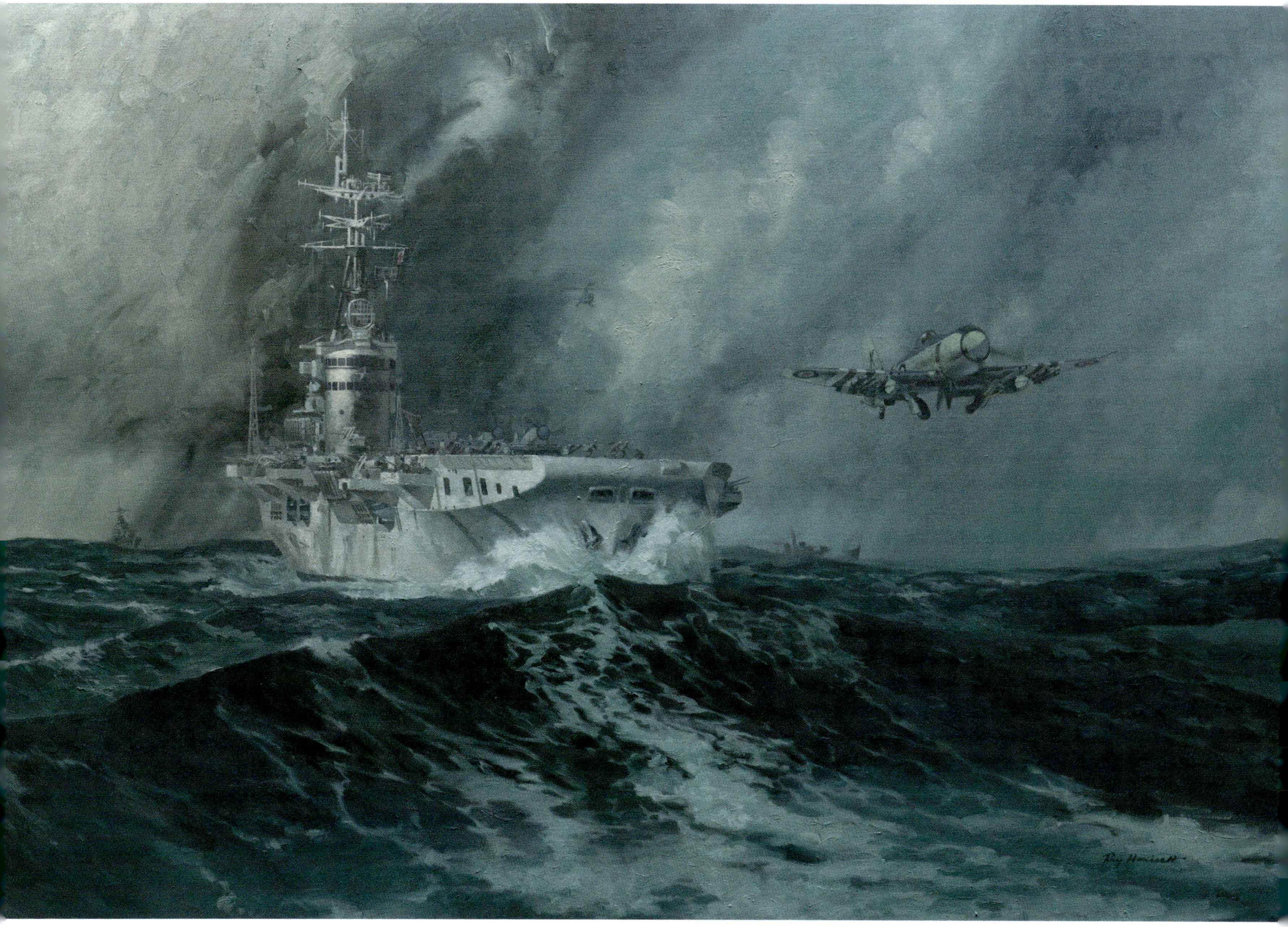

Success with the first two AWM commissions immediately led onto a third, the single subject identified for Korea. *HMAS Sydney in Korean Waters* was a difficult subject to realise and the artist eventually settled for a dramatic head on view of the ship in heavy seas, with a Hawker Sea Fury being launched directly out at the viewer in a moment reminiscent of the Beaufighters painting. At 4' x 6' the picture was somewhat smaller than the earlier paintings, anticipating a restricted display space in the new Korean War gallery. After the strong colours of the Lincolns in Malaya, the Sydney was a much more subdued work, the dark and moody colours perhaps an apt metaphor for a conflict few understood or even remembered.

HMAS Sydney in Korean waters, 1951-52 *(1973-75)*
Oil on canvas, 121.9 cm x 182.9 cm
Australian War Memorial ART28077

Sir Hughie Edwards VC

The Hughie Edwards painting had a rather more difficult road to completion than Ray Honisett's earlier War Memorial commissions. With three significant paintings under his belt, he was in the position to take the initiative and make some suggestions of his own as to possible new subjects. With the 'official list' complete, he proposed a trio of new topics he felt warranted a place in the collection — the three RAAF VC winners not already covered with major works. The Septimus Power painting of McNamara aside, there was only a drawing of Newton VC in his Boston bomber, a small, rather odd painting of the Middleton VC action and a posthumous portrait of Middleton by Harold Freedman. Edwards was the subject of a small sketch portrait by Stella Bowen, done in the UK in 1944. Ray made three sketches on card, which were submitted to the AWM Art Committee in 1977. In earlier times things might have moved on reasonably quickly, but with a new Director, Noel Flanagan, in place and, perhaps more importantly, professional curators now appointed for the collection areas, new commissions were a much more complex procedure. He was informed that the Art Committee 'liked' the Edwards picture, but were less keen on the Newton and Middleton compositions. The concepts were batted back and forth within the Memorial, with the idea of any commission receiving stubborn resistance from the new Curator of Art. The matter became so obscured with argument and counter-argument that it became unclear as to whether the Memorial was negotiating on the purchase of the sketches, or the final paintings. The curator was keen to invest the limited acquisition funds on paintings by significant artists who had worked during the war — individuals like Albert Tucker and Margaret Preston, rather than works by someone seen as only having a 'talent for documentary illustration'. By the middle of 1979, the commission was still not formalised as the curator had made the commission a stand of principle, suggesting that such commissions ... *"make my position here one of great embarrassment and frustration: my peers would assume that I am responsible for the more unfortunate recent commissions".* [AWM File 895/3/105, 26.10.1978] A year later a new Art Curator was in office, by fortunate chance one who happened to be rather more sympathetic to Ray Honisett's way of working. The commission was formalised and the picture was completed, and accepted, in 1982.

Having read several descriptions of the action, my broad concept had been of the aircraft flying very low over Bremen, preferably the dock area, as it was a major port containing several obvious targets. The plan was agreed to during my first meeting with Sir Hughie who was back in Australia at the time. From his logbook and a crystal clear memory he furnished a myriad of essential details: his aircraft's code letters and number, time of day, weather, the disposition of squadron aircraft over target, barrage balloons and his personal feelings at the time – heightened responsibility, fear, and concentration.

With insight from the central character, a wealth of photographs of Blenheim IVs in correct colours and markings from my own library, my research seemed well under way, the only remaining gap being the target itself. Where on earth was I to find accurate photographic reference of the Bremen docks in 1941? In this instance the War Memorial was unable to assist and a cursory search of local libraries yielded next to nothing, certainly nothing with a pictorial component. German wartime publications seemed to have harboured a reticence about displaying close-up pictures of what would have been plum targets. RAF photo-recon shots would be mostly verticals, any obliques would be most unlikely to have been taken from the extremely low level at which the attack took place and the existing few taken by 105 Squadron aircrew were only of details of saw-toothed factory roofs. It began to dawn on me that I may have a substantial reference problem ahead.

I took my problem to long-time friend Arthur Hyland, whom many will remember as having a prodigious knowledge of technical and specialised books and of knowing every enthusiast who had ever entered his store. Arthur was unable to fulfill my need with any current publication or recall any past possibility. However, never without a helpful suggestion, he put me in contact with several groups and individuals whose particular interests were naval and maritime history. None could find anything approaching what I had in mind, although all offered future help with anything falling within their orbit. The offer was later taken up, being of substantial help, but still no joy with Bremen docks.

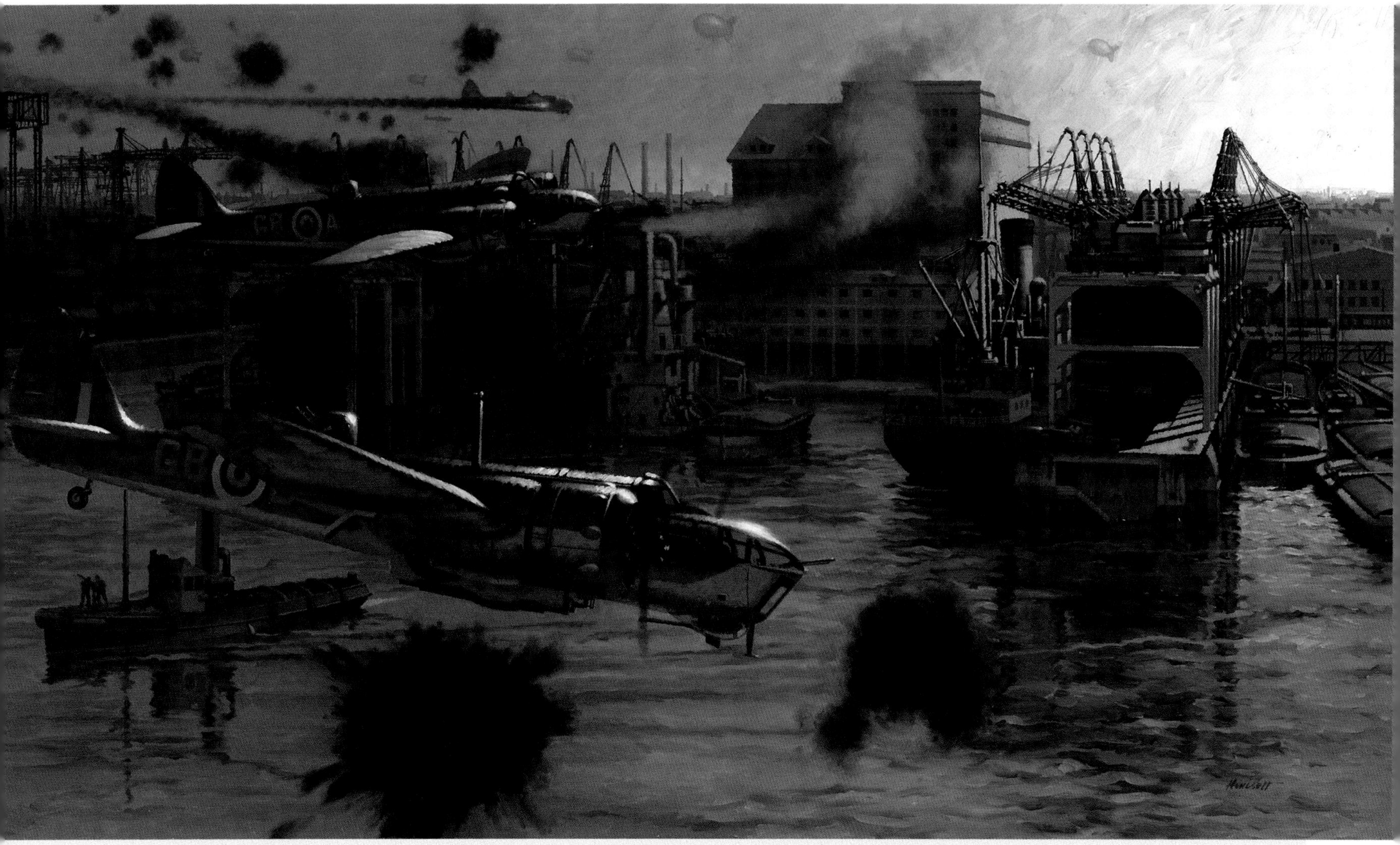

It was time to move from those sources that had seemed likely to those with only remote possibilities. Several times each week, on my way to and from work at RMIT, I would pass the Victorian State Library, often wondering just how many thousands and thousands of books, etc., reposed under that gold dome; how much history and knowledge. Gradually I came to believe if ever I was to find the Bremen reference in Australia it could well be somewhere on those vast shelves or archives. This appeared to be my final source before having to resort to Bremen itself – find some obscure historical society that may have the reference material that I required! This was speculative in the extreme, considering I neither spoke nor read a word of German.

The episode in which Wing Commander Hughie Edwards won his VC *(1982)*
Oil on canvas 74.8 cm x 120.3 cm x 5.7 cm
Australian War Memorial ART28455

Wing Commander Hughie Edwards leading a bombing raid on Bremen Harbour 4 July 1941. Bristol Blenheim light bombers of 105 Squadron can be seen in very low level attack on Bremen Harbour. Edwards was awarded the Victoria Cross and, as Air Commodore Sir Hughie Idwal Edwards, VC, KCMG, CB, DSO, OBE, DFC, was knighted in 1974, the year he was appointed Governor of Western Australia.

I've lost count of the number of hours subsequently spent with patient State librarians who would huddle over computers, then disappear for a time, returning with arm loads of books for my examination. Nothing! The few containing anything pictorial were not even close. It had become embarrassing and I was ready to surrender. But not the librarians – the exercise had become a challenge and they cast their nets in further, more obscure directions, even tapping into the contents of State Libraries in other Australian states. All results were the same until a librarian emerged from the archives carrying not the usual load of books but what appeared to be a thin, about two thirds tabloid-sized newspaper and a broad smile. What I had assumed to be a newspaper was, in fact, a semi-promotional, in-house magazine for the Bremen – Amerika Shipping Line. Promising. The front page carried a photograph of three distinguished looking ships' captains dressed in dark uniforms with many rings of rank on the sleeves. The caption was in German, of course, but, from their stern expressions, I gathered they were about to retire, or be promoted, or to be shot!

Page two, though, was the pot of gold. It carried a half page photograph of the Bremen docks with two foreground crane-encrusted piers, ships, tugs, barges, waterfront buildings and an industrial background. The shot had been taken from the opposite side of the harbour, At a height of around 20 metres – probably from another building or even a crane at the ideal height for my viewpoint. The paper was dated late 1938. It wasn't 1941 but it seemed reasonable to assume such substantial and permanent looking structures would be little changed in three years, even with the transition from peace to war. The librarian was as delighted with the find as was I and accepted my effusive thanks graciously, adding that the paper simply appeared at the top of a pile of uncatalogued pieces awaiting sorting – "all a bit of a mess, really", he commented.

From my clear copy of the precious photograph, the naval and maritime historians confirmed it to be of a Bremen terminal of the Bremen-Amerika Line and even identified some large background gantries as being part of the shipyards of A.G. Weser, shipbuilders and constructors of 160 U-Boats. And, honouring their earlier offer of assistance, the historians supplied helpful advice and photo reference for Bremen tugboats, German cargo ships of the era and a Sperrbrecher that would have preceded any shipping movement down river.

Finally satisfied that all necessary reference was to hand, I felt confident in pulling it together into something resembling my earlier, vague concept. The result was a down-sized study in Acrylics for approval by Sir Hughie and the War Memorial. Sir Hughie examined the study for quite a few minutes before slowly nodding and saying, "Yes, it could well have been like that". With a positive response from the AWM, too, I then executed the final painting, larger, of course, and in Oils but with similar content and feelings of the study. Unfortunately, Sir Hughie Edwards died before the final painting was completed.

In my experience the research and reference gathering are almost as pleasurable as painting, itself. An adventure with a degree of luck added. Steeping myself in the subject first seems to heighten the anticipation of coming to terms with canvas and paint. However, of greater importance is the fact that any (even vaguely realist) painter can only be as good as his or her reference – whatever form that may take!

Ray Honisett
Newsletter of the Australian Society of Aviation Artists
February 2013

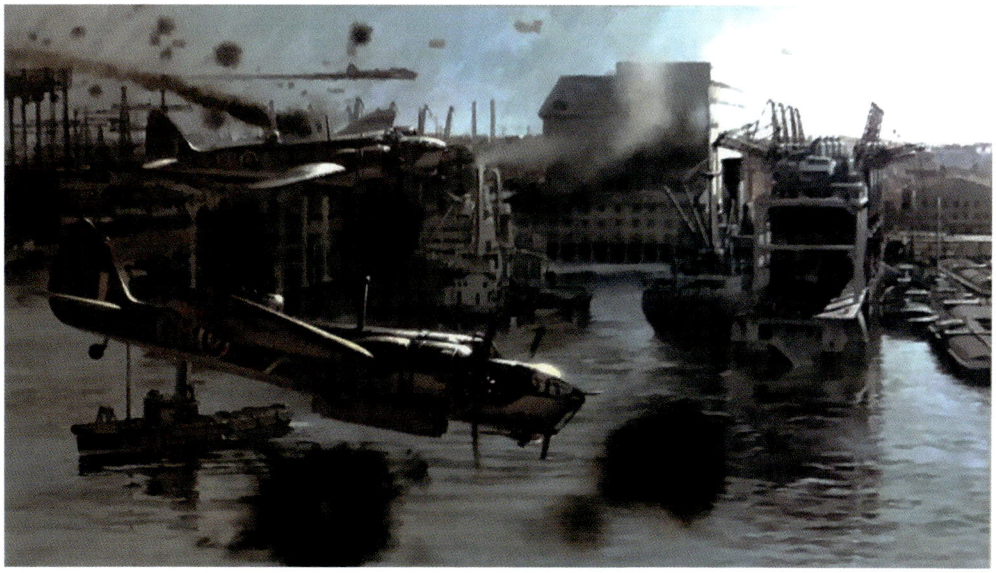

In its original form the sketch for the Hughie Edwards painting was what the artist termed a 'rough' — finished enough to get a good idea of the composition and colours, but not considered a finished work. When the sketch was eventually returned after some years at the Memorial, Ray decided to bring it up to a level of completion that satisfied him as a finished work in its own right. The painting later passed to a friend where it remains a fascinating link to a controversial moment in the artist's career.

In painting this work the artist sought to strike a very delicate balance between two requirements: an expressive and artistically satisfying painting and an accurate and historically verifiable illustration of a specific event. Draughtsmanship of a very high order was required to realize convincingly a complex set of relationships between the action portrayed and the land and seascape background of the work. Colour and tone had both to serve the expressive needs of the work and to clearly explain the technical detail of the subject. Before any preliminary drawings could be commenced considerable research had to be undertaken. Exact details of the colouring and markings of Edwards's No. 105 Squadron RAF Blenheim Mark IV had to be established. Finding accurate photographic reference for the Bremen dock area as it was on the morning of 4 July 1941 was an even harder task. The artist was able to consult with Sir Hughie to establish details of the raid not clearly evident from squadron reports. Weather and lighting conditions over the target were obtained from contemporary meteorological reports as well as from the recollections of the aircrew.

To produce a convincing image of the action and the scene the artist had to combine these historical facts with his imaginative interpretation of the story of the raid. He has visualized the action as though seen from one of the aircraft on the raid, using a high horizon to stress the low-level flying which was crucial to the success of the operation. The drama of the scene has been heightened further by placing the aircraft against the threatening jungle of cranes, dock machinery and ships which fills the upper half of the picture. The lightness and vulnerability of the small aircraft is contrasted with the strength of the target buildings as the aeroplanes fly from the dark flak-filled sky over the shipyards towards safety, represented by the clear early morning light to the right of the picture. As in his fine painting of the aircraft carrier HMAS *Sydney*, Honisett has restricted his palette to greys, blues and dark greens, allowing a very subtle use of tone to build up the complex detail which is such an important part of the painting. This careful control of tone, combined with excellent drawing and a most accurate rendering of detail, has produced a painting of great complexity yet the action is very simply and clearly expressed. As a piece of historical reconstruction it is a great success. It also succeeds admirably as a highly evocative work of art, capturing the spirit and danger of wartime flying. Sir Hughie approved enthusiastically of Honisett's preliminary study for the composition, but unfortunately died just before the final painting was completed. It was unveiled at the Australian War Memorial on the day his medals and decorations were presented by Lady Edwards for inclusion in the Hall of Valour.

Gavin Fry,
Australian War Memorial Journal, No.4 April 1984

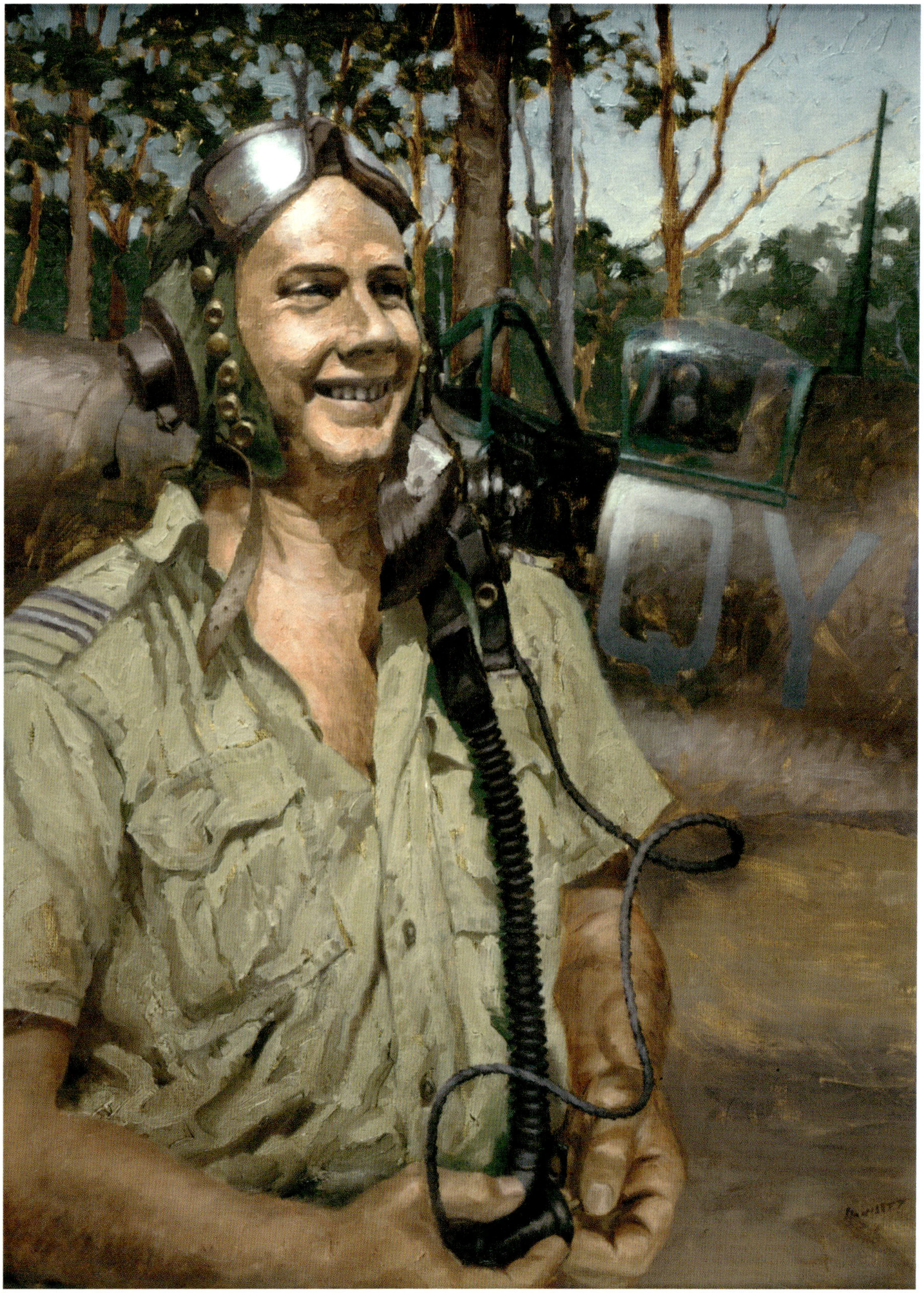

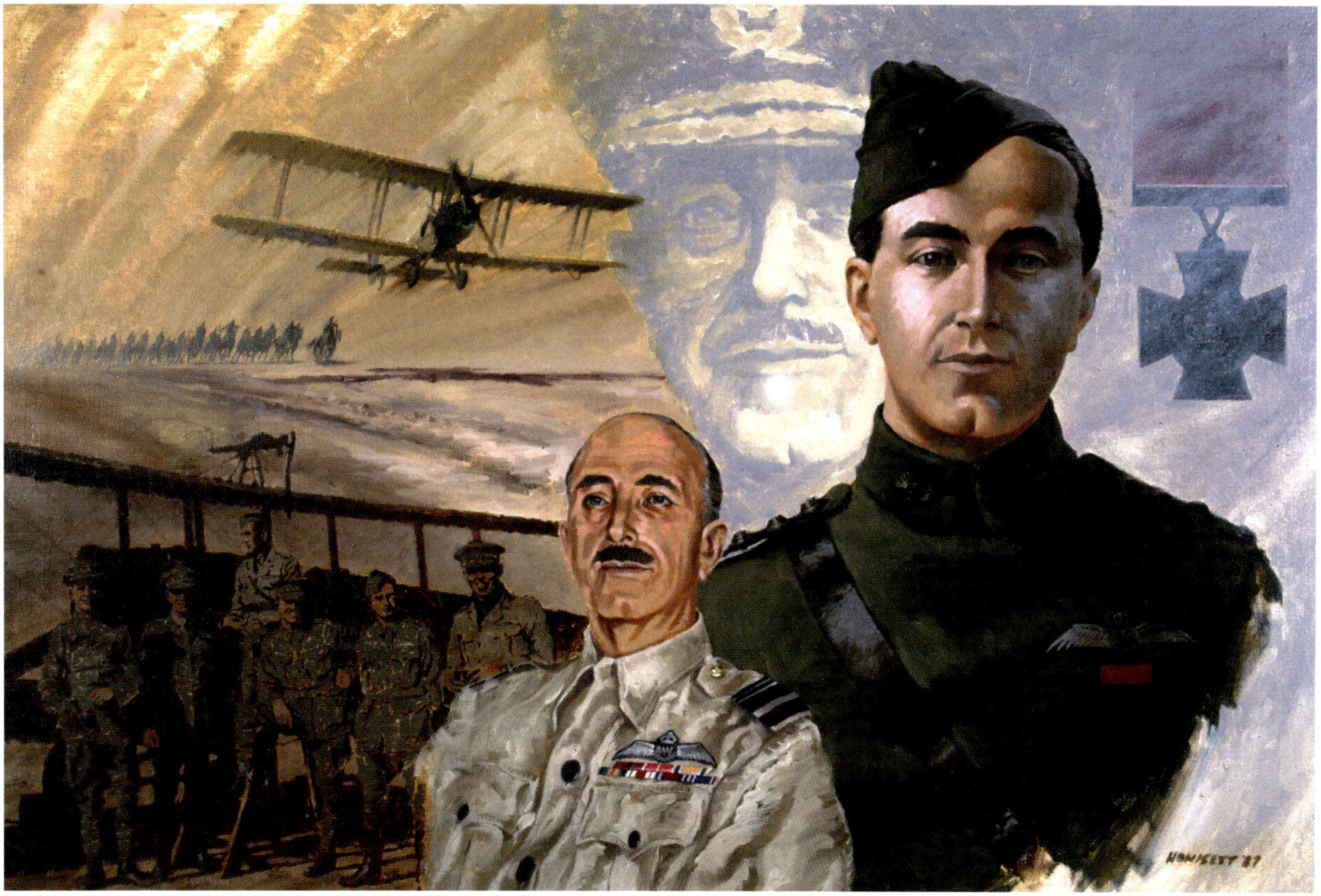

In 1983 the Australian War Memorial's fifth Director, Noel Flanagan, retired and was replaced by recently retired Air Vice Marshall James Flemming. Unsurprisingly, Flemming continued the push for a more equal representation of the RAAF in the Memorial's displays and Ray Honisett immediately became the beneficiary of that enthusiasm. A portrait of Tim Goldsmith DFC was commissioned, along with a new painting commemorating the first Australian Flying Corps VC winner, AVM Frank McNamara. While the Memorial already had a large painting by Septimus Power depicting the incident in which McNamara had won his VC, many thought the painting, despite its size, was somewhat underwhelming and told little of the man. Ray produced a montage showing McNamara at different times of his career as well as graphic images of the action. The work was considerably smaller than the Power painting so it could hang in the new Hall of Valour, opened by Prince Charles in 1981. At a personal level the painting was a stylistic precursor to the Australian sports montages the artist produced in the 1990s.

Ray Honisett, ***Air Vice-Marshal Frank McNamara****,*
1987
Oil on canvas, 80.5 x 122.3 cm
Australian War Memorial ART29094

Opposite, Ray Honisett
Flight-Lieutenant Adrian ('Tim') Goldsmith *1987*
Oil on canvas, 83.2 x 65.9 cm (framed)
Australian War Memorial ART29256

A FEW NOTES ON PAINTING THE SEA

RAY HONISETT

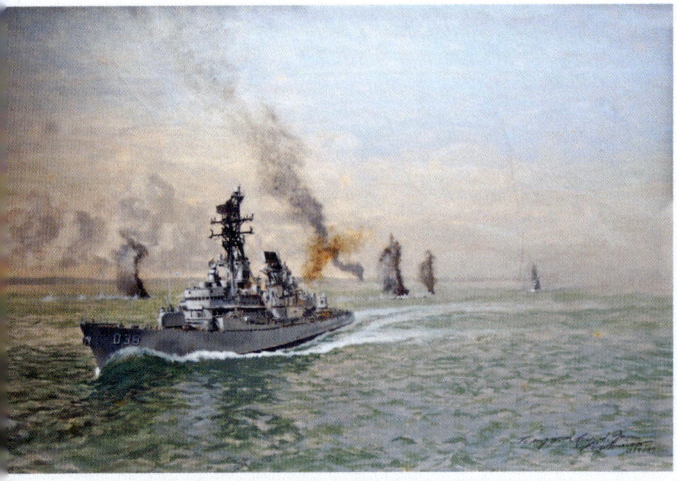

Sketch for HMAS Hobart, 1984
Acrylic on card, 40.3 x 60.5 cm
Private collection, Sydney

Ray Honisett's capacity to produce preliminary sketches of high quality was again seen with the Vietnam commission. A more sympathetic management at the Australian War Memorial meant that the push by some to simply acquire the rough instead of proceeding to the final commission was no longer an issue.

The following notes are the product of understanding gained from more erudite writings on oceanography which explain physical structure of the sea, some comprehension derived from many years of observation and the actual experience gained while attempting to portray the sea's countless moods and continuous restlessness.

Right at the beginning, let me assert what I believe is an unarguable point ... that neither you nor I have any hope of convincingly portraying the sea unless we are prepared to spend time drawing and painting it directly from life; either before it, in it, under and/or over it, in order to try to grasp the sheer vastness and power of it. Searching out a few of the physical 'signposts' that will communicate what we want to show and say ... this is the reality of painting. The following is mere theory.

The most obvious 'signposts' are related to the sea's movement — what I would call the 'short term' surface movement that is manifest in waves and ripples of all sizes, caused by the wind. 'Long term' movement would be larger yet less obvious influences such as tides, currents and perhaps underwater volcanic activity. The later category we will leave for another time.

Across the sea there are few obstructions to the wind, therefore it is almost always disturbing the water's surface. The wind creates a texture of ripples — small, scalloped, wave shaped forms, constantly evident even when there are also large waves. The ripples cover the entire surface of wave and space between and only smooth out when the wave front curves, just prior to breaking.

Waves, too, are the product of wind, wind much stronger than the breezes responsible for localised ripples; wind often of violent storm velocity. While the exact process is not as yet fully understood, it is believed that wind pressure on the surface, as much as across the surface, contributes to the initiation of water movement, which creates waves.

Wave sizes are determined by wind strength at their place of inception, the length of time that the wind has been blowing and the extent of open water across which the wind has blown. This last factor is known as "fetch." When waves move out from under the wind, which gave them birth they are referred to as "swells", a relatively regular surge, which may carry for thousands of kilometres.

It is interesting to note that a wave is not water moving forward but rather a disturbance that progresses in a consistent direction. The disturbance causes the water particles to rotate in an ellipse, carry forward a little, than drop down and back - almost to where they started. Flotsam (and the quicker moving Jetsam) floating on or near the surface will eventually wash ashore — but a far slower rate than the prevailing wave (or disturbance) speed that carries it.

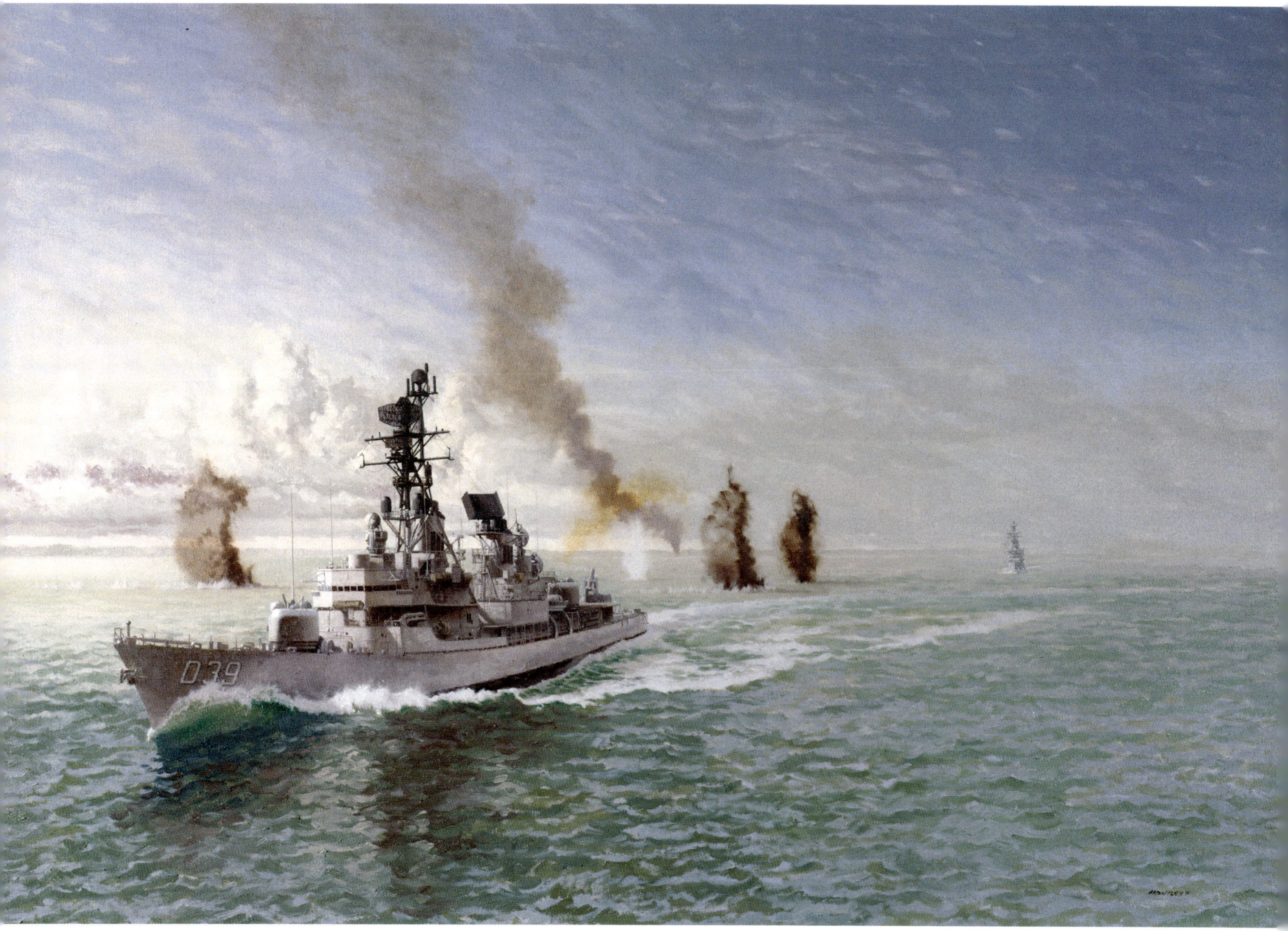

By the mid 1980s sufficient time had passed since the end of the Vietnam War for a more analytical approach to the interpretation of the conflict. There were still considerable tensions between supporters of the traditional 'military' story and those seeking a more rounded view that included some of the social impacts on the home front. One thing lacking in the exhibition was a signature artwork for the Royal Australian Navy and Ray Honisett was commissioned to paint HMAS *Hobart* operating with the US Seventh Fleet off North Vietnam.

Ray Honisett, ***HMAS Hobart with the US Seventh Fleet off Vietnam****, 1968 (1985)*
Oil on canvas, 121.6 x 182.5 cm
Australian War Memorial ART45009

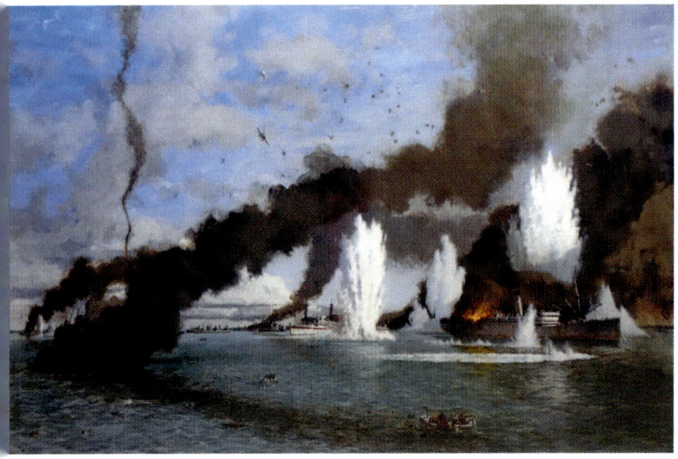

Ray Honisett, Sketch for the attack on Darwin, 1942 (1983)
Acrylic on card, 39.5 x 60 cm
Private collection

The Raid on Darwin, 19 February 1942

The institutional motivations for awarding a new commission are many and varied. In the case of Ray Honisett's painting of the Japanese air raids on Darwin in 1942 it was, as in the case of the McNamara commission, a feeling of unease about the way the event was represented in earlier works. The approach of the 40th anniversary of the Darwin raid led to an examination of the Memorial's current holdings. A large painting of the raid had been acquired from Queensland artist Keith Swain in 1975, but that work had few friends among the professional staff of the organisation. It was seen as overblown, almost a comic-book rendition, with strange changes of scale and perspective that produced an unconvincing impression of the action. One of the attendants, a veteran of the action, would often quip as he passed the picture — 'glad I wasn't there when that was going on!' It was decided at Council level to acquire a more appropriate rendition to mark the event and Ray Honisett was again the chosen artist. He produced a highly finished sketch for final approval and the completed work was delivered in 1984.

Wave movement is perhaps best understood by creating an imaginary model, as follows:

(1) If we were to cut a small cardboard mailing tube in half (lengthways) and place the resultant semi-circular half-turn ends down on a bed sheet, then move our half-tube across the sheet in the direction that resembled a wave, we would be stimulating wave movement as most of us initially perceive it.

(2) BUT, if we place an uncut mailing tube on the bed sheet and then placed a top sheet across the bed and tube, rolling the tube in a constant direction, our tube becomes a disturbance beneath the top sheet (or surface) lifting it as the tube rolls along, thereby creating a more accurate simulation of what actually takes place.

While waves at sea are less regular than those produced by a passing ship, they do have loose patterns and a constancy, which is generally maintained until shallow water, usually associated with a coastline, is encountered. Waves travel freely as long as the 'wave length' (i.e. the distance between crests) remains less than double the depth of the water. When a wave of 100 metres length reaches water of 50 metre depth (measured from the crest to the sea bed) it is slowed by the drag of the bottom, the circular moving water particles are squeezed together by the following crests crowding up behind. Also, the back of the wave travelling faster than the front, overtakes it forcing it to rise in a peak. Finally, the peak, still moving, leans further and further forward until the crest eventually topples over … and our wave breaks. It breaks when the peak height reaches a ratio of 3:4 to the water depth, i.e. 3-metre wave in 4 metres of water, (measured from surface to bottom).

What we call "white horses", i.e. waves breaking well away from the shore, are the effect of local wind pushing the surface into peaks until they overbalance. But for the most part the wind simply helps the waves along.

Waves, then, obviously come in an almost infinite variety of shapes and sizes, dependent upon the strength of the local wind, the conditions prevailing at their birthplace, how far they have travelled and the depth and topography of the sea-bed. While it is true that in any one area of ocean it is possible to experience all manner of seas, it is equally true that areas of ocean have their own characteristics, as do areas of countryside. They are recognisable to those who frequent them. The formation of seas in the Southern Ocean, for instance, is discernibly different from those in the North Atlantic, even when weather conditions are similar. The characteristics are more obvious the closer one is to a coastline. But even hundreds of miles from land most 'salts' know what ocean they are on.

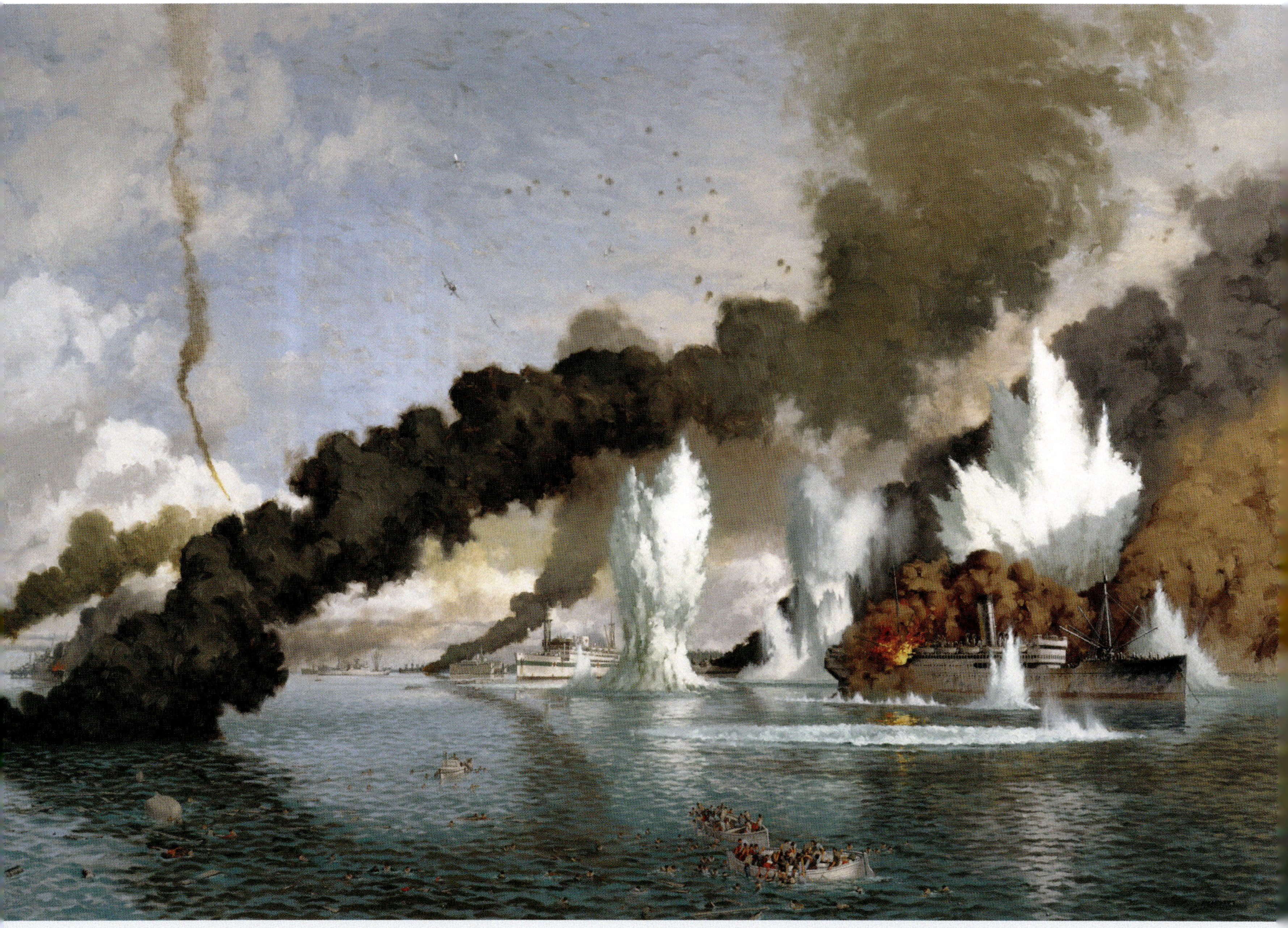

Ray Honisett, ***Darwin, 19 February 1942****, (1984)*
Oil on canvas 122 cm x 182.8 cm
Australian War Memorial

The Battle of the Coral Sea Triptych 1992

The major actions, contained in the two large canvases, are separated to symbolise the opposing fleets being distanced from each other. They reflect the contrast of attitudes at the time — the desperate attack by the U.S. Navy dive bombers on the Shoho as against the steady, confident approach of the Japanese torpedo bombers towards the Lexington and Yorktown. The middle panel, showing the Hobart, Australia and Chicago combing the torpedo tracks from the Japanese land-based aircraft, symbolises the beginning of the Australia-US alliance and is also central to the action in that the cruisers are moving to cut off the actual invasion fleet. Overall, the balance of the large pictures either side of the pivot is to be consistent with the balanced outcome of the action whereby the Japanese probably inflicted greater material damage but the Allies achieved their prime purpose of preventing the Port Moresby landings.

Ray Honisett, 10 January 1992

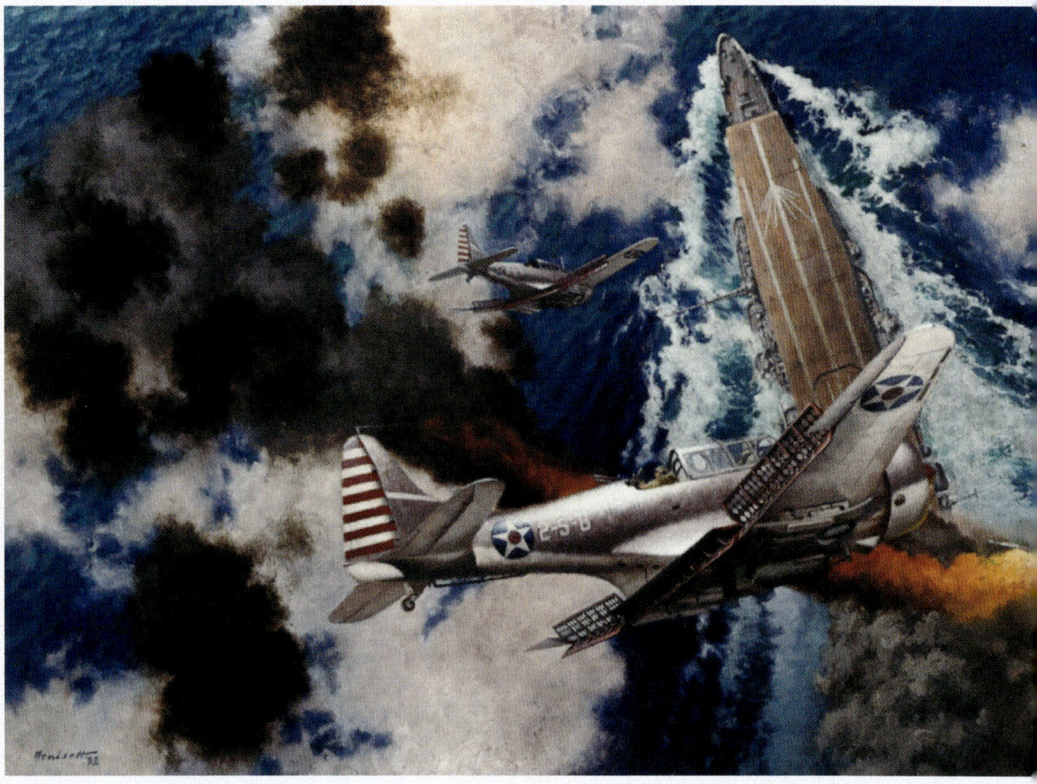

Whatever one's proximity to the sea it is being affected by wind, tides, currents, prevailing weather or sudden squalls. It is seldom without some movement, be it subtle or downright awe-inspiring. When we paint the sea, its continuous, restless, motion is an essential quality.

It is impossible to consider the sea without simultaneously considering the sky, which is to say, the weather. The sea's form, its colours, its very mood are largely determined by the condition of the sky above it. While it is possible for a strong sea to be running beneath a cloudless sky, most often the link between sea and sky conditions is patently obvious. Nature is seldom anomalous and never in visual conflict — contrast, but never conflict.

In my own paintings I always work into the sky first, to establish light, weather, temperature, time of day, etc. If the picture contains sea (or any water, for that matter) then the predominant sky colours are continued into it to form the key under-colours. Whatever form the water may eventually take, the chromatic link is logical, as it takes account of both the painting's overall cohesion and the fact that the sea's surface will reflect the sky.

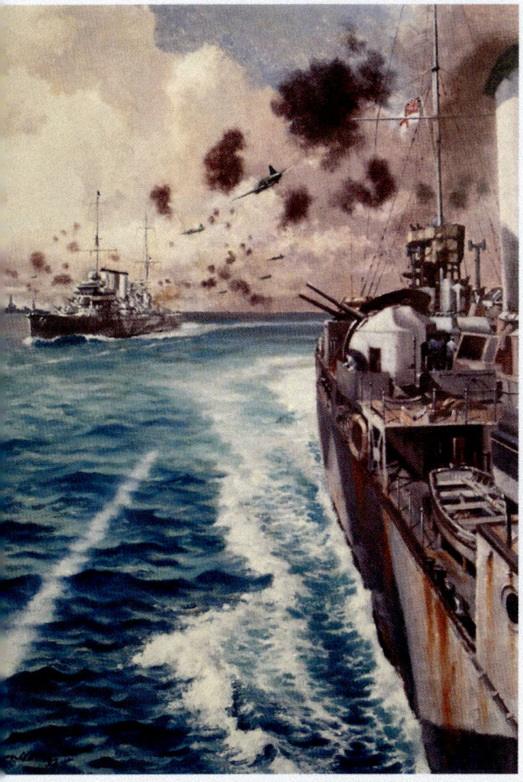

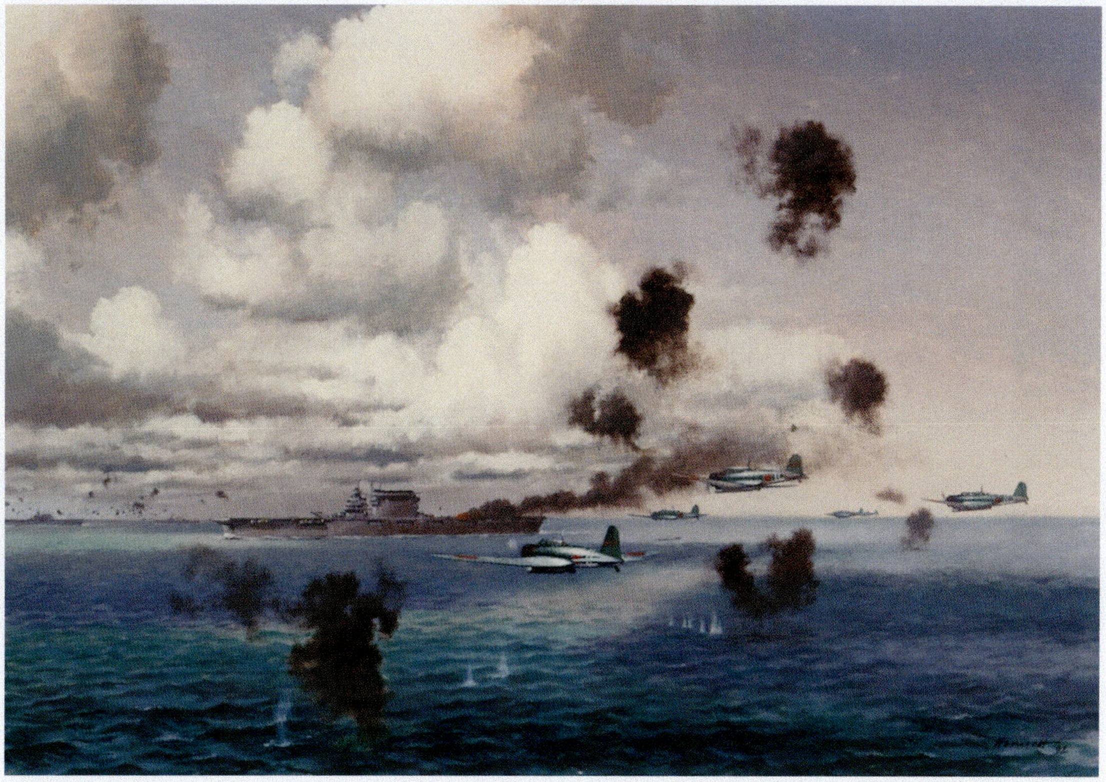

This brings us to the point where we should examine the question of colour.

Setting aside for the moment the whole fascinating area of reflected light and colour, the sea itself when viewed directly and en masse does have its own definitive colours and different bodies of water are reputed to be of a predominant colour. Through time we have heard much of the 'Blue Mediterranean' and the 'Blue Pacific'. The Persian Gulf waters are referred to as a 'beautiful green' while those of the northern Indian Ocean are reportedly black. In fact, a whole range of colours including white, sapphire, blue, olive green, red, purple, etc., are said to be known in various oceans. Water depth, level of salt content, type of bottom and microorganisms are obvious factors contributing to these localised colour saturations. Whatever the accuracy of these descriptions, they provide yet another valid reason for having a close look at the body of water you wish to depict.

If one must use a formula, probably the only one that can be used with reasonable safety is that deep water is darker than shallow water, often tending to progress (on a fine day) from dark blue through turquoise to light emerald when the influence of sandy beach colours move in under the water's edge.

*Ray Honisett, **Battle of the Coral Sea**, 1992*
Left, (US Dive bombers attack the Shoho) 1992
Oil on canvas, 83 x 121 cms

Centre, (HMAS Hobart, HMAS Australia and USS Chicago, Coral Sea), 1992
Oil on canvas, 83 x 60 cms

Right, (USS Yorktown and USS Lexington under attack from Japanese torpedo bombers) 1992
Oil on canvas , 83 x 121 cms

In this we assume a viewpoint on or near the surface, looking into the water at a fairly acute angle. As we lift out gaze beyond 45 degrees we become progressively aware of the reflection of the sky colours bouncing off the surface of the sea. Our vision is moving from looking into the water, to looking across it.

Reflections of the sky are only apparent on the flat or near flat surfaces such as areas between waves and ripples and the top facets of waves and wavelets. As we look at the water between 45 degrees and the horizon the slivers of sky reflection tend to stack progressive closer together, so that the chromatic effect becomes more that of sky influence than sea colour influence. Wherever the water builds up to become waves the sea colours intrude once again because we are looking with the water, seeing it without the overlay of reflection.

Along the horizon itself there tends to be darkening of the sea; just a strip, grading back towards the viewer, often with a purple or violet tinge – especially on a sunny day. This darker horizon often presents a problem in that it contrasts with the usually lighter sky, disturbing the recession of distance and cutting the painting. One must find ways of breaking this intrusive line – birds, boats, masts or foreground details.

Painters are frequently drawn to depicting the sea where it meets the shore, for this is where occurs most activity and variation of colour, form and texture. Fortunately, this is also the most convenient place to set up for observation and painting. Because the sea is so interrelated to the sky and the sky itself is of such prime concern to the aviation artist, it is natural that may of us also gravitate towards seascapes. But, as with the land, significant changes occur to the appearance of the sea as our viewpoint lifts above the surface to any great degree. Distance simplifies formal detail; tone and colour variations are brought closer together and a smoothing out becomes evident. Sky reflections remain obvious, if more generalised.

Although it is amazing how much surface texture one is able to discern from heights of thousands of metres, I quickly reach a point (or height) at which I am not prepared to attempt the depiction of the sea's surface texture, which often resembles a pattern similar to that of coarse hide. It seems almost geometric in its regularity and I void it, preferring to lay in convincing colour with short, self-texturing strokes that imply an underlying order and sense of movement.

When dealing with ingredients that impinge on the sea it is important to ensure that they do not end at the surface or float on it - a ship or flying boat should be painted in to include that which is below the surface. Even if this part is later obliterated by the surface itself, such subjects float in the sea, not on it. Similarly, land — rocks, beaches or whole islands should subtly seem to be continuing on below the surface all the way to the bottom. Aircraft, in flight over water, are a much easier task in that the generally cool and dark colours of the sea provide a satisfactorily backdrop to their warmer and/or lighter colours and highlights, thus pulling them of the surface.

Considering their vastness, it is perhaps puzzling that both the sea and the sky have received comparatively casual attention from painters of the past and present. Few have painted either with skill, understanding or emotion. Of those who have succeeded, Monet, Sisley and Boudin are undoubtedly worth studying in terms of restricted water – rivers harbours, lakes, etc – while Norman Wilkinson, the English naval war artist and Winslow Homer both created a great affinity between paint and the sea. It is interesting and illuminating to note that all five of the above-mentioned painters worked extensively from life, feeling the pull of the subject at very close quarters.

What becomes patently obvious when we look at so many maritime paintings executed down the years is that artists know their ships (be they sail, steam, large or small) inside out, but the actual environment in which the ships lived is so often put down with guesswork, clouded memory and a deplorable lack of sympathy and understanding. The same can easily be happen to the aviation artist. Beware!

On a not too distant date we could take a closer look at smaller, more enclosed bodies of water. These have different structures, patterns and influences than those of the sea, often carrying a great deal more reflection and shadow and presenting different problems. However, as is easily understandable, these have been worked through by more people and with greater success, so we have many, many more good examples to study and be inspired by.

Ray Honisett
Newsletter of the Australian Society of Aviation Artists,
June/July 1995

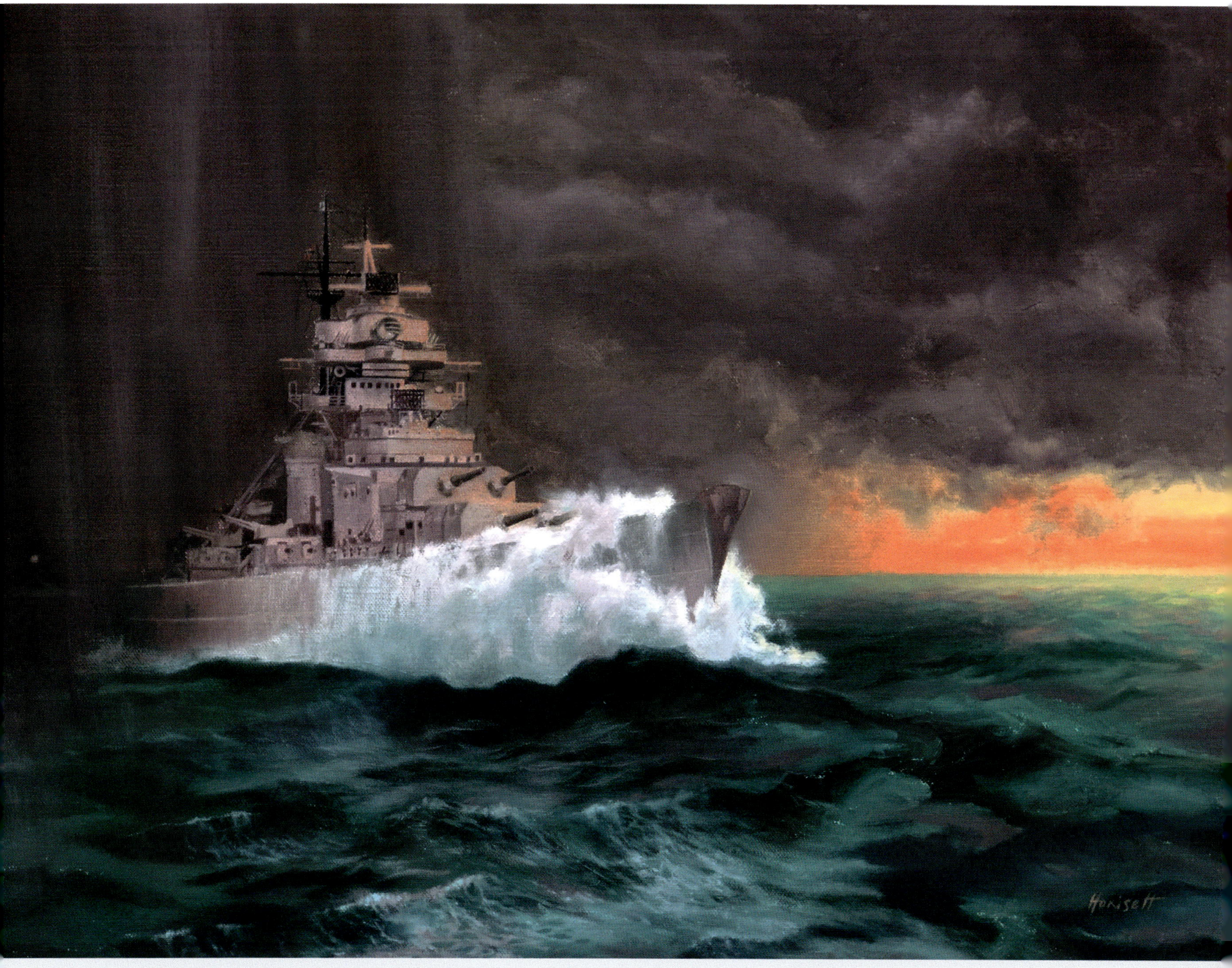

The Bismark; Denmark Strait, 23 May, 1941 (2016)
Oil on canvas, 65 x 97 cms
Private collection, Melbourne

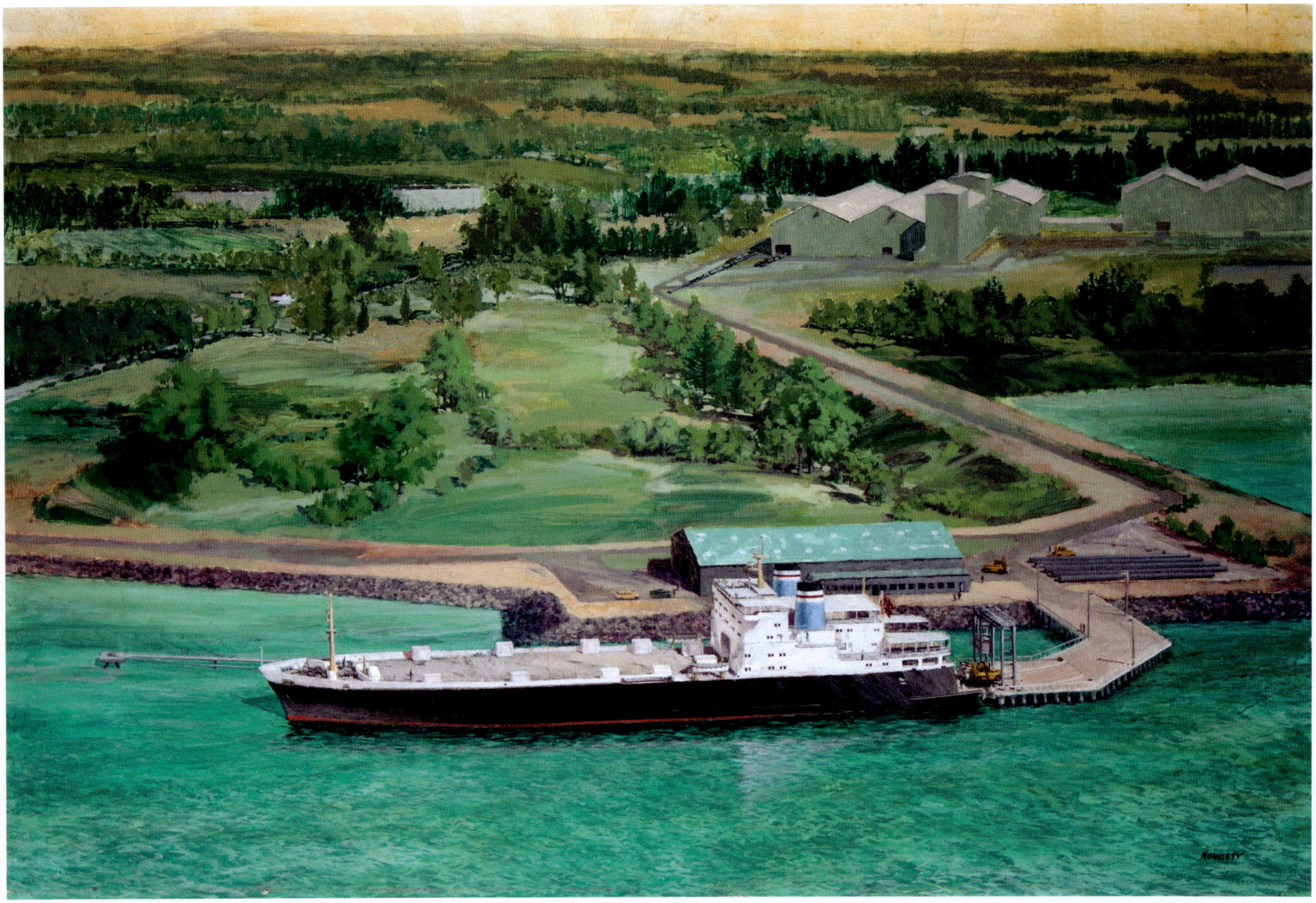

Ray Honisett, (Lysaght Endeavour loading at Hastings) c.1995
Acrylic on hardboard, 61 x 91.5 cm
Collection of the Mission to Seafarers Victoria, Melbourne

Lysaght Endeavour and its identical sister-ship Lysaght Enterprise were built in Newcastle in 1973 specifically to serve the regional steel industry on the route Port Kembla - Westernport - Adelaide. Even after being lengthened by 17 metres, the two ships had remarkably short working lives, with both being broken up by 1988.

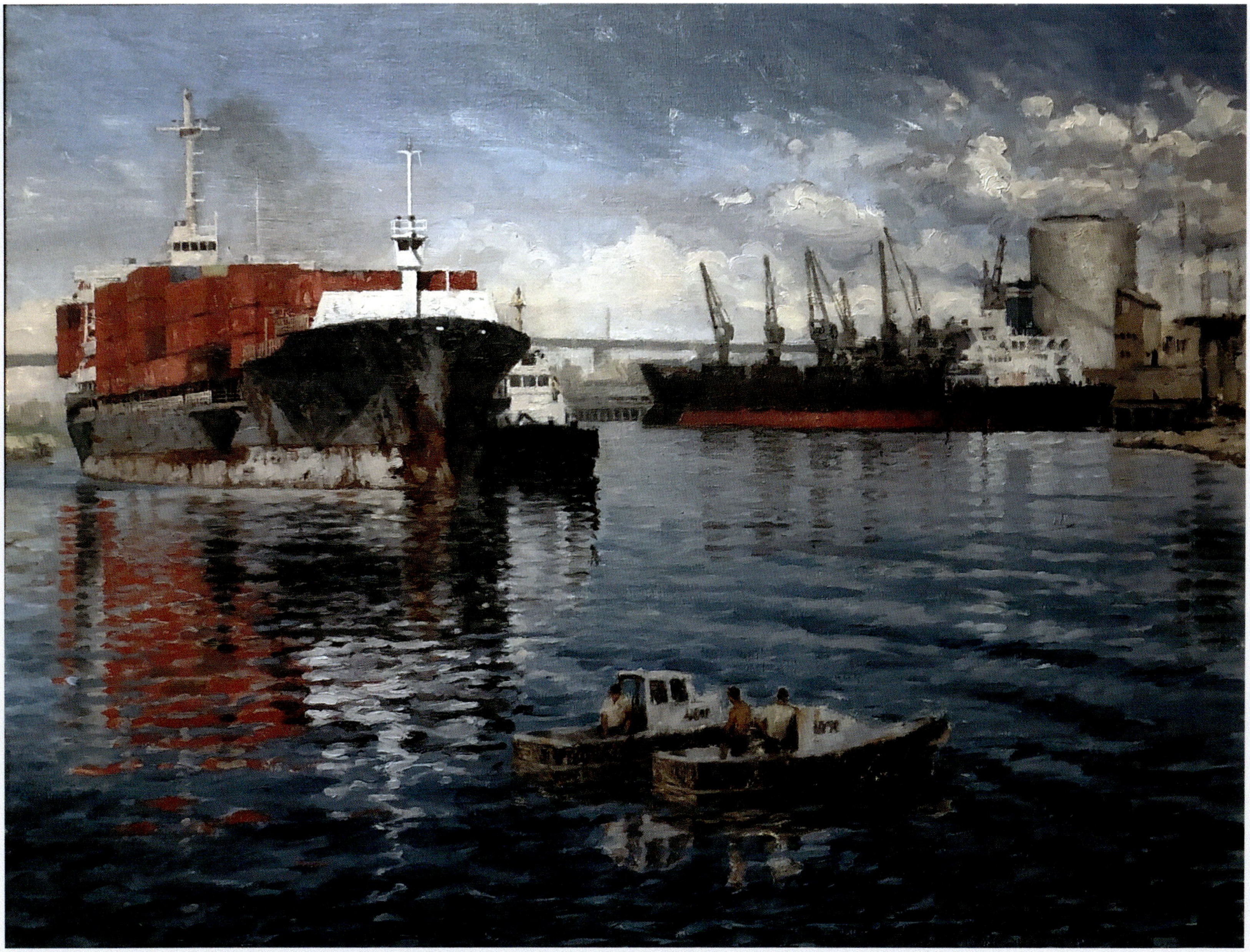

Ray Honisett, ***Autumn Morning,*** *c.2000*
Oil on canvas
Private collection, Melbourne

AUSTRALIAN NATIONAL MARITIME MUSEUM

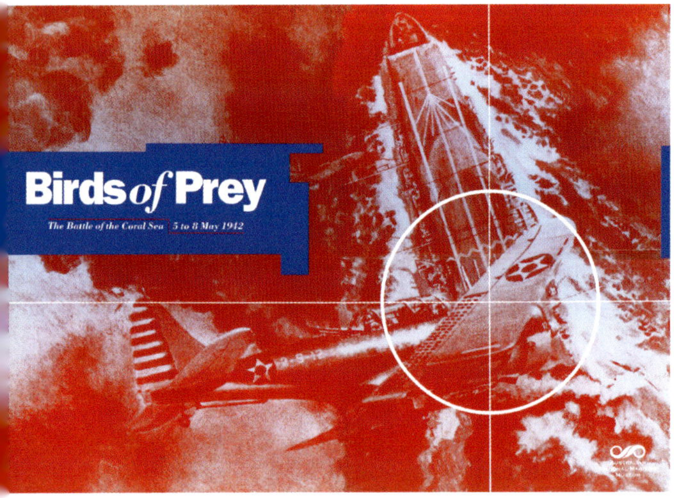

Promotional brochure for the 40th Anniversary Battle of the Coral Sea Conference at the Australian National Maritime Museum in May of 1992. The artist later completed two more charcoal sketches for a planned Coral Sea triptych.

The new Australian National Maritime Museum in Sydney was a prominent marker for the Australian Bicentennial in 1988. While initially it was to follow on from the establishment of the National Museum of Australia in Canberra, internal politics within the Hawke Government saw it become the priority for development. New South Wales Premier Neville Wran was also President of the Australian Labor Party and pressed hard for a major Commonwealth contribution to the state government's flagship development at Darling Harbour. While originally slated for the large finger wharf at Woolloomooloo Bay, a last minute decision was made to locate the museum at Darling Harbour on a site immediately to the north of the old Pyrmont swing bridge. The project commenced in 1986, initially under the Council of the National Museum of Australia. The Commonwealth committed over $100 million to the project and the museum itself would become the beneficiary of a number of international gifts, including restored vintage yachts from Norway and New Zealand.

One of the more controversial gifts was a cash donation of $US5 million from the United States of America, tied to the development of a special gallery documenting the maritime relationship between Australia and the USA. While there was a good story to tell, there was precious little in the way of major museum artefacts to support the development of the exhibition. With a number of AWM staff having made the move to the new museum it was not surprising that Ray Honisett's name would be put forward when it was decided to commission some appropriate works to interpret the naval relationship between the two countries. Vietnam was a necessary inclusion and an incident was selected where Australians came to the aid of a US Navy ship.

On 29 July 1967, a fire broke out on board the aircraft carrier USS Forrestal after an electrical anomaly caused a Zuni rocket on a F-4B Phantom to fire, striking an external fuel tank of an A-4 Skyhawk. The flammable jet fuel spilled across the flight deck, ignited, and triggered a chain-reaction of explosions that killed 134 sailors and injured 161. At the time, Forrestal was engaged in combat operations in the Gulf of Tonkin. HMAS Hobart, then operating with the US fleet, provided fire-fighting equipment and transferred Surgeon Lieutenant L. Barnett, RAN to the carrier to help with the care of the injured.

A second commission to produce a work for the celebrations for the 40th anniversary of the Battle of the Coral Sea was commenced, but completed elsewhere. Ray produced a fine charcoal sketch of an attack on a Japanese carrier, but the painting was not proceeded with at the time. The drawing was used on the program and publicity material for a major national conference on the Coral Sea, while the final painting was later completed for the Naval and Military Club in Melbourne as part of a triptych celebrating RAN and US actions in the Battle of the Coral Sea. *(pages 56-57)*

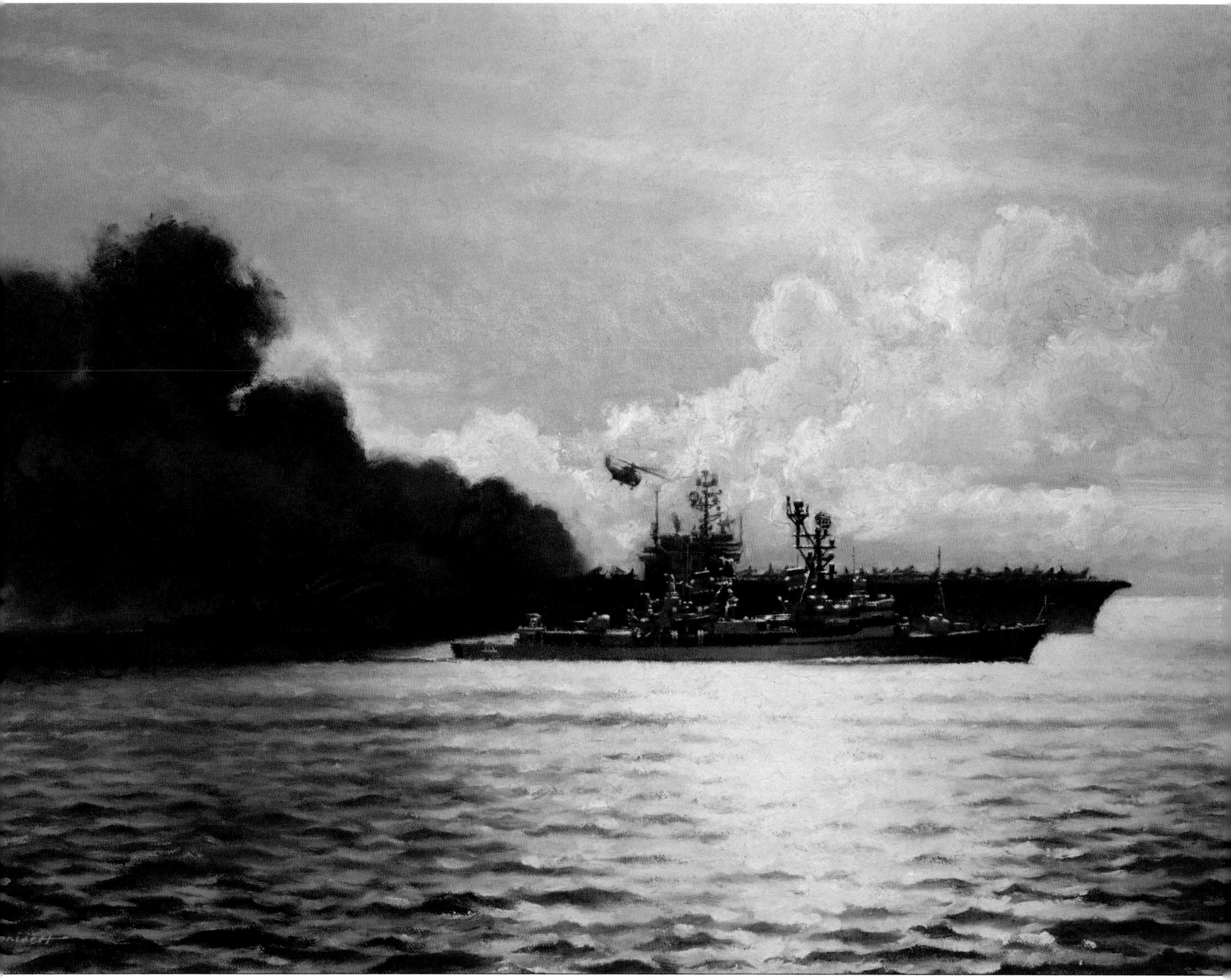

*Ray Honisett, **HMAS Hobart and USS Forrestal** (1991)*
Oil on canvas 60 x 80.cm
ANMM Collection Purchased with USA Bicentennial Gift funds

A CHANGE OF SCALE

THE PHILATELIC WORLD

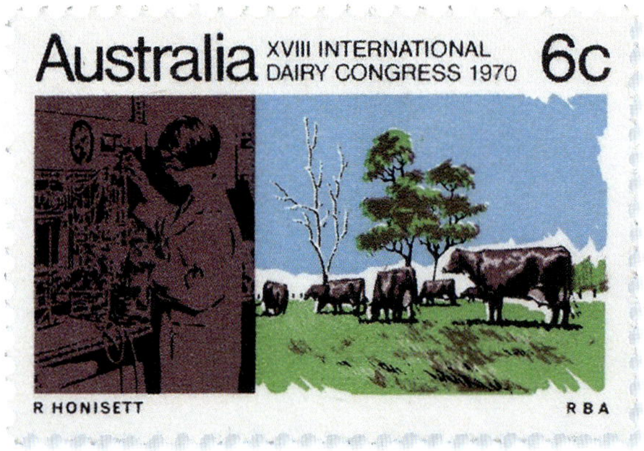

Above. 7 October 1970 - 18th International Dairy Congress

Below, First Day Cover for Dairy Congress, 1970

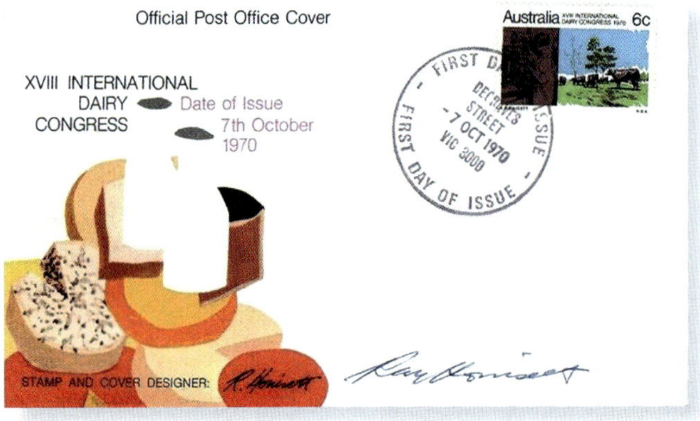

The initial success of Ray Honisett's Australian War Memorial commissions in 1969 did not go unnoticed in the world of illustration and commercial art. Even before the second painting was complete he received another commission that would lead to an important body of work created over the next twenty years. In the Beaufighter commission he had proved he could work at a grand scale, the largest painting he had ever attempted. One year later he was offered a commission for something at the opposite end of the creative spectrum, the design of a stamp for the Australian Post Office.

Until the 1960s all stamps had been printed by the traditional intaglio process, restricting designs to a single colour and images that could be realised using finely engraved lines and dots to produce a half tone pattern. The only exception to that constraint was the special commemorative issue for the Melbourne Olympic Games of 1956, which featured 1 and 2 Shilling stamps featuring coloured photographs of the city. These stamps were printed in London and Switzerland, making them both highly collectible and also a clear marker for the way ahead in stamp design and production. In 1962 the change was made to photogravure stamp printing, allowing multicolour printing and photographic rather than engraved imagery. Because of the high technical specification required most of the first four colour stamps were printed by the Note Printing Branch of the Reserve Bank of Australia — identifiable by the RBA initials opposite the artist's name at the base of the stamp. There were still tight restrictions in what could be reproduced and designers were constrained by the size of the finished item. Type had to remain legible at an extremely small point size as the artist's name below the image is less than 1 millimetre in height and detail had to be sufficiently clear not to smudge or blur in the printing process. Ray's broad experience as a designer and artist in the UK came into play and he quickly made a specialty of the philatelic discipline.

Given Ray Honisett's prowess in the field of aircraft illustration is was perhaps unsurprising that his first commission would involve the depiction of aircraft. The Australian Post Office had for some time produced stamps for other jurisdictions, most notably the newly independent Pacific Islands that did not possess the sophisticated printing technologies required for stamp and bank note printing. Ray was commissioned to design a set of four stamps for Samoa, the first of the South Pacific islands to gain independence, having come out of New Zealand administration in 1962. The stamps showed four aircraft of great importance in modernising communications with the tiny island nation — the Douglas DC3, Britten Norman Islander, Samoa

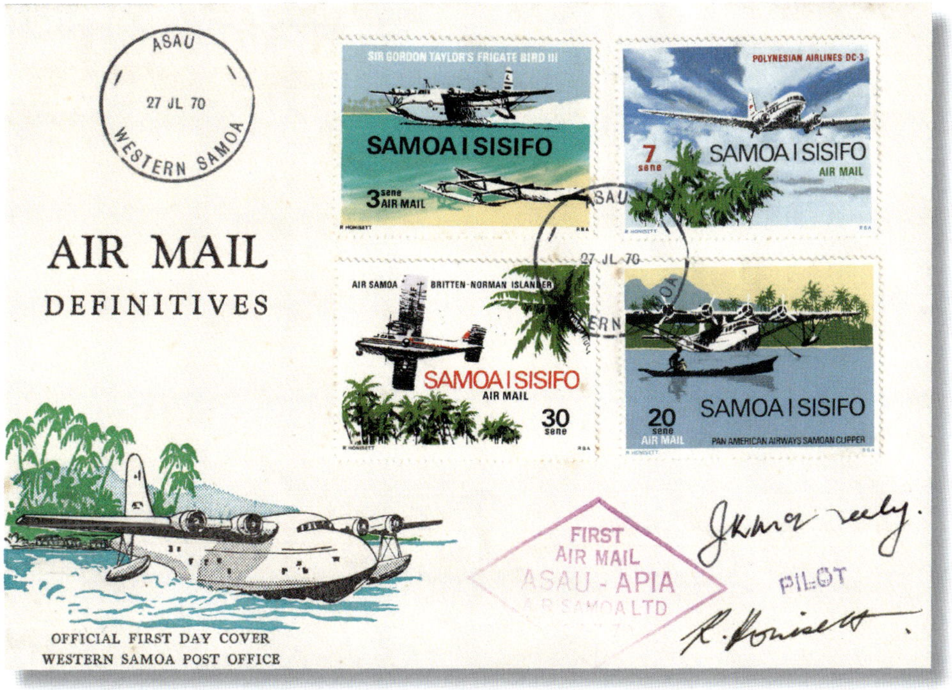

First Day Cover for the Samoan aircraft stamps, 27 July 1970. The production of colourful first day covers became an increasing lucrative market for Post Offices in many jurisdictions. They also gave the artists scope to enlarge upon the designs incorporated in the stamps themselves. This example was given added value by the signatures of the artist and the pilot of the aircraft that carried the officially stamped covers back from Samoa.

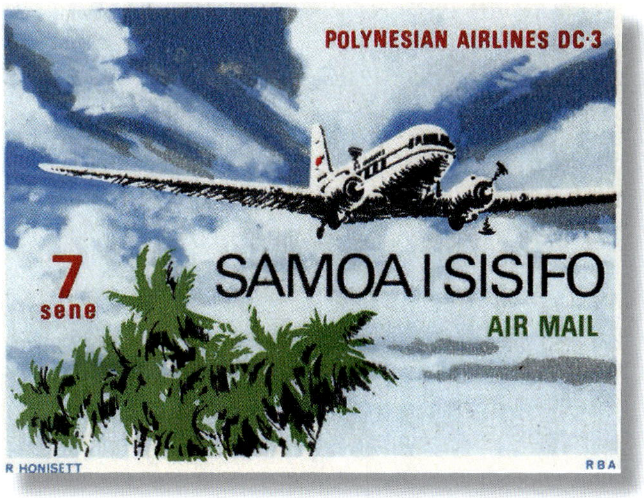

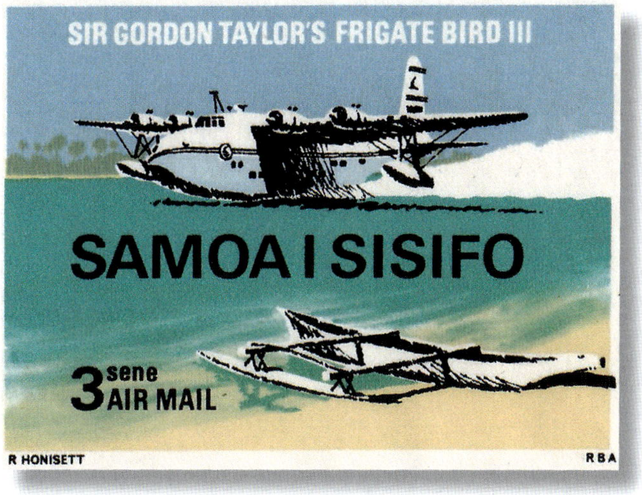

Clipper and the Short Empire Flying Boat. The design exploited the colour printing possibilities of the gravure process with bold graphic renditions of the four aircraft, each seen in an island context with palm trees and azure seas.

Ray's first stamp for local consumption followed just a few months later. In October 1970 Australia played host to the eighteenth International Dairy Congress in Sydney. The event was considered sufficiently prestigious to warrant a special stamp issue, still a relatively unusual event at that time. He chose to design the image in two halves, one showing the modern scientific advances in dairy technology while the other depicted dairy cows in a paddock – significantly he chose the Illawarra Shorthorn, the only dairy breed developed in Australia. The tone drop-out photographic technique had been highly successful in the Samoan stamps and was employed to good effect again in the dairy design.

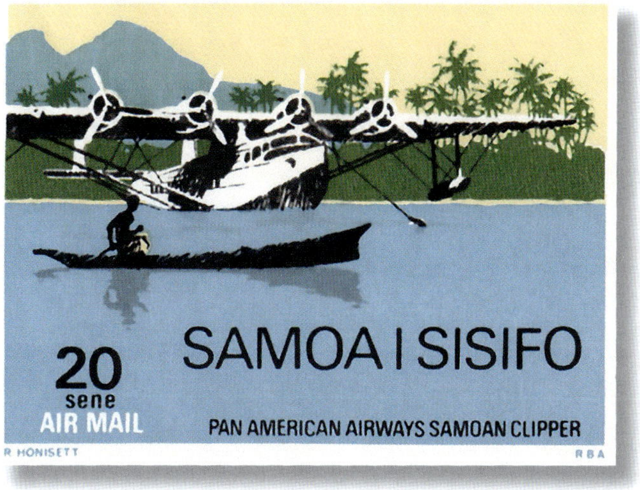

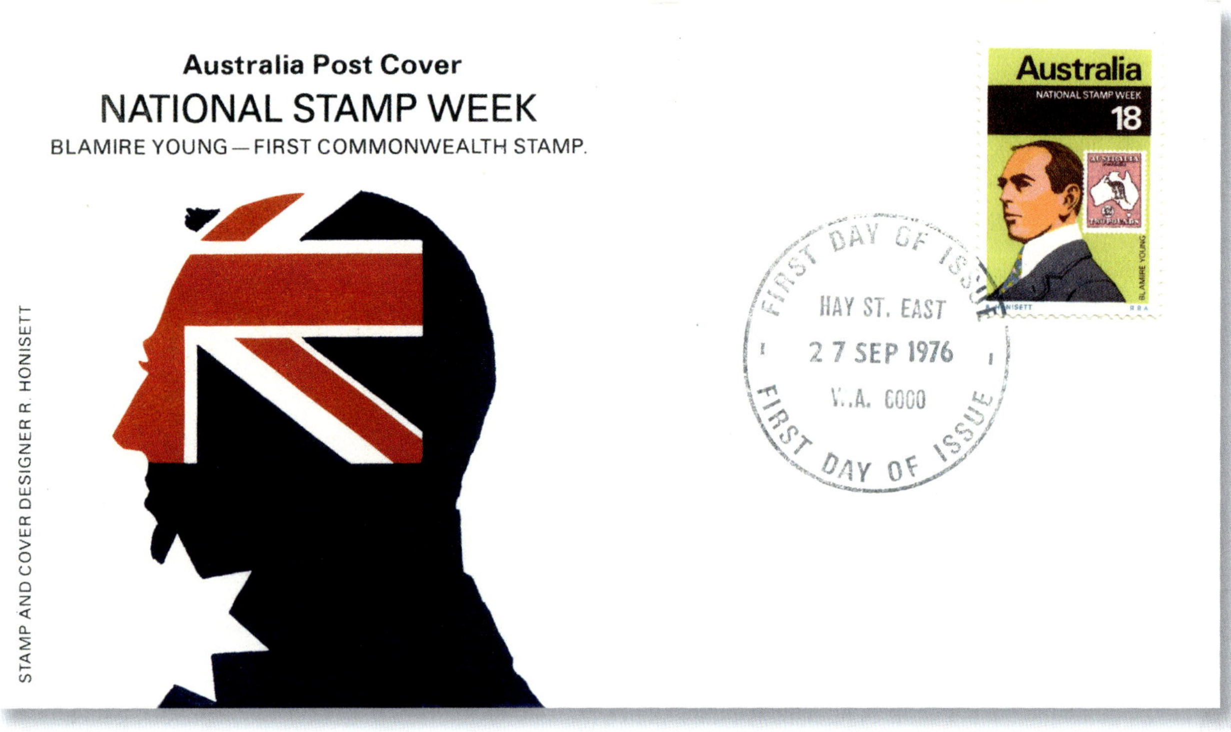

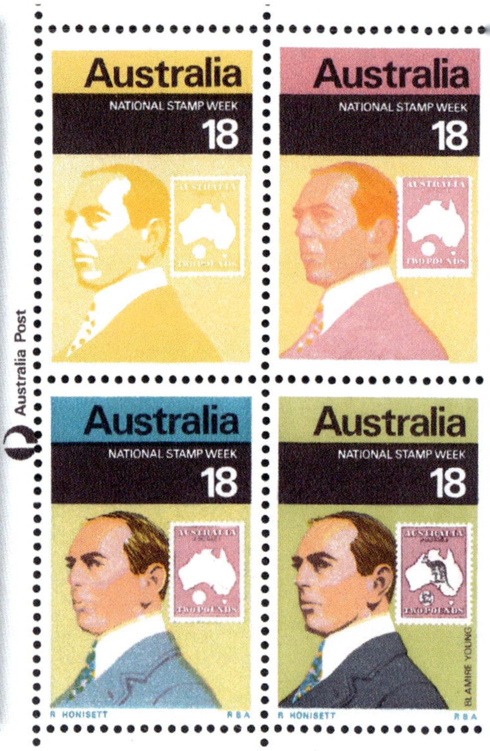

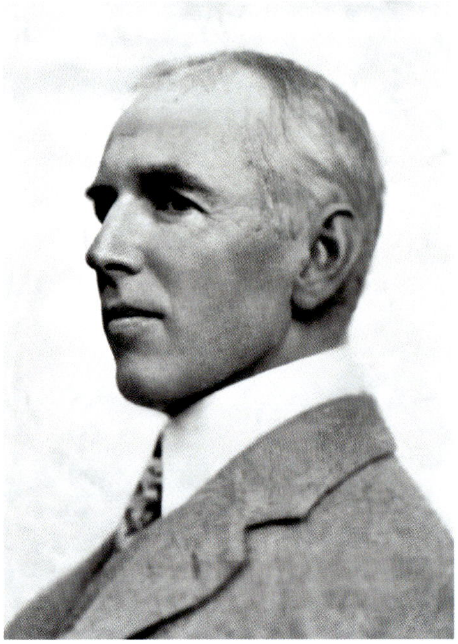

A third commission issued just eight months later put Ray squarely back in his comfort zone, a commemorative stamp for the 50th anniversary of the Royal Australian Air Force. A striking flight of four Mirage jet fighters led the eye up to a technical profile of a World War One Avro 504, making a memorable and dynamic design.

Over the next 20 years Ray Honisett would design 84 stamps for the Australian Post Office, later to become Australia Post. Of the total, 32 would be for the Australian Antarctic Territory, ten for Christmas Island, 29 for the Cocos (Keeling) Islands and four for Samoa. It was not surprising that he would receive the commissions for the various island states and territories, for their very isolation meant that aircraft and boats would be the most relevant subjects. Over that period Ray undertook just nine designs for 'local' Australian stamps out of the total of 84 designs, or just over 10% of the total. His interests and specialisations meant he was particularly suited to work for the territories and Pacific Island nations, with the Australian Antarctic Territory becoming his favourite domain, accounting for more than one third of all his work.

The design for the Christmas aerogramme (opposite left) further consolidated Ray's place with the Australian Post Office, for the following year it was a prize winner in the 1972 San Gabriele International Prize for Philatelic Art, given annually for the best stamp or postal item featuring a religious subject. His design incorporated a stamp image of three stylised heads, based on a design by John Lee, balanced by a bold abstracted Christmas star.

Ray Honisett's increasing status within the cohort of Australian philatelic artists was recognised in 1976 when he was commissioned to design a special issue for National Stamp Week. His subject was the pioneering Australian artist William Blamire Young, who had designed what was effectively the first true Australian stamp, rather than one which was essentially a re-branding of those from the United Kingdom. Young's design was disarmingly simple, just a Kangaroo placed within a map of Australia, but the issue marked our increasingly independent place in the world in the years before the Great War of 1914–18. Honisett's design block of four stamps was a technical expression of the four colour printing process – first yellow, then magenta, cyan and finally black. Young's image had been built up from an original black and white photograph. Ray also designed the strongly graphic First Day Cover, which featured Young's profile cut into an Australian Flag. The issue was also notable as being under the new banner of Australia Post, created on 1 July 1975 after the dissolution of the old Post Master General's Department, which also saw the creation of Telecom out of the old Post & Telegraph Office.

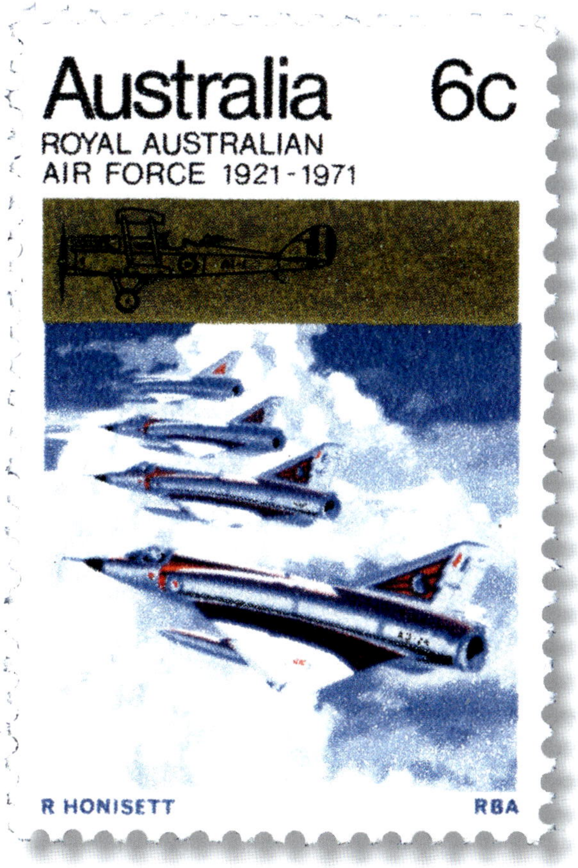

Above, 9 June 1971 — 50th Anniversary of the Royal Australian Air Force

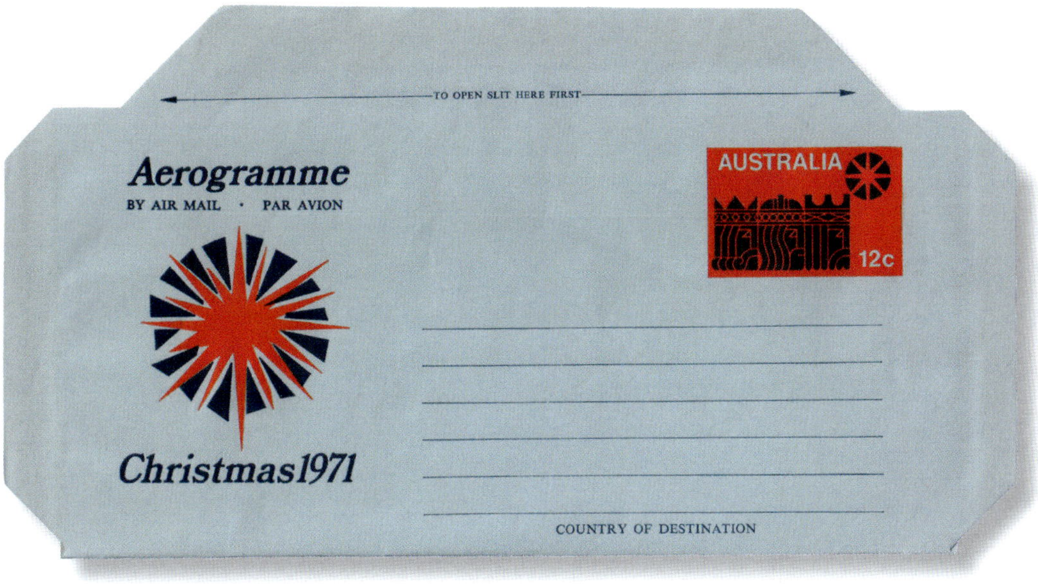

A different project that was released soon after the RAAF stamp, a special Christmas aerogramme. As international air mail was an expensive option, the prepaid aerogramme was an economical option for overseas letters.

"Going from the sublime to the ridiculous, I have recently been engaged on the design of the stamp and first-day covers marking the RAAF 50th Anniversary. It's quite a shock to drop down from working on a 45 sq.ft. canvas to a 2 square inch stamp."

Ray Honisett, 19 March 1971

ANTARCTICA

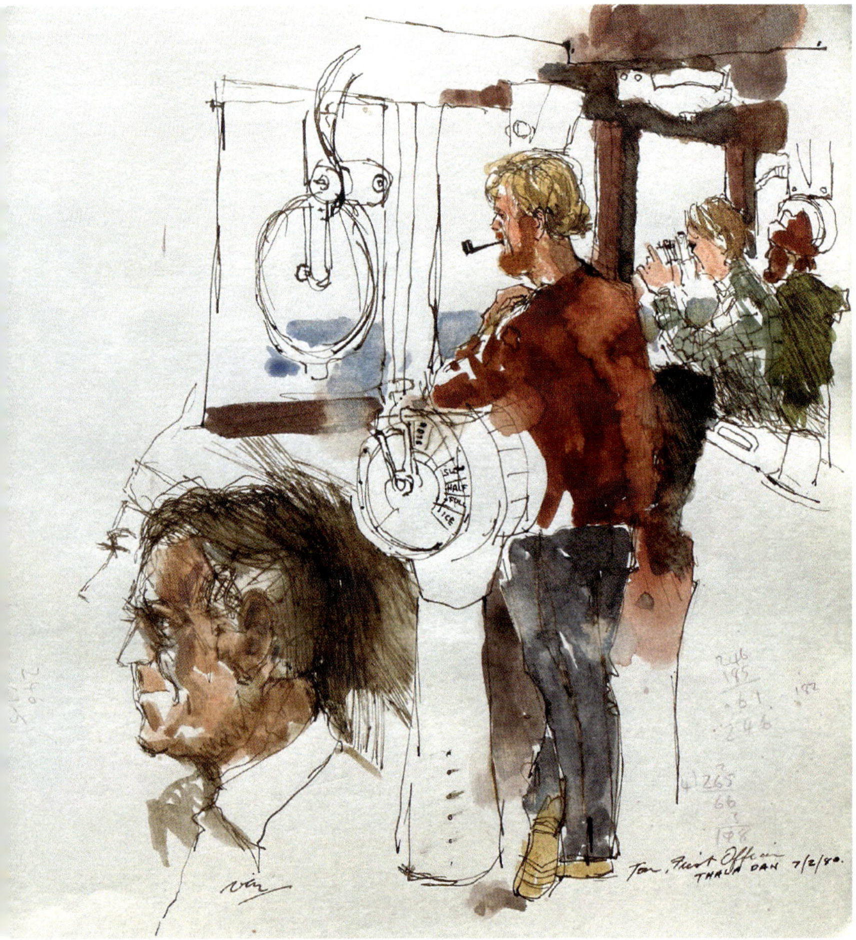

Above, page from Ray Honisett's sketchbook:
on board ***Thala Dan*** *on the way to Antarctica, 1980*

The Royal Melbourne Institute of Technology, formerly known as Melbourne Technical College (and now RMIT University) was the premier art and design school in Victoria, with a broad base of subject areas and highly qualified and experienced staff. When Ray Honisett joined the school in May of 1968 he was, like many of his contemporaries, very experienced but without any formal qualifications. The Illustration Section was headed by Harold Freedman, formerly the Senior Official War Artist for the Royal Australian Air Force in World War Two and a formidable character with some 20 years service at the college. The two did not always get on, but fortunately there were other senior staff with whom he was able to form excellent working relationships. British born and trained John Mason would have a profound effect on Ray, both at the personal and professional level. Mason had studied at the Royal Academy in London and worked in many disciplines, including design, sculpture, painting and drawing. Mason had a close association with the Australian Post Office and had designed 15 stamps for the Australian Antarctic Territory between 1966 and 1972. In that year John married a fellow artist, Gwendoline Mannerheim, whose husband Frank had been killed in a farming accident two years before. John recommended Ray for a challenging new Antarctic stamp commission, which required six designs based on historic aircraft used in exploration and supply. The work was based on photographic and documentary research — it would be another six years before he could make the journey south to experience Antarctica for himself. He would travel to Antarctica three times (twice to the Antarctic mainland and once to Macquarie Island) and in the process become the most prolific designer of Antarctic stamps for the next ten years. Tragically, John Mason died a little over a year after his marriage, not living to see Ray marry Gwen's daughter Sherryl Mannerheim, on 20 November 1978. His position at the college had been formalised with the granting of a Fellowship Diploma in Art in 1977, recognising his considerable experience and, in particular, his contribution to the philatelic discipline.

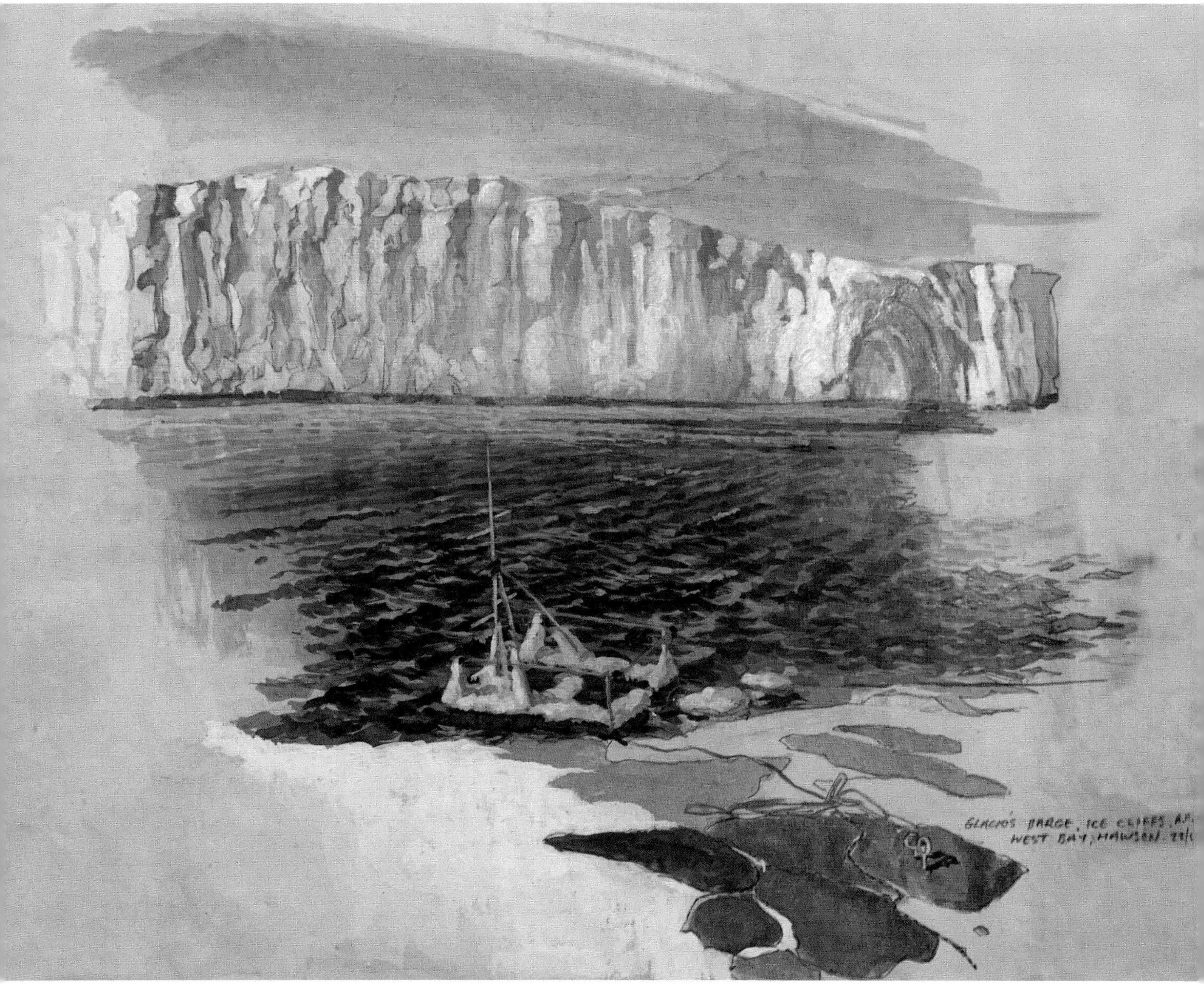

Glacio's Barge, Ice Cliffs, AM, West Bay, Mawson, '80
Watercolour and gouache on paper, 1980
Private Collection, Melbourne

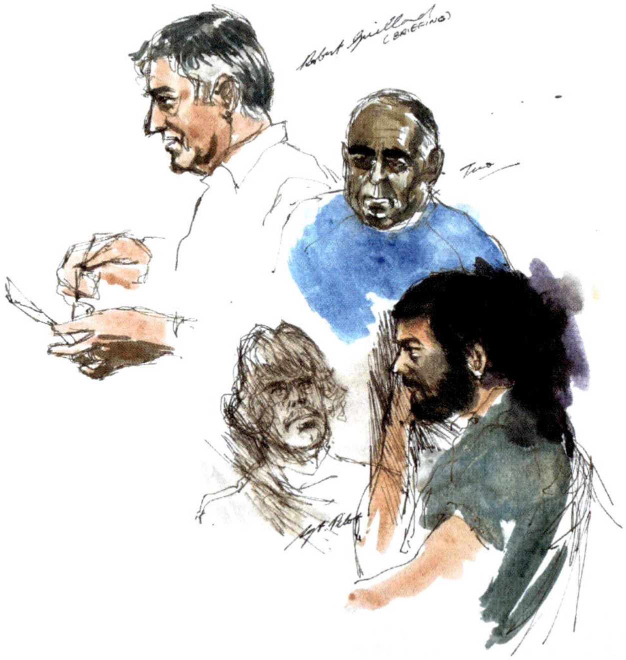

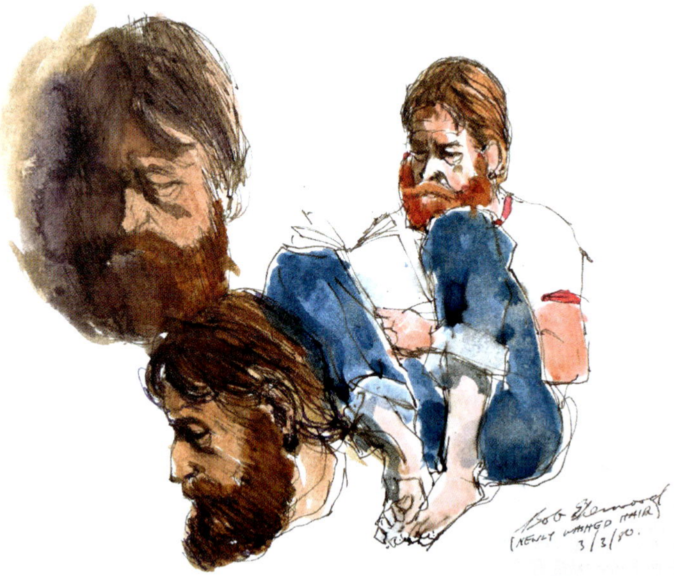

Pages from the Thala Dan sketchbook, February 1980

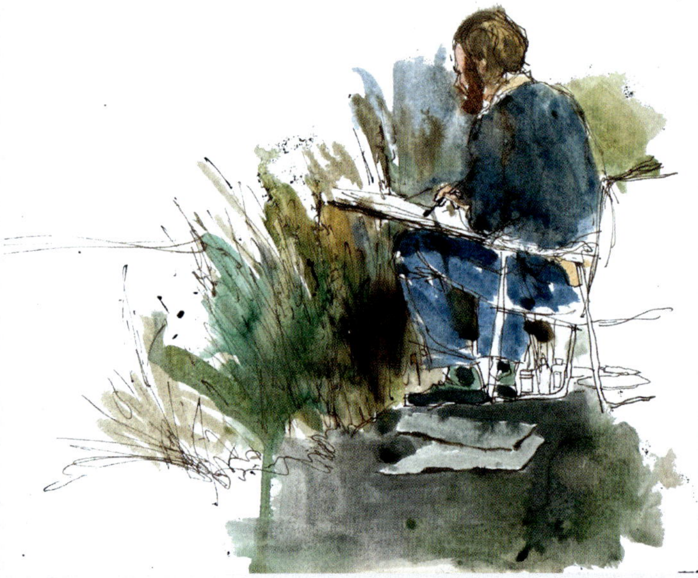

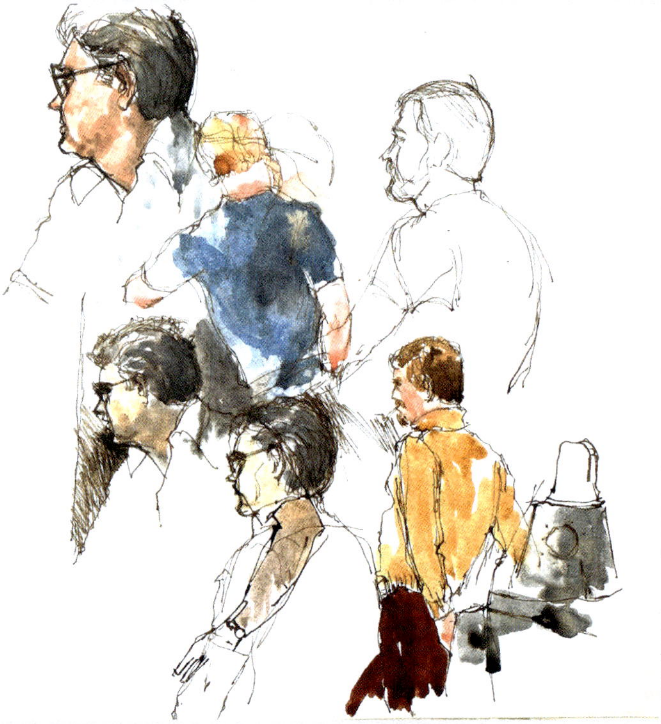

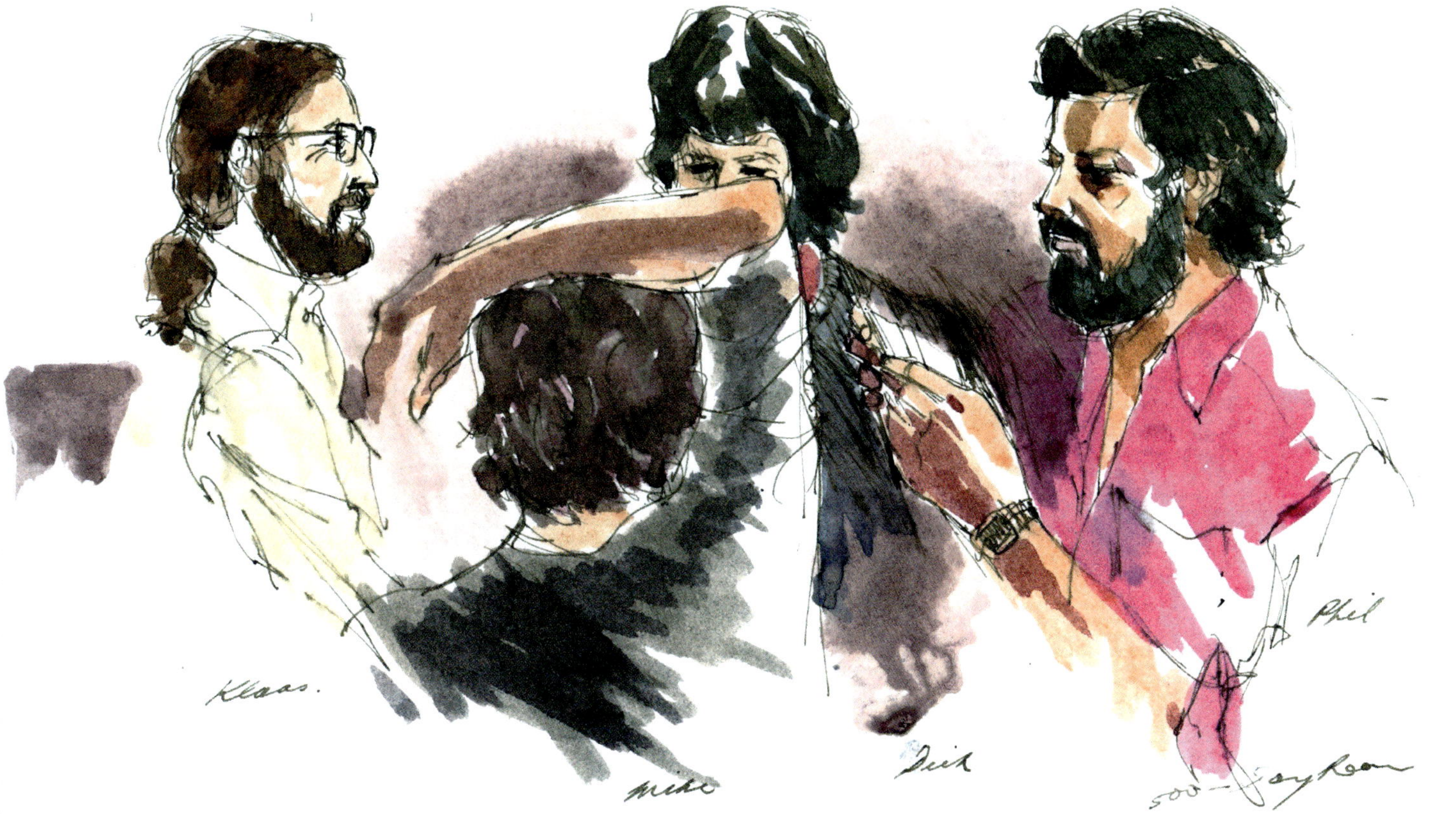
Klaas.
Mike
Dick
Phil

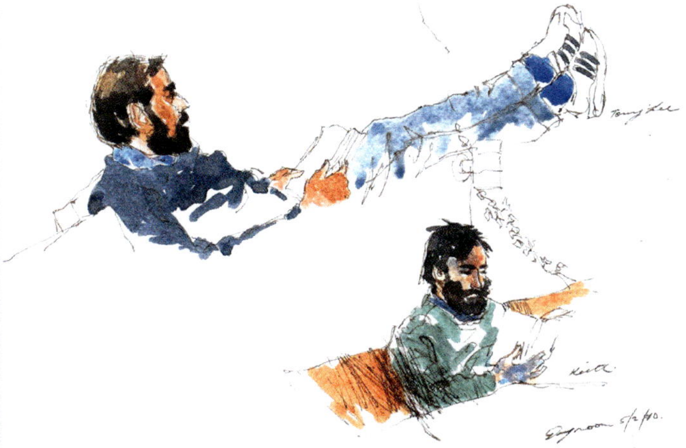
Keith

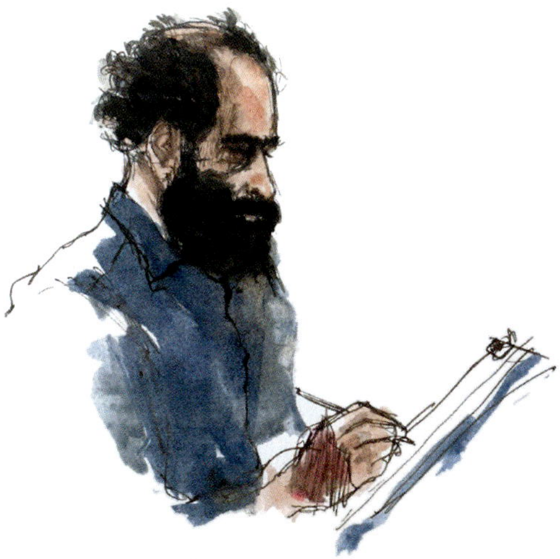

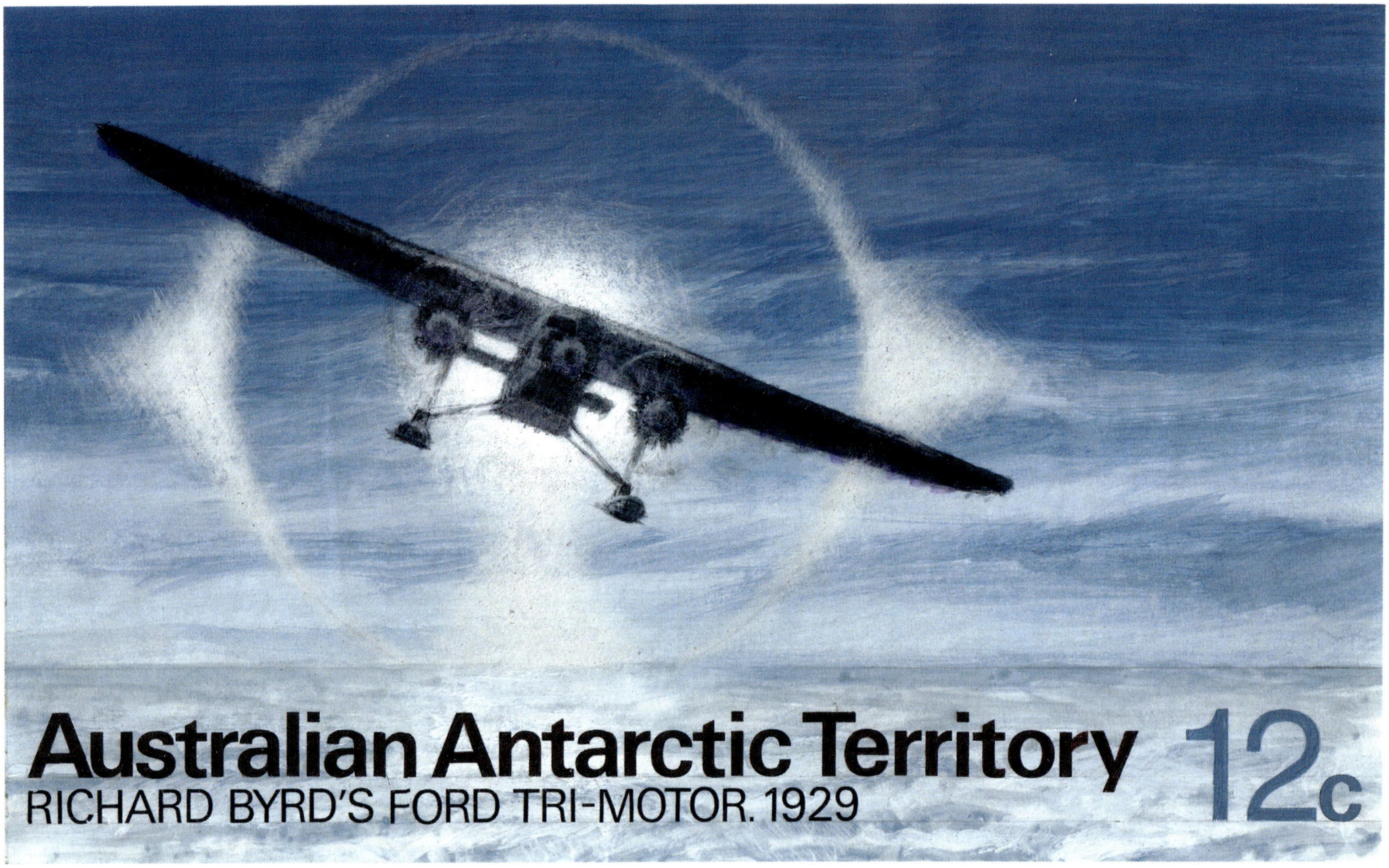

*Unadopted artwork, **Byrd's Ford Tri-Motor, 1929**,*
Acrylic on card
Australia Post Archives, Melbourne

All the paintings reproduced in this section were directly scanned from the original artwork, held in the Australia Post Archives in Melbourne. A standard stamp commission required the design to be prepared six times the size of the final imprint. While stamps tended to vary in size, the standard for most of Ray Honisett's commissions was an image size of 34.5 x 21 mm. The painting above has been reproduced at an exact 1:1 scale of the original work. The image opposite shows how the paintings were usually presented, with a wide bleed and a border that showed marks such as the 'test' brushstrokes made by the artist. The lettering was applied using Letraset, the dry transfer system introduced in 1961. The Letraset process was revolutionary for the commercial art and design world, allowing relatively inexperienced designers to produce high quality camera ready artwork, once the preserve of highly skilled hand lettering experts. This was particularly useful for the stamp designer, allowing the application of small scale type, which in earlier times would have to have been printed separately by conventional letterpress or engraved by a skilled engraver. The design selection process required that designers submit two distinct sets of artwork that fulfilled the brief. A selection committee of art and philatelic experts would then choose the design to go to print. The formal designation of 'Adopted' and 'Unadopted' has been used in captioning these works.

RESEARCH FOR A STAMP SERIES — EXPLORERS' AIRCRAFT

BY RAY HONISETT

One might reasonably assume that a graphic designer's first work on a new commission is to make preliminary pencil sketches; thumbnail "doodles", gradually increasing in dimension and content as his ideas crystallise and develop into something worth full-scale colour interpretation. This process, or something similar, is followed by most designers I know, and by me, of course. However, it is never the first step — it occurs at anything up to one third along the way, only after much other work; in effect, for every six days of work on an assignment, only four are spent in drawing or allied tasks.

What, then, occupies such a large proportion of time? Gazing out of the window, until the mood or a bolt of divine inspiration arrives? This important first stage can be described in two words — research and planning.

Research is gathering as much pictorial and written information necessary to understand all aspects of the subject. Planning is philosophical — deciding what the design should communicate, what emotions should be evoked. I regard research as most important — the foundation for all subsequent work. Indeed, the quality and depth of reference is directly related to the quality and conviction of the final design. As a designer whose approach is illustrative, I find this particularly true, and with the commission for the Explorers' Aircraft stamps it would have been disastrous not to have held and implemented this belief.

This commission required all the care and time of normal research, with the added difficulty of all subject matter being historical. Reliance had thus to be placed on material published many years before. Fortunately, the events involved men of science who always recorded faithfully and in detail. However, their records cannot be found in just any bookshop or library — mostly, they are limited-edition books published by the learned bodies to whom they were originally presented as scientific papers.

Of the aircraft concerned, three were thought no longer to exist. The remaining three are in museums overseas. Their existence made research easier, but even then, one has no guarantee that they have been preserved in their original colours; they may have been modified later in their working lives. Crosschecking is necessary, and photographs taken at the time of the relevant event must be studied. For the aircraft not now in existence, the problems multiplied.

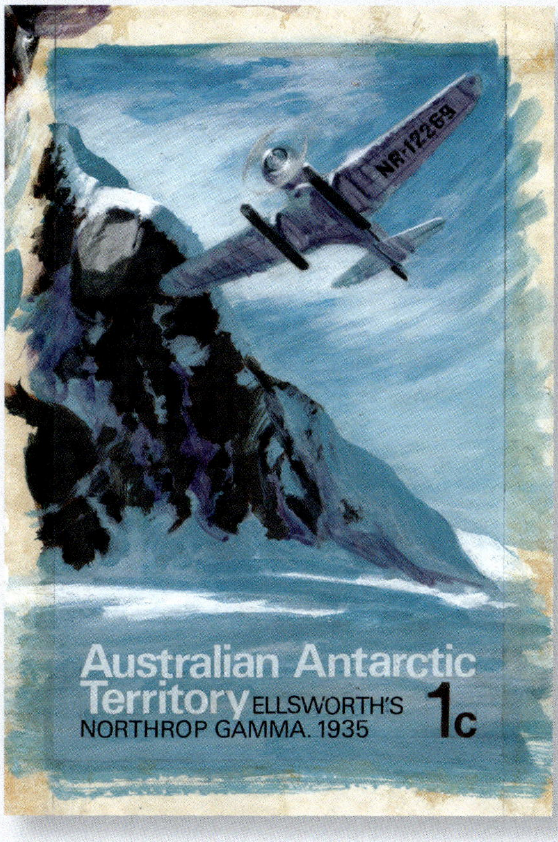

Adopted artwork, ***Lincoln Ellsworth's Northrop Gamma,***
Acrylic on card
Australia Post Archives, Melbourne

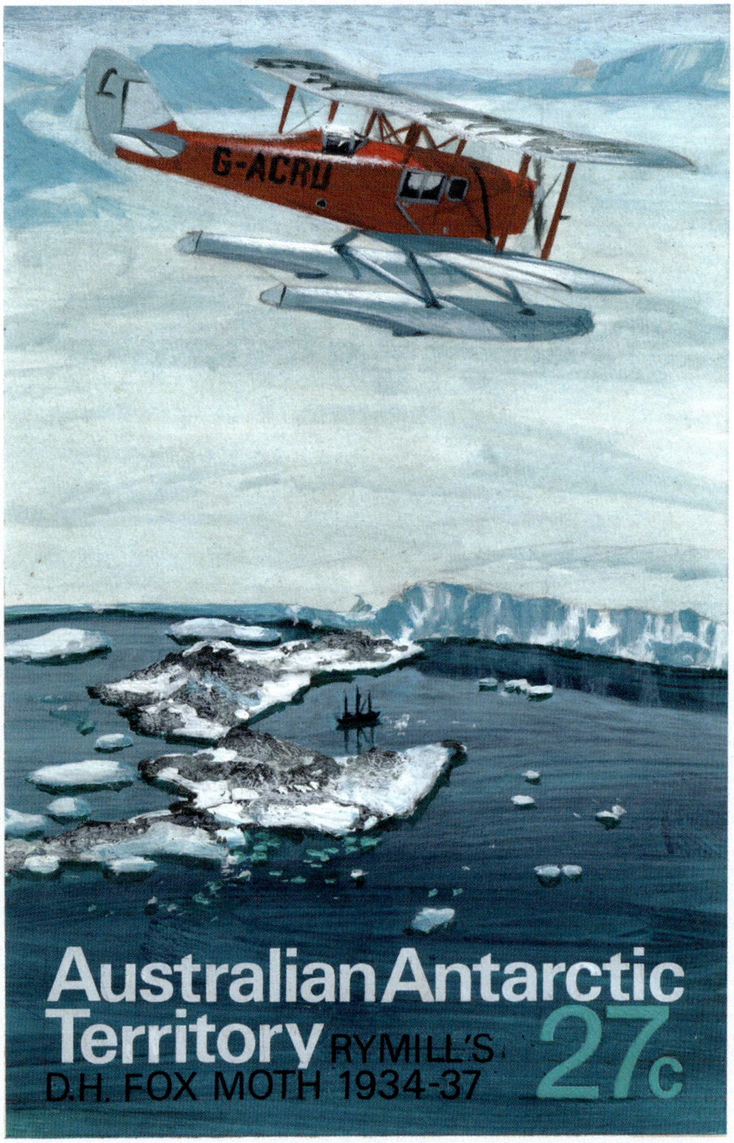

Adopted artwork, ***Rymill's DH Fox Moth,***
Acrylic on card
Australia Post Archives, Melbourne

Of the three major ingredients of the proposed designs, the event, the aircraft and the environment, only the last can be considered to be unchanged. But practical considerations prevent one from taking sketchbook and camera to the South Pole! So one is indebted to the thoroughness and sense of history of the Antarctic scientist-explorers, and their cameras and journals.

Why, if there are so many problems, accept such a commission? For me the answer is that it is one of the most challenging and interesting facets of the work, one which not only gives a welcome change away from the drawing hoard, but which adds to my store of knowledge.

Unlike most artistic work, done in studio solitude, research is accomplished by an expanding team; with the artist acting as a member and final filter. His studio, untidy at any time, becomes littered with increasing piles of books, magazines, reports, photographs, maps and vital notes which magically hide in unrelated material. The sum total is the result of hard work and patience of many people, most not directly involved with the reasons for the research, and with no motivation other than a desire to help in ensuring historical accuracy.

The starting point of this commission was the Post Office's brief: … "two alternative designs of each of six proposed stamps for Australian Antarctic Territory, featuring the following: Sir Hubert Wilkins' Lockheed Vega, Admiral Richard Byrd's Ford Tri-motor, Sir Douglas Mawson's De Havilland Gipsy Moth, Lincoln Ellsworth's Northrop Gamma, John Rymill's De Havilland Fox Moth, and Lars Christensen's Avro Avian." Immediate assistance came from the Philatelic Section, P.M.G.'s Department; its staff includes two members interested in aviation history, who were able to supply many photographs, plans and documentation on the Lockheed Vega, the subject of the first designs. *(Editor's note: irrespective of our personal interests, we always need to do a "feasibility study" on any proposed stamp series, to determine whether sufficient diverse material is available, before seeking Ministerial approval in principle).* In later months, these officers unearthed more references from their files, or tracked down, from other sources, further references or information.

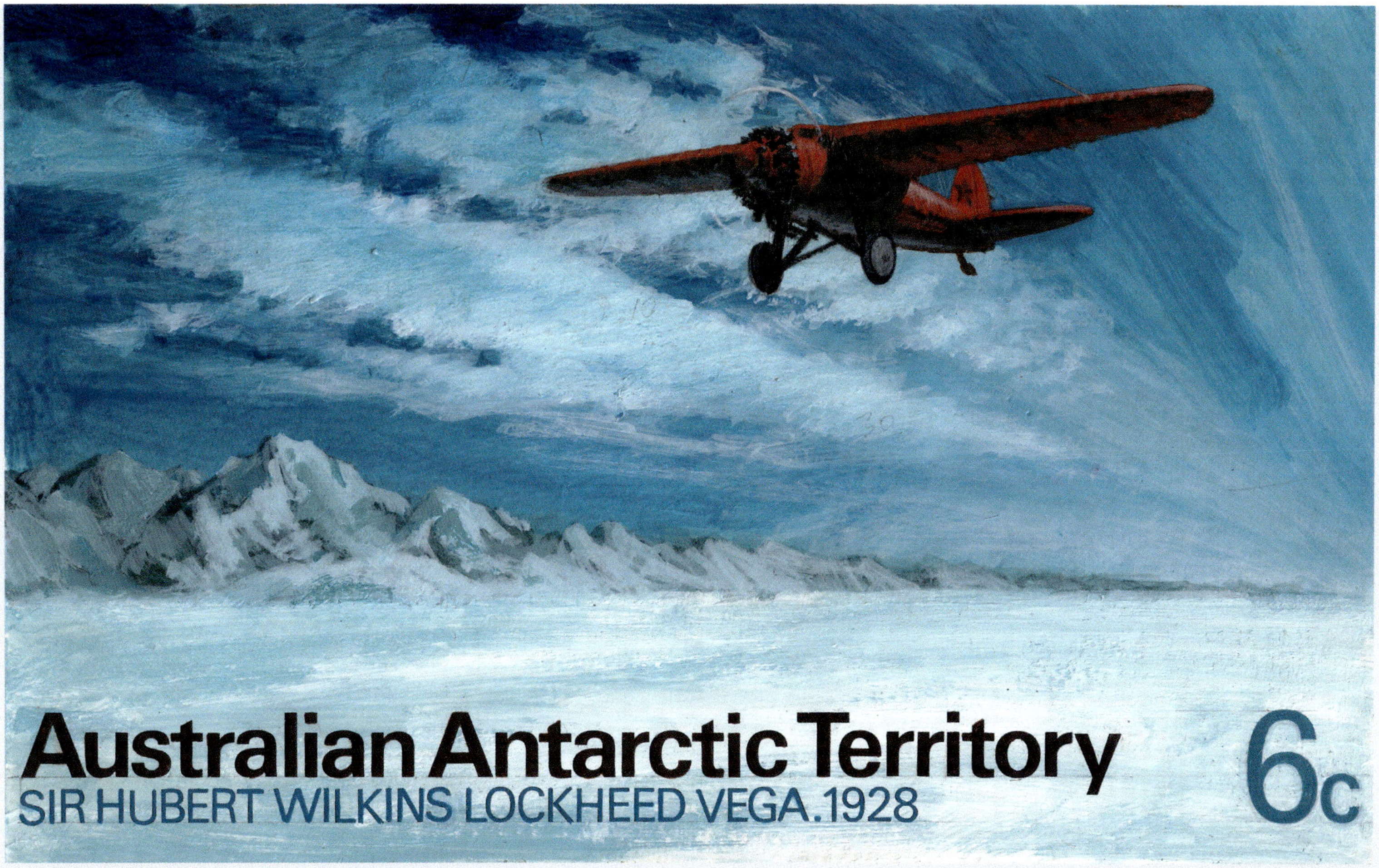

Soon, many people of the Australian National Antarctic Research Expeditions office in Melbourne added their considerable efforts, especially Mrs Ida McMahon, the chief librarian. For clarity, I shall deal with the jigsaw by taking each stamp separately, although as time passed, earlier research began to fit with later subjects.

Sir Hubert Wilkins' Lockheed Vega

This, the first design, was the most crucial; the design philosophies which would be constant throughout the series had to be established, and the format of caption and denomination had to be incorporated flexibly, to work for this design and for the remainder. I was also aware that as the first of a series, the Vega designs would be closely scrutinised by the Stamp Advisory Committee, who rightly demand a high standard. I was thus determined that authenticity would not be queried at Committee level.

*Adopted artwork, **Sir Hubert Wilkins Lockheed Vega**,*
Acrylic on card
Australia Post Archives, Melbourne

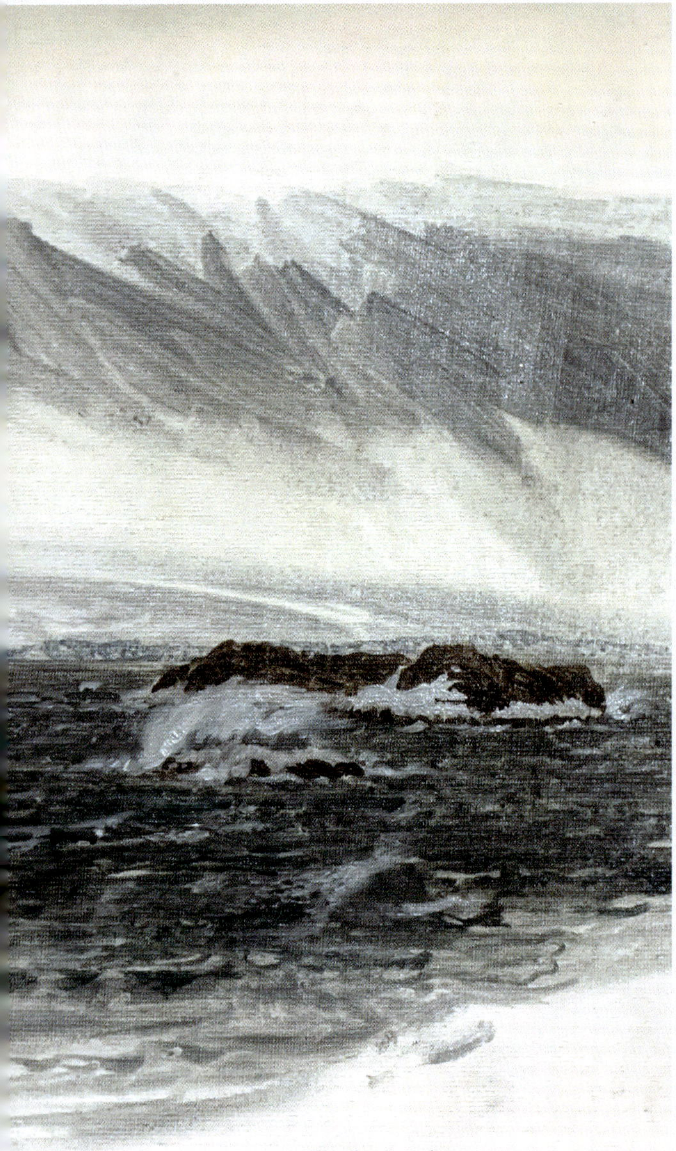

To research the event and terrain I made the first of many visits to A.N.A.R.E. headquarters, whose library has a wealth of books, scientific papers, maps, etc., covering every aspect of Antarctica. Additionally, one can consult personally with eminent scientists, most veterans of several Antarctic expeditions.

The historical aspects of Wilkins' epic flight proved fairly simple — as a "first", it was well documented, and we established that wheels were used for the flight; hardly the best type of undercarriage for emergency landings, but the only form possible for the enforced take-off conditions.

The terrain was more of a problem. After reading accounts of the flight, I had decided on two basic approaches: one with the aircraft flying along the east coast of Graham Land's jagged mountains, and the other setting course from the Deception Island base.

Although few photographs of this area of Graham Land were found, it is similar to some other areas of which colour photographs are plentiful in the A.N.A.R.E. photographic section. Subsequently, the topography was confirmed by the albums of the late John Rymill.

Deception Island, a small volcanic outcrop in the South Shetland group, proved difficult. The only frequent human activity there in 1928 was that of whalers. Photos of that period are rare, and Wilkins' descriptions are insufficient a basis for full-colour illustration. The only worthwhile reference was in a geographic magazine, but unfortunately it reported 1967-69 volcanic upheavals, which radically changed the shape of the island. Many hours of careful analysis of description and photographs were needed to piece together an image of its 1928 aerial aspect.

Three alternative versions of this subject were submitted, and in the event the laboriously researched Deception Island design was not chosen, although not for any reason connected with research.

Admiral Richard Byrd's Ford Tri-motor

As Byrd was the first to fly over the South Pole, it seemed that this event would be simple to research. The Tri-motor was a very successful design (some 200 examples being built); many remain in flying condition even today and the Byrd Ford, "Floyd Bennett" (named for his companion of the earlier North Polar flight) is preserved in the Henry Ford Museum at Dearborn, Mich., U.S.A. The Chief Curator kindly supplied three photographs, and further material was slowly accumulated from elsewhere.

Documentation in the A.N.A.R.E. library described how, over the South Pole, Byrd dropped an American flag weighted with a stone from Floyd Bennett's grave at Arlington, Virginia. I decided to portray this moment for its emotional quality.

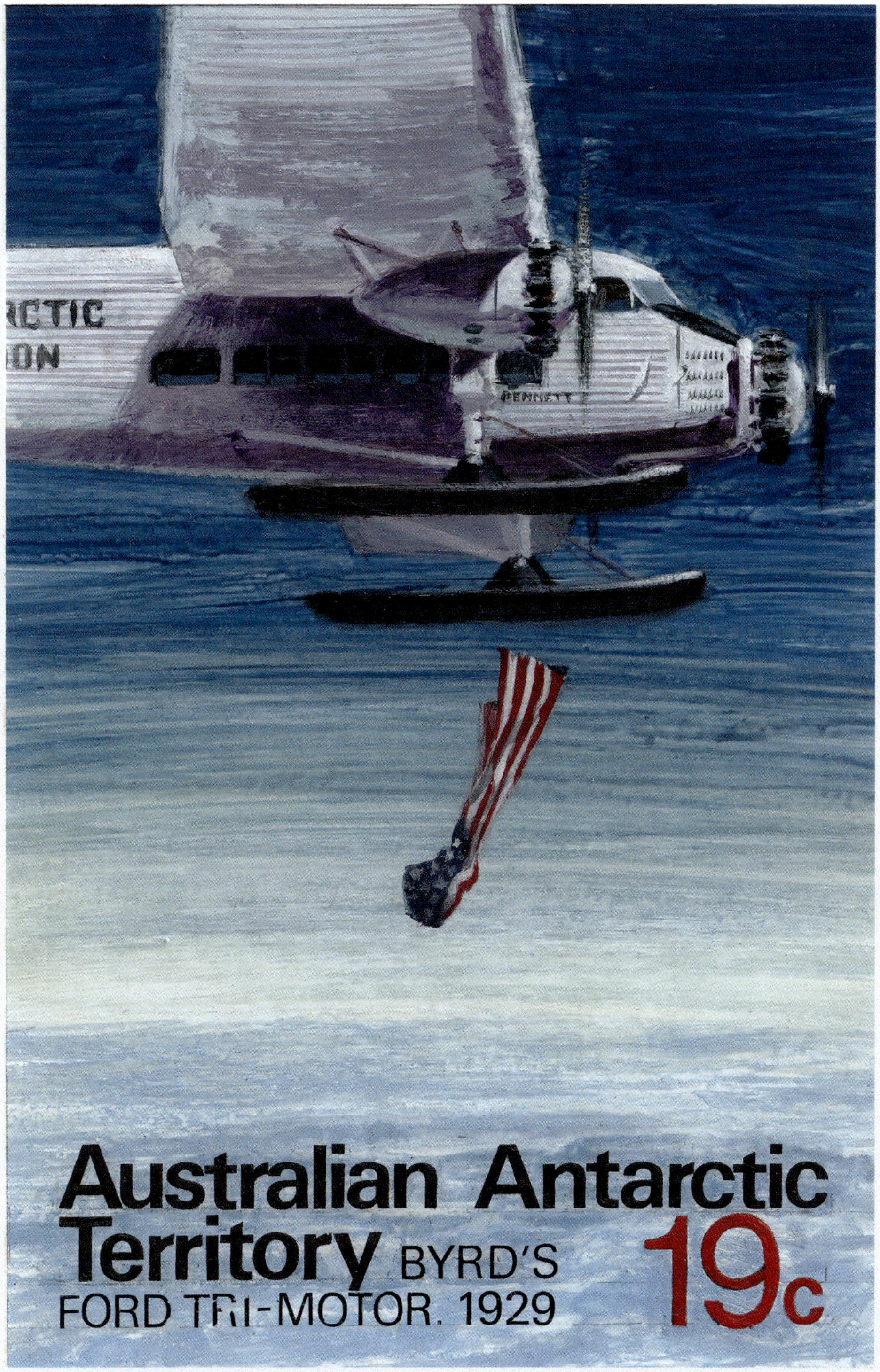

The alternative design proved a problem — I could have shown the same action from a different angle, but that hardly provided a choice. Finally, an A.N.A.R.E. meteorologist mentioned "halo phenomena" — optical effects of rings, arcs, pillars or bright spots, produced by refraction of light by ice crystals in the atmosphere. After studying photographs, mental images of the Ford silhouetted against a halo began to form. Then a chance question brought the type of unbelievably precise response for which I was indebted time after time to the scientists.

"Roughly how high above the horizon would the sun be at the South Pole at that time of year?" I asked. Twenty minutes later, an A.N.A.R.E. man gave me a computer print-out, listing the degree of angle of the sun to the zenith, at 30-minute intervals, for the whole 24 hours of 29th November 1929! Ironically, as with the Wilkins stamps, the design which had taken the greater amount of research and mental exercise was not the eventual choice ... no matter.

*Adopted artwork, **Byrd's Ford Tri-Motor, 1929**,*
Acrylic on card
Australia Post Archives, Melbourne

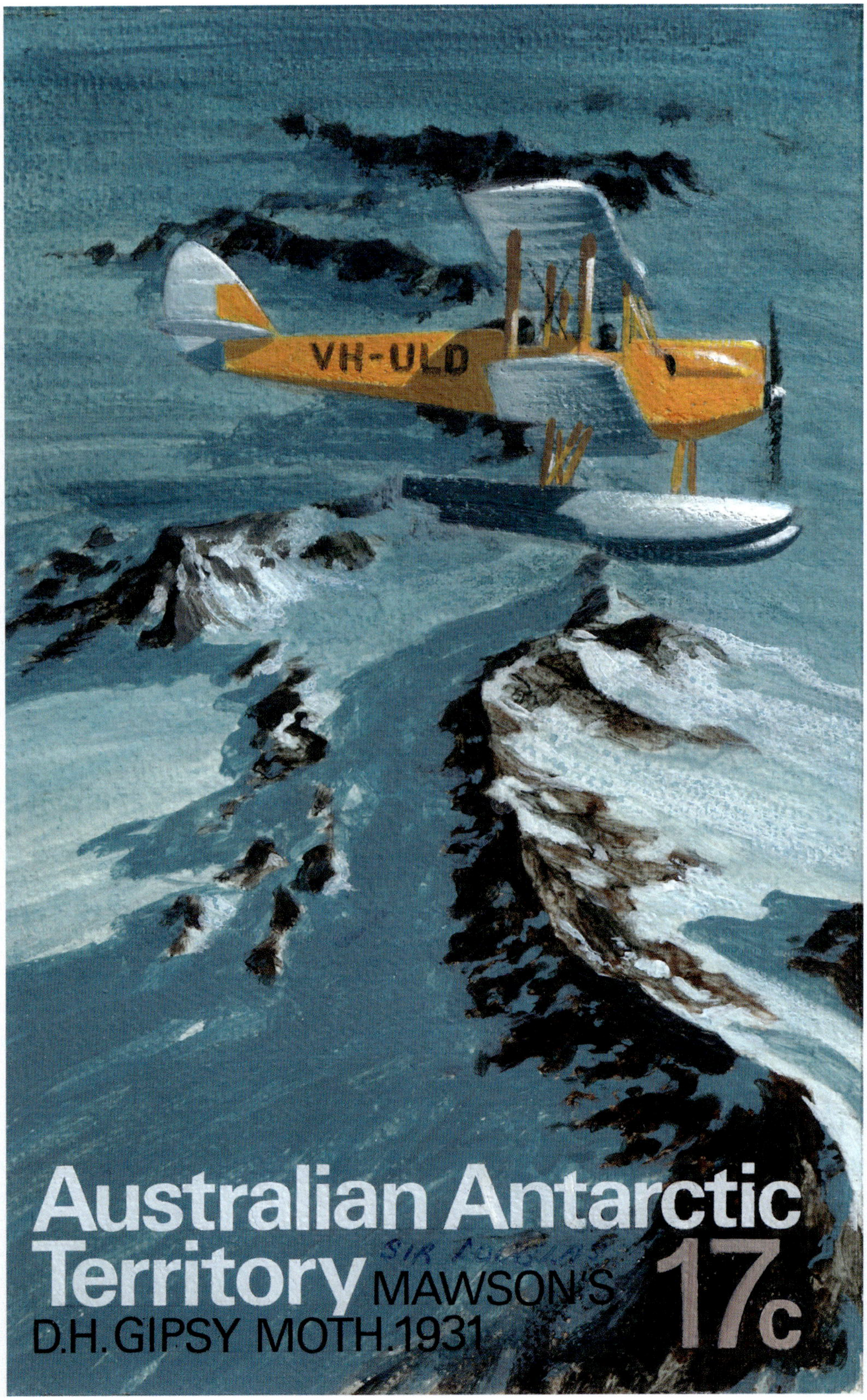

Adopted artwork, ***Mawson's D.H. Gipsy Moth 1931****,*
Acrylic on card
Australia Post Archives, Melbourne

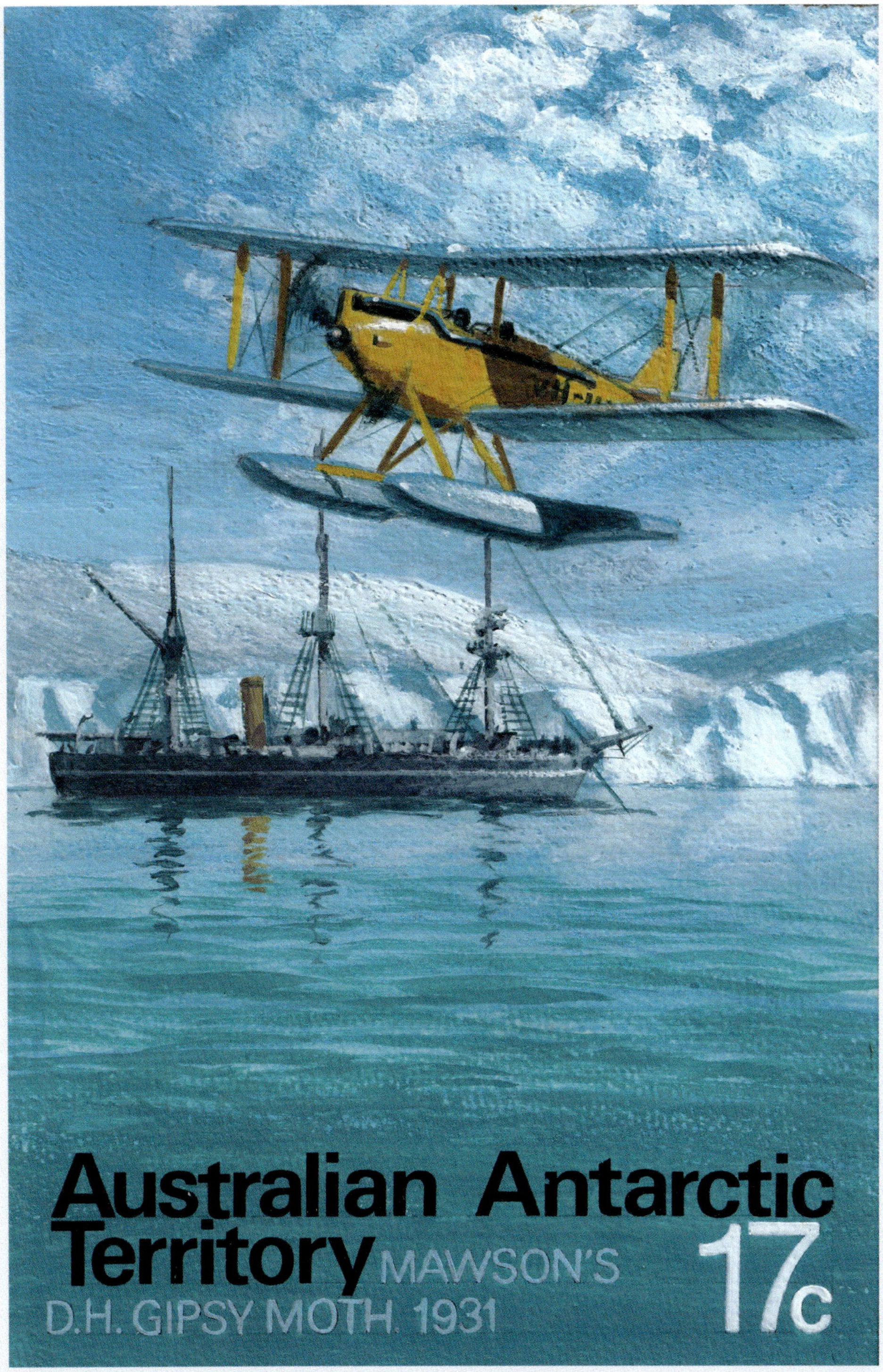

Sir Douglas Mawson's De Havilland Gipsy Moth

The British–Australian–New Zealand Antarctic Research Expedition of 1929-31, led by Mawson, took a small De Havilland Gipsy Moth as deck cargo on *Discovery I,* Captain Scott's old ship. This machine was the next subject probed.

Since its appearance in 1928, the Gipsy Moth has flown to almost every corner of the globe and many examples still survive in flying condition. Thus, I was able to gather much reference on the type generally, and on Mawson's Moth, VH-ULD, specifically.

Colour is always a problem in interpreting black and white prints. Despite the 1931 expedition's detailed reports, no clue to the colour of VH-ULD could be found; thus, the tonal greys of black and white prints must be equated with the renditions given by film of the period. Sometimes, these can be related to a known colour on the same print, for instance the red, white and blue of an Air Force roundel. A certain amount of guesswork is also involved.

Finally, we tracked down the son of the expedition's pilot, who supplied the necessary details.

Many of the Gipsy Moth's flights were over areas surrounding modern A.N.A.R.E. bases, thus colour shots were available. One, of a rugged mountain range snaking off into the distance, afforded the exact balance of drama, spectacle and typicality to offset the frail-looking biplane.

The alternative design showed the aircraft with *Discovery I* behind it, but for once I had correctly anticipated and was well content with the Committee's choice.

Unadopted artwork, ***Mawson's D.H. Gipsy Moth 1931****,*
Acrylic on card
Australia Post Archives, Melbourne

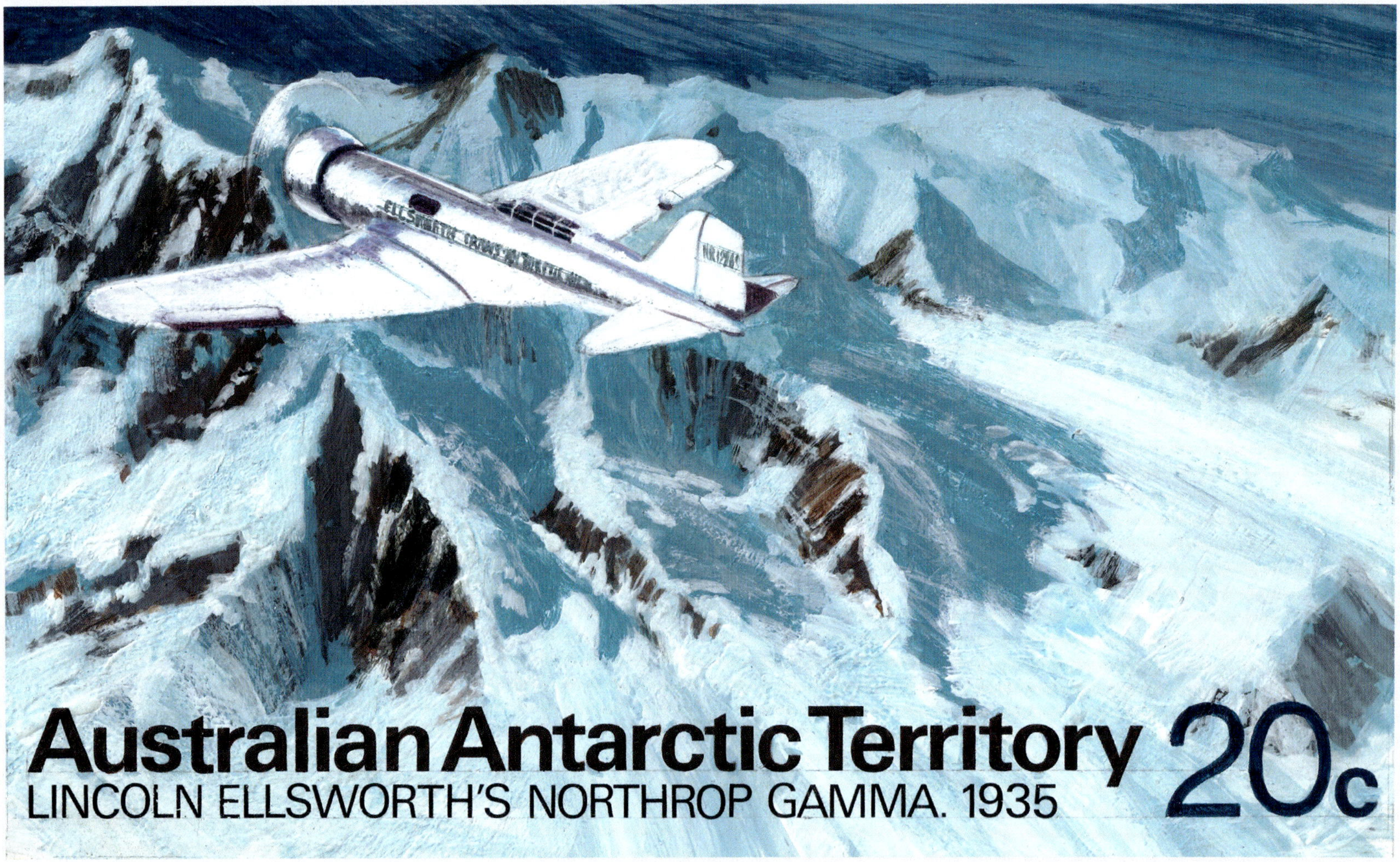

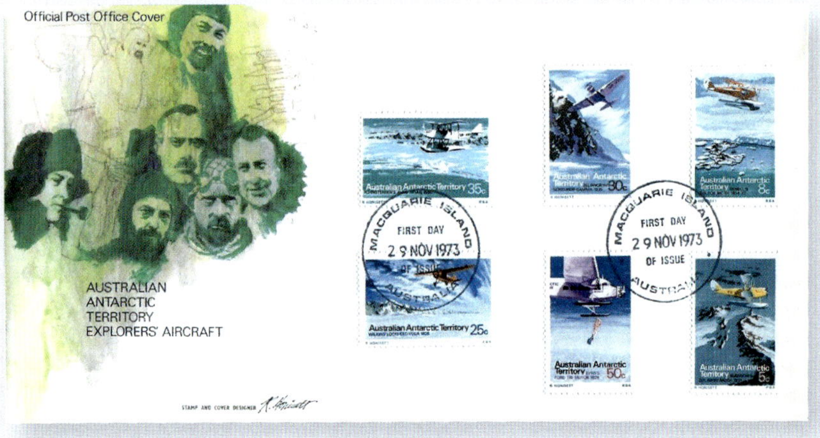

Unadopted artwork, ***Lincoln Ellsworth's Northrop Gamma****,*
Acrylic on card
Australia Post Archives, Melbourne

Left, First Day Cover, Antarctic Aircraft
29 November 1973

Lincoln Ellsworth's Northrop Gamma

By the time we reached this subject, it was possible to make cross-use of existing references; for instance, Ellsworth's flight began from Dundee Island, some 220 kilometres west of Deception Island. His flight path, therefore, must have been close to that of Wilkins. Also, the Rymill family had made available some magnificent photo albums, adding to the knowledge of the Graham Land area of Ellsworth's first 800 kilometres. The design ultimately selected was based upon this reference. The unadopted alternative showed part of the Ellsworth Mountains, from American sources.

Like other aircraft in the stamp series, the Northrop Gamma was used by a number of record-breaking pilots of the period, many of whom required modifications built into their machines. Ellsworth's aircraft is now in the Smithsonian Institute, Washington, D.C., but is it in its original state? The archives of an American aviation journal eventually established the 1935 configuration of Gamma NR 12269.

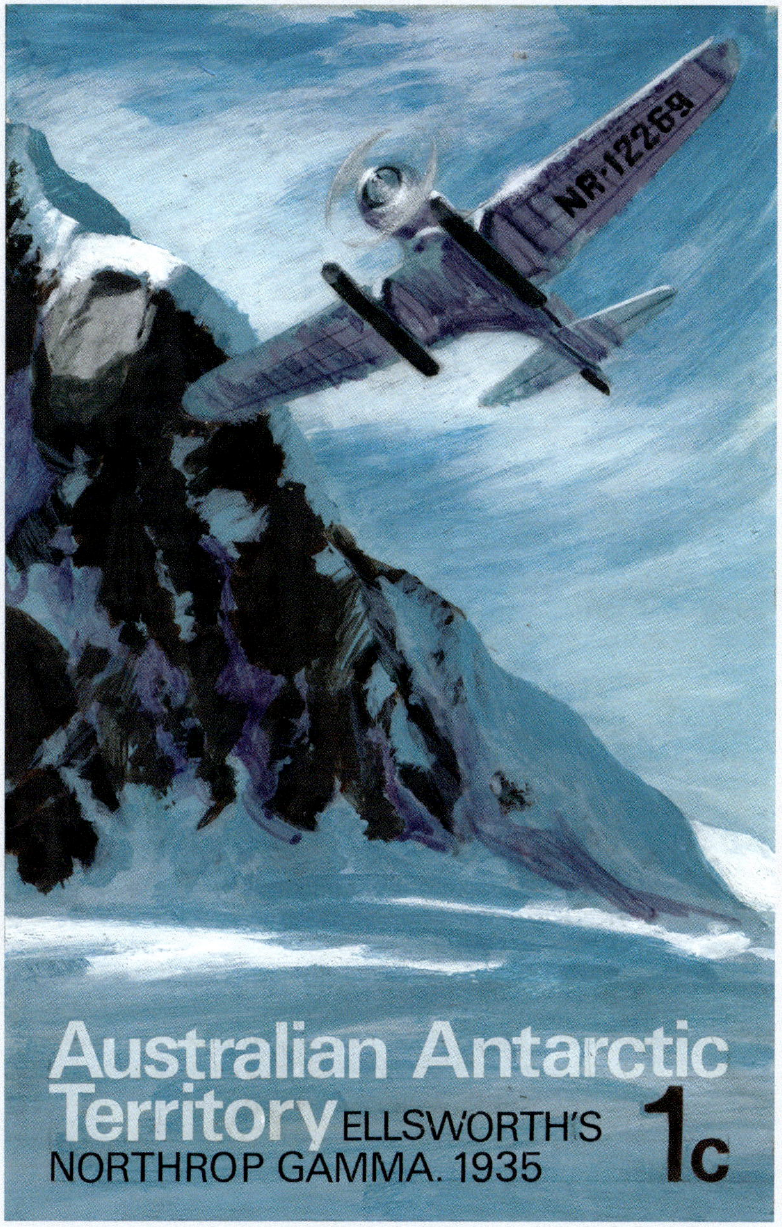

Adopted artwork, ***Lincoln Ellsworth's Northrop Gamma,***
Acrylic on card
Australia Post Archives, Melbourne

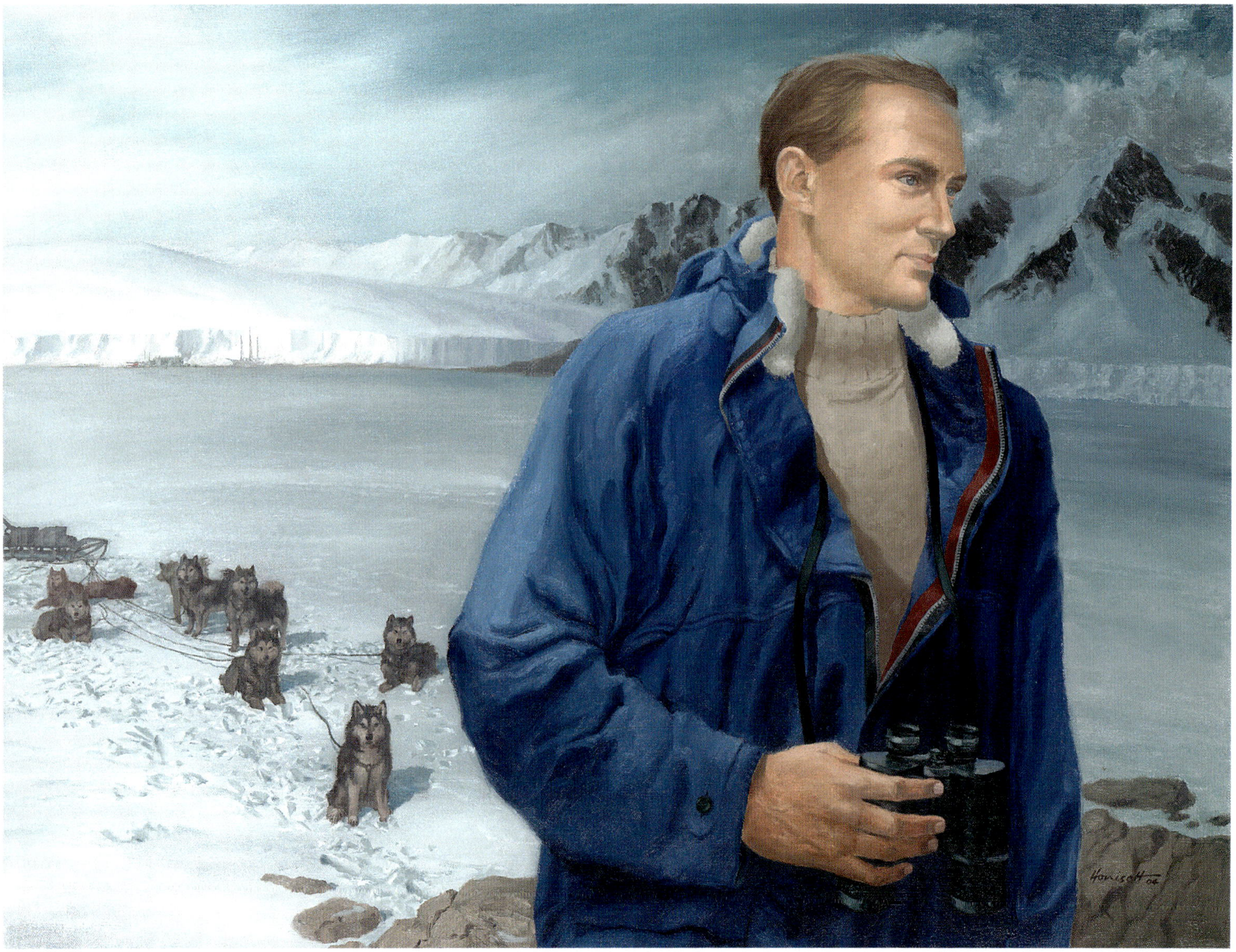

Ray Honisett, ***John Rymill and the dog team****, 2004*
Oil on canvas
Private collection

Thirty years after the stamp designs Ray Honisett revisited the Rymill story, producing a portrait based on a period photograph of the young explorer, with the body modelled by Melbourne sporting identity Ken Jacobs.

John Rymill's De Havilland Fox Moth

As noted above, from the family of the late John Rymill we received two huge photographic albums and the complete reports, maps, etc., which he submitted to the Royal Geographical Society. However, although these supplied all but a tiny fraction of the required information, it was that fraction which consumed more time and energy than all other aspects. Once again, it was the question of the aircraft's colour.

Monochrome photos indicated that early in the expedition, when the Fox Moth was fitted with floats, the paint scheme was two-tone — the fuselage and fin had photographed a dark tone (possibly red or deep orange) while all other surfaces seemed much lighter (possibly pale blue or silver).

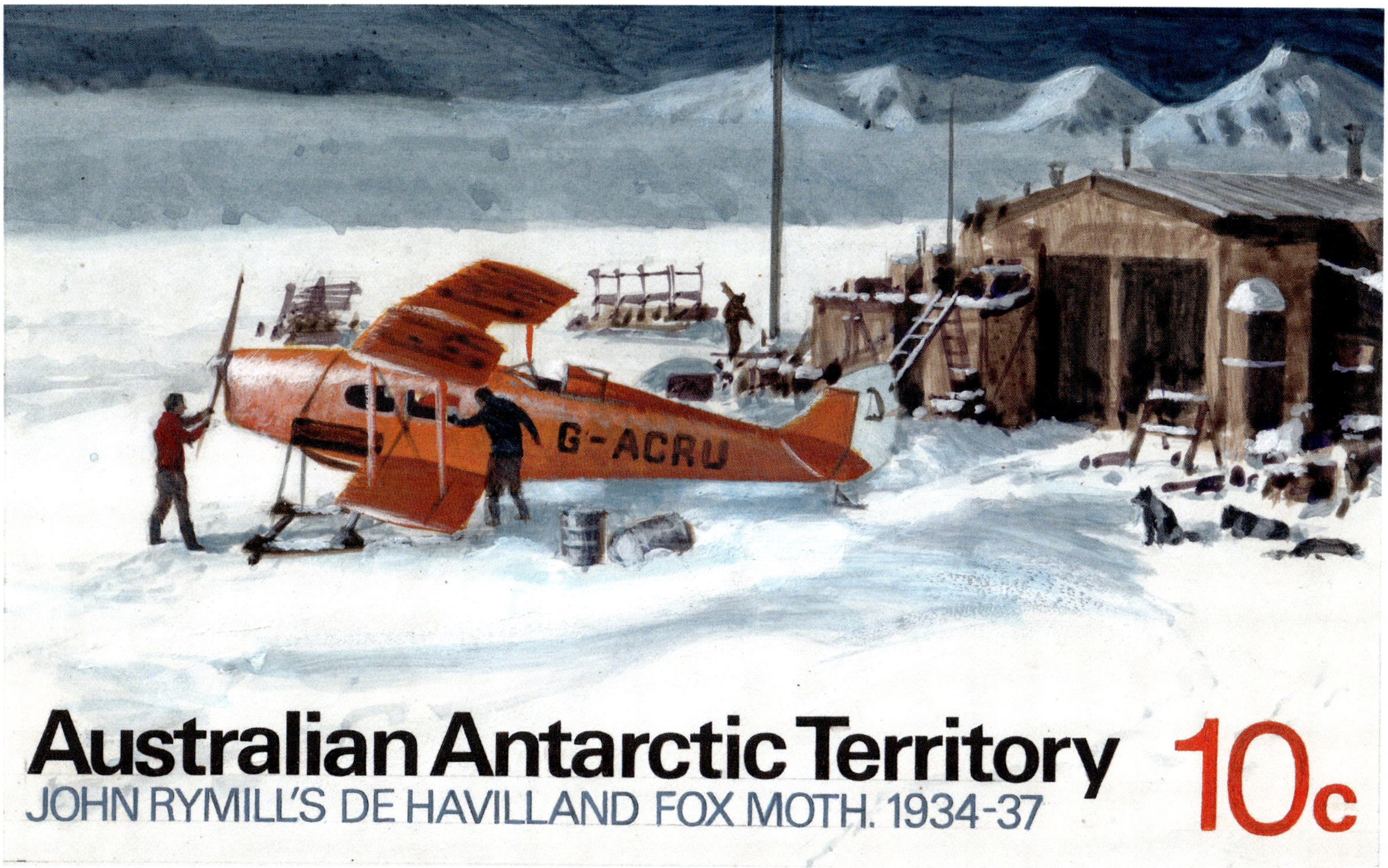

Unadopted artwork, ***Rymill's DH Fox Moth,***
Acrylic on card
Australia Post Archives, Melbourne

Later photos, taken during winter, showed a light overall colour. I wanted to solve both problems, as I wanted one design showing the Fox Moth in summer with floats, and the alternative with ski undercarriage for winter operation. So began several months of correspondence with the Royal Geographical Society, which advised that several expedition members recalled the aircraft was all-over orange, and that the apparent differences in photographs could be caused by light conditions. Eventually, Mr. Wilfred E. Hampton, who had cared for and flown the Fox Moth throughout the expedition, wrote that it "was painted yellow-orange, both fuselage and wings, with tail silver… I am sorry that I am not able to give you the actual paint specifications as I cannot trace the original invoices." One could not ask for more than this, coming from a pilot who was so careful about facts that he had attempted to find a 38-year-old paint specification.

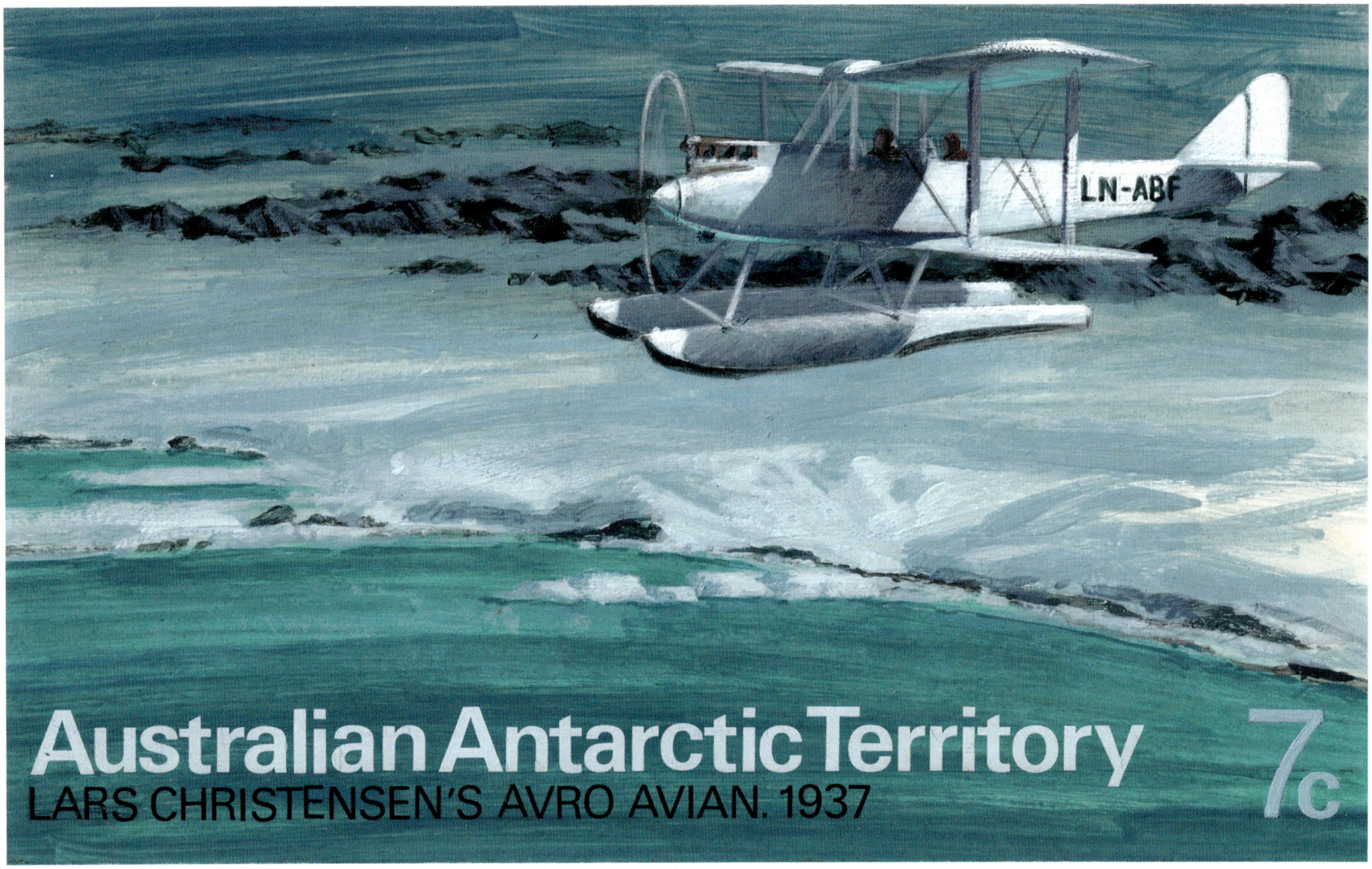

Lars Christensen's Avro Avian

References for this Norwegian's many voyages to the Antarctic were found in a pre-war autobiography, with photographs of the numerous aircraft types he employed at various times. Additionally, he had travelled in much the same area as Mawson.

Nevertheless, as for the other aircraft, I obtained other photographs of the Avian from as many viewpoints as possible. This familiarisation with an aircraft is for me an essential prerequisite. The environment in which an aeroplane moves assures some degree of fundamental aesthetic beauty of form — the nature of aerodynamic shape. Each aircraft has its "essential" viewpoint, which one must find by becoming familiar with every line. This, then, is an example of how a designer spends the early days of a commission. If he finds it fascinating, as I do, well and good. If he finds it a chore it is nonetheless vital.

Ray Honisett

The Australian Post Office gratefully acknowledges the assistance of all those who contributed to the new series — the Australian National Antarctic Research Expeditions Division of the Department of Science, the Rymill family, geographical societies, museums, aviation historians, and others who went to much trouble and lent irreplaceable historical material to ensure authenticity.

Published in the *Australian Post Office Philatelic Bulletin*, September 1973

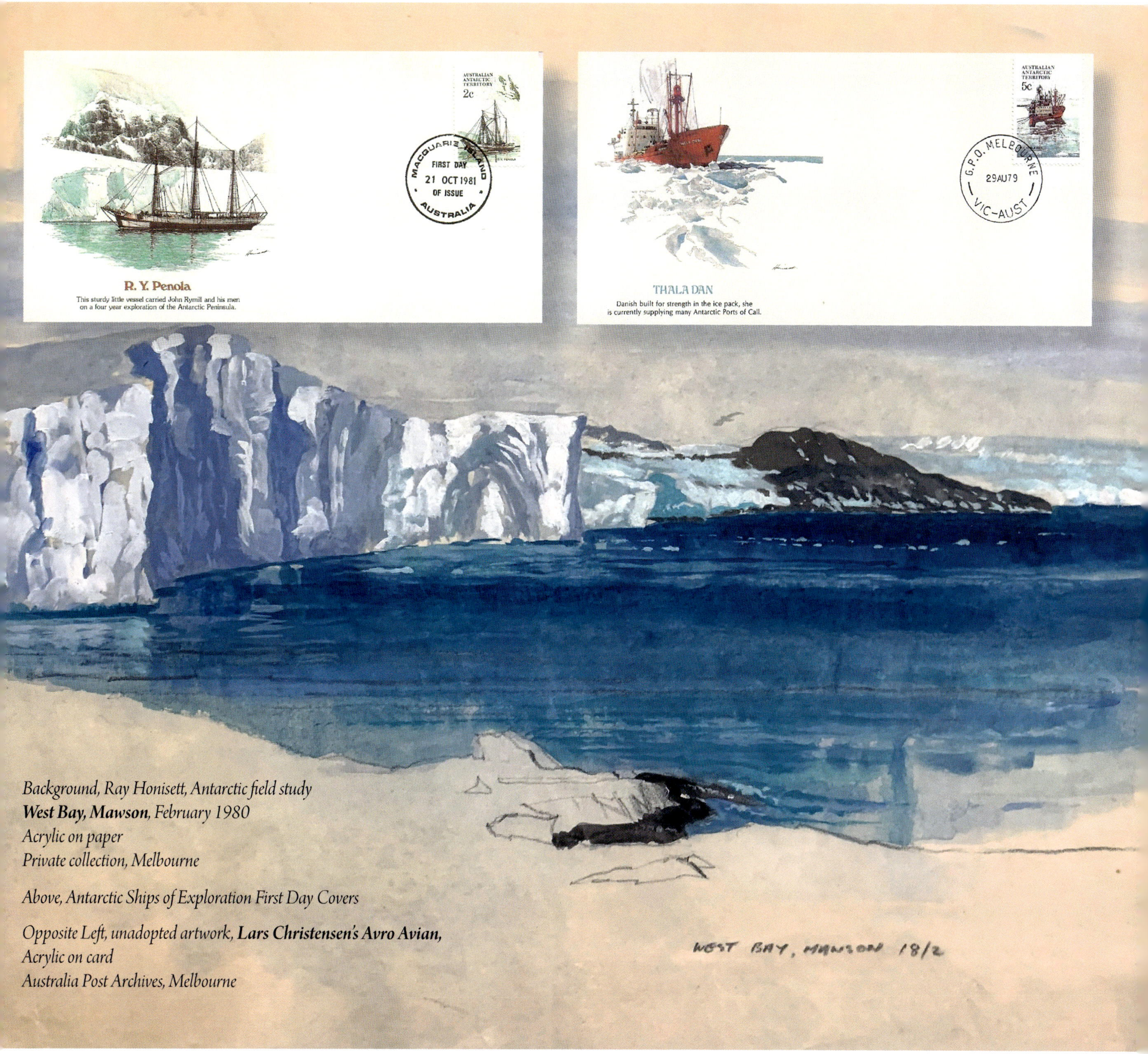

Background, Ray Honisett, Antarctic field study
***West Bay, Mawson**, February 1980*
Acrylic on paper
Private collection, Melbourne

Above, Antarctic Ships of Exploration First Day Covers

*Opposite Left, unadopted artwork, **Lars Christensen's Avro Avian,***
Acrylic on card
Australia Post Archives, Melbourne

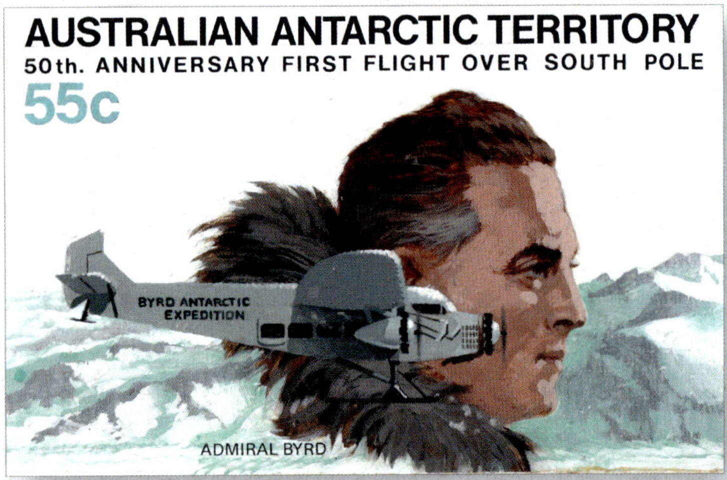

Left above, Adopted artwork 55c, ***50th Anniversary Flight over South Pole****, v on card*

Left below, Adopted artwork 20c, ***50th Anniversary Flight over South Pole****, Acrylic on car*

Background, unadopted artwork, ***50th Anniversary Flight over South Pole****, Acrylic on car*

Opposite, Souvenir cover, 50th Anniversary Flight over South Pole
Australia Post

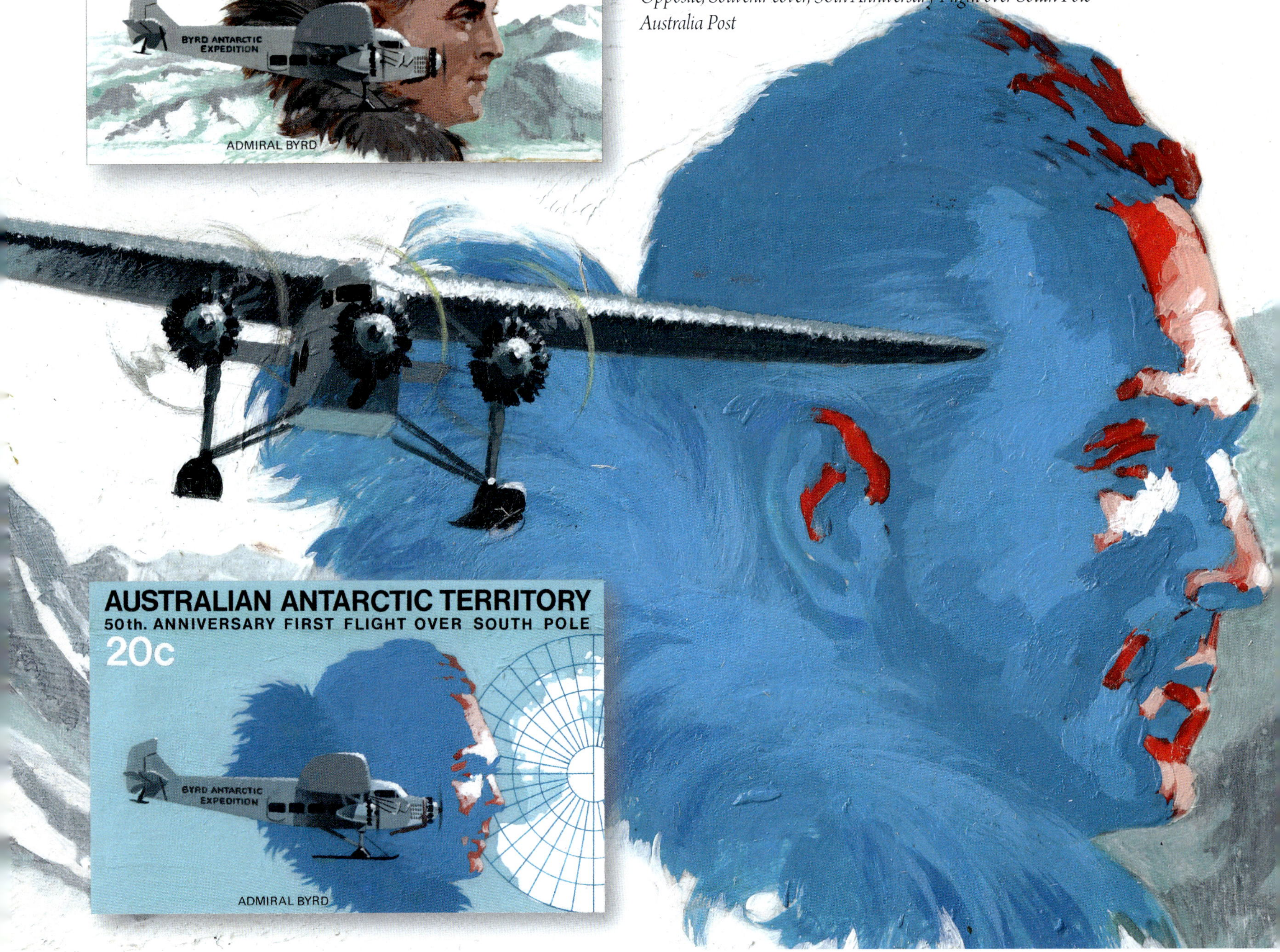

Many directions and themes were explored during the design development of these stamps with the roughs paying several visits to the Stamp Advisory Committee before receiving the eventual approval. In part the reason lay in the variety of thematic options open to me —the man, the aircraft, the event, the place, the climax, the return, the back-up, the reliance on reports of earlier explorers and so on. Unlike the Mawson Centenary designs, the portrait core was not, at first, an obvious choice because Richard Byrd's inherent modesty quite permeated his record of the expedition, stressing the team aspect, minimising his own strengths. But, while all of the expedition members no doubt made fine contributions, the inescapable fact is that without Byrd there would have been no expedition, no achievement.

Eventually the profile portrait of Byrd became the design feature, highlighting the remarkable man whose adventurous spirit, organisational drive and leadership was to earn him the title, 'Father of Polar Aviation'. And, closely linked to the portrait, the simple, robust Ford Tri-motor aircraft that bravely carried Byrd and his companions to the South Pole and back to base. As with the later Mawson stamps, it was necessary to make two separate, related statements. I chose to depict Byrd, the crisp leader, organiser, master navigator on the 20c design with a predominance of blue overall and touches of orange reflecting in the sharply lit profile. The 55c stamp carried a softer, more sympathetic rendering of the modest man behind the decisive leader. To this I added the aircraft and a view of the Trans Antarctic Mountains, setting the motifs on a white background to contrast with the blue of the 20c design.

The stamp held together as a pair through the common profiles, aircraft and sans serif typography. Reference for the portraits came from photographs in Byrd's book, *Little America* and from a painting in the National Portrait Gallery of the Smithsonian Institution, while that for the aircraft had been previously accumulated when researching for the Explorers' Aircraft stamp series. The Trans Antarctic Mountains were assembled from the expedition's photographs. The above illustration depicts Rear Admiral Richard E. Byrd, USN, as he looked in 1947, some eighteen years older than the stamp image. Hair greyer, face fuller but still gazing steadily into the unknown.

50TH ANNIVERSARY FIRST FLIGHT OVER SOUTH POLE

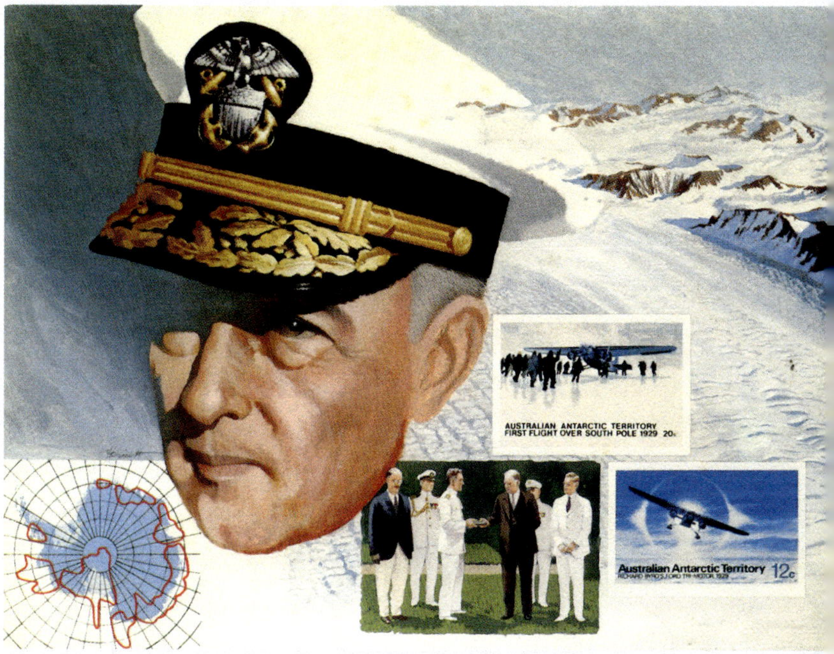

The map shows the areas of the Antarctic photographically surveyed by aircraft of the 1946-1947 US Navy Antarctic Expedition which he led. The figure group depicts Byrd receiving from President Hoover the Special Gold Medal of Honour conferred by the National Geographic Society on 20 June 1930. The stamp designs are the unadopted alternatives for the Polar Flight and Explorers' Aircraft. The former shows the Ford arriving at base after the South Pole flight while in the latter the aircraft is flying over the South Pole with a sun-halo phenomenon behind. The background for these ingredients is the mighty Liv Glacier which provided the aircraft with a pathway through the mountains to the Polar Plateau.

Ray Honisett

SIR DOUGLAS MAWSON CENTENARY — 1982

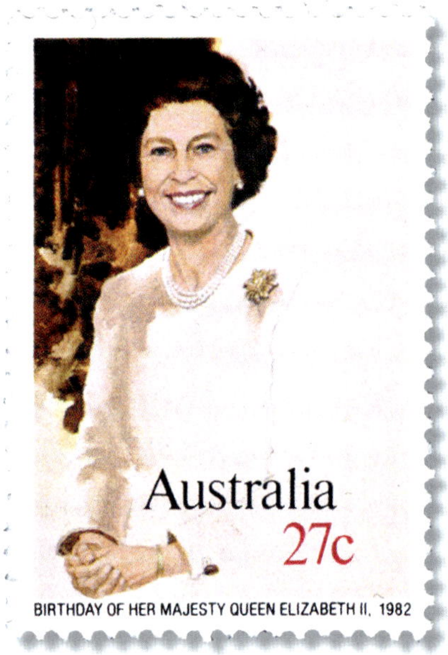

A second 'portrait' stamp was issued on 21 April 1982, just two weeks before the Mawson centenary. It was of a very different nature to Ray Honisett's previous work, a commission to design the traditional Queen's Birthday stamp for that year. Using photographic reference, he produced a striking design, with nearly two thirds of the image left blank to balance the detailed face, jewellery and background treatment. In style and content it could not have been more different to the rugged, weather beaten face of the Antarctic explorer Sir Douglas Mawson, executed at the same time. Ray Honisett would become known for his portrait work, especially historical reconstructions, such the Tim Goldsmith and Frank McNamara portraits he would do for the Australian War Memorial two years later.

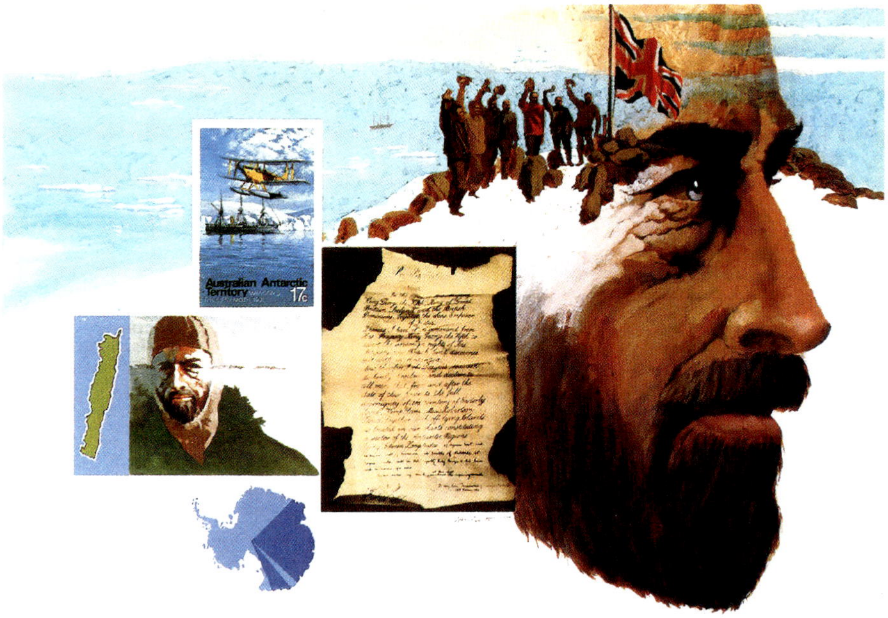

Usually the first and greatest difficulty in designing stamps is that of finding a tangible theme on which to work. The Sir Douglas Mawson Centenary pair, however, presented the opposite problem — how to select a theme, achievement or facet from the long list of outstanding Antarctic contributions made by this great scientist/explorer, any of which could have been developed into an interesting stamp design, relating an absorbing story. Obviously a portrait was essential but, how to set it, how to relate it and translate the whole into meaningful communication? The answer, I decided, lay in grouping or generalisation. One design should echo Mawson's practical achievements while the second told of Mawson, the dreamer. Thus, the 75c design with its conventional, cartographic background groups together Mawson, the scientist, explorer, geographer and leader. A colourful and ordered proclamation of Australian Antarctic Territory which owes its existence to his endeavours. The 27c design depicts Mawson, the visionary and man of the Antarctic. Mawson — Antarctica; each synonymous with the other. This was, to me, a vital message. Certainly the philosophical and spiritual sides of the man had to be as important as his achievements for they were the motor-force, the catalyst that set him on the path to his visioned goals. I had arrived at the conclusion some two years earlier that the visual blending of the head and the Antarctic was the essence in portraying Mawson. This was when I designed a large mural proposal for the Australian Antarctic Division. In fact the 27c design is virtually the same as the central panel of that mural and it is, therefore, interesting to note that the final 34.5 x 23 mm stamp image started life as the major section of an 18 x 4 metre work!

The area of Australian Antarctic Territory in which I chose to set Mawson is within sight of the station that carries his name, where the Casey, David and Masson ranges thrust through the drift-polished ice plateau. The large area of windswept sky typifies the importance of weather and is designed to impart a wild, free contrast to the structured 75c stamp. The above illustration (opposite right) features the angular, noble profile of Sir Douglas Mawson with his piercing blue eyes staring into the unknown distance. Set around and into the portrait is the flag-raising occasion atop the summit of Proclamation Island, Enderby Land. The weathered but still legible proclamation, retrieved in recent years from the cairn, is adjacent. Grouped around these are a map of Antarctica with the areas of Australian Antarctic Territory delineated in darker blue; a map of Macquarie Island, an important wireless link for Mawson's expeditions; the portrait panel from the above mentioned mural design. Also shown is the unadopted, alternative design submitted to Australia Post for the Explorers' Aircraft stamp series. The design centres on the de Havilland Gipsy Moth aircraft, carried on Mawson's 1929-1931 voyages, low over the water with the Discovery in the background.

Ray Honisett

Opposite right, , Mawson Centenary Stamp Issue souvenir cover, Australia Post 1981

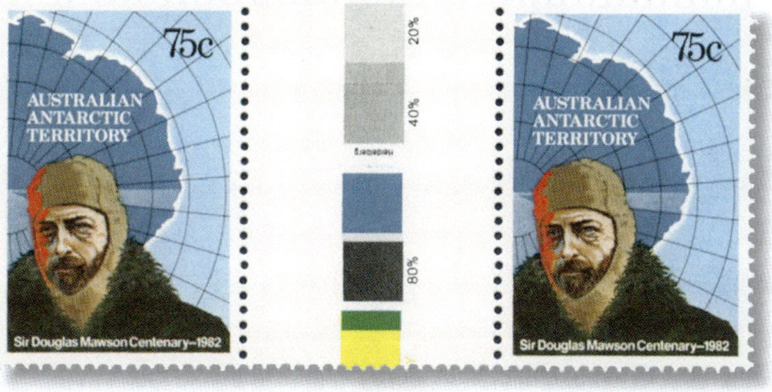

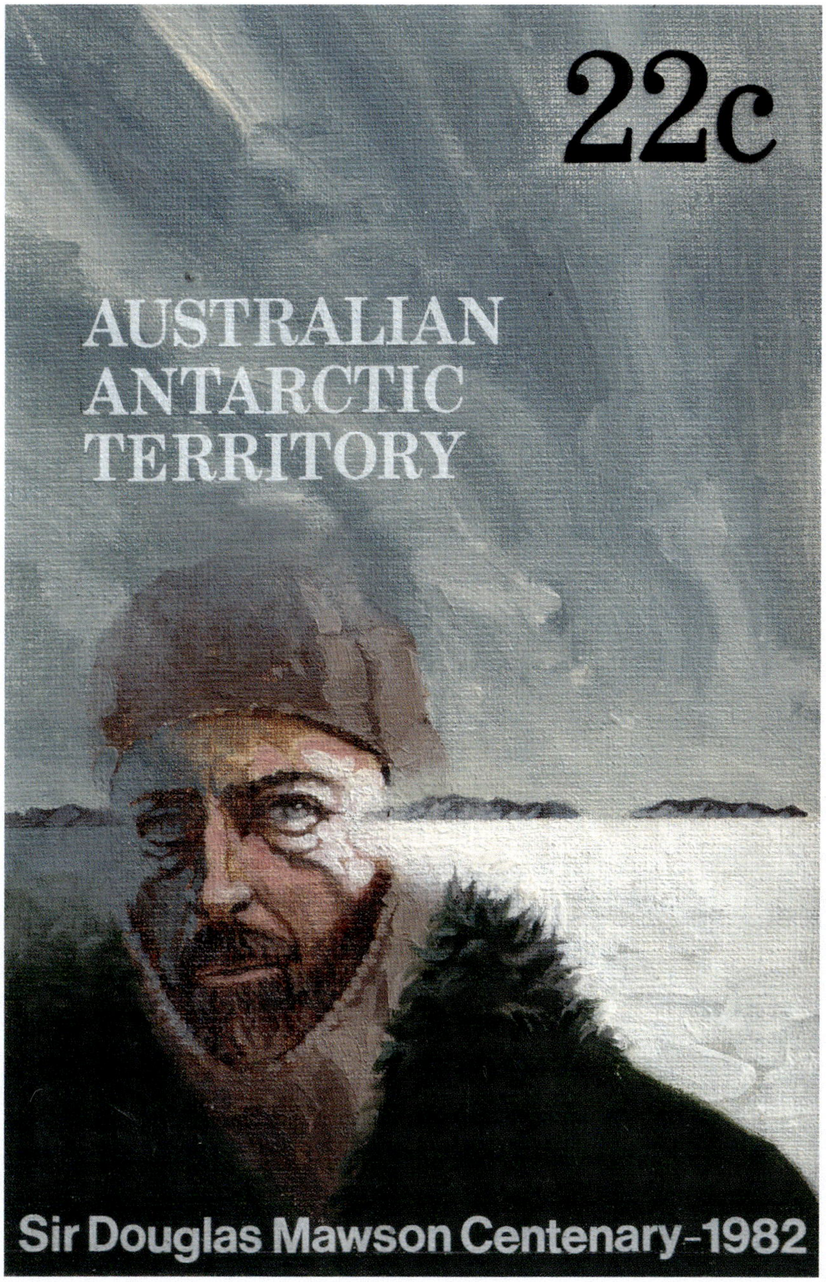

Adopted design, ***Mawson Centenary, 1982***
Acrylic on card
Australia Post Archives, Melbourne

ANTARCTIC SHIPS

SERIES I - III

1979 - 1981

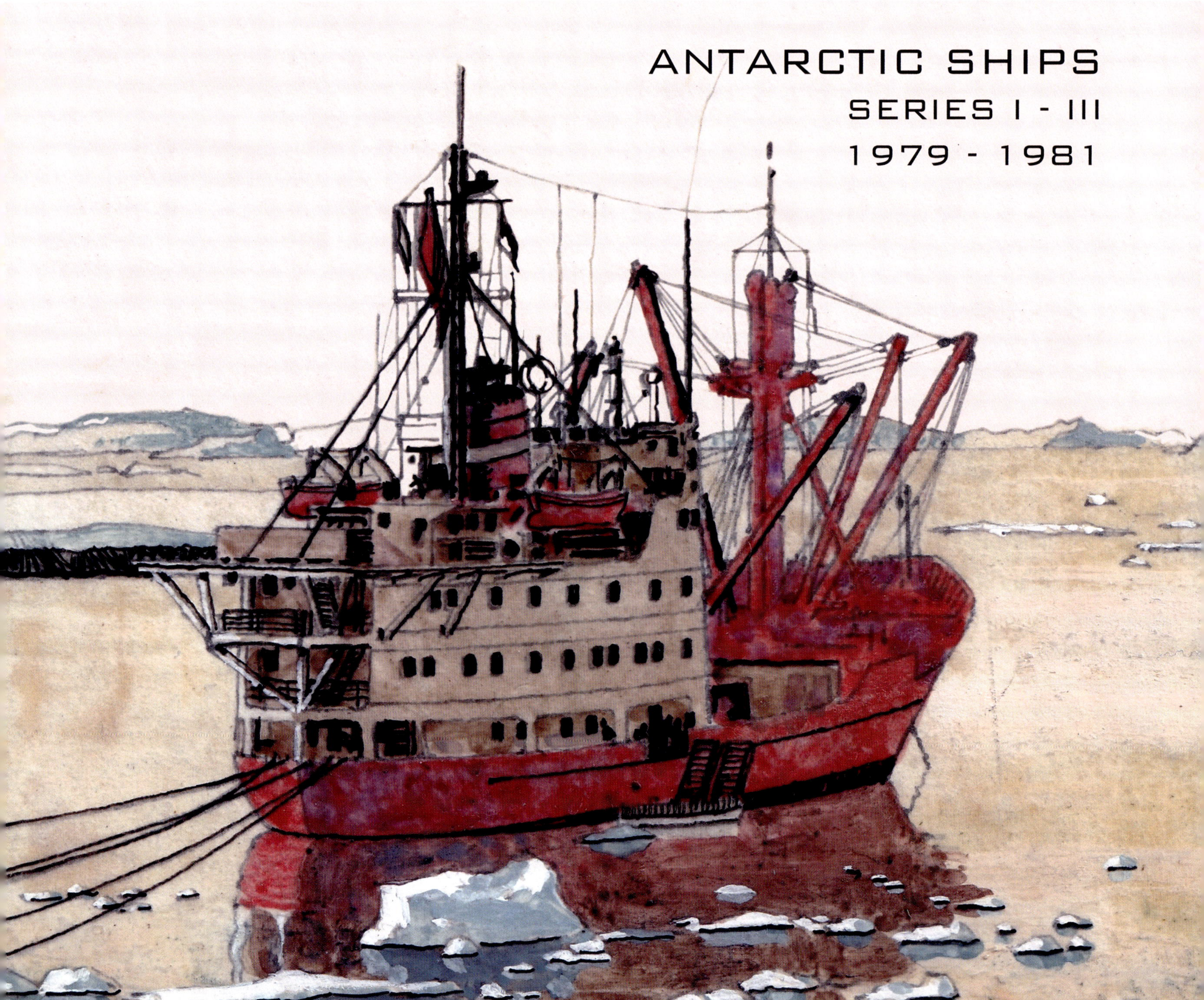

Between 1979 and 1981 Ray Honisett would produce his most significant work for the Antarctic Territory, three sets of stamps depicting the ships that were inextricably linked to the exploration and development of the frozen continent . The three groups of five stamps each ranged from the earliest ships of discovery to the modern icebreakers and supply ships that service the Australian bases. In one unusual aside, it was found that the original reference for *Nimrod* in 1980 had been incorrect and the stamp was reissued in 1981 with the correct vessel. As with the Australian steam locomotive series of 1979, the Antarctic ship issues were accompanied by a lavish series of first day covers and supporting material published by Wind River Studios in the United States. The specialist philatelic company undertook the publishing of a series of supporting paintings which complemented the original stamp designs. These works, mostly in a square format, were at a larger scale and able to incorporate considerably more detail than the official stamp paintings.

The fourteen months between 31 January 1979 and 18 March 1980 would be among the busiest of Ray Honisett's long career. He sailed from Melbourne aboard the MV *Nella Dan*, arriving at Davis Base on 13 February, spending two days there before heading for Mawson, where he had a week on the ground before heading back to Melbourne, arriving on 10 March 1979. Exactly a year later he again shipped from Melbourne, this time on the sister ship, MV *Thala Dan*. They arrived at Casey on 9 February, spending six days there before heading for Dumont d'Urville. Ray spent six days on the ground before heading to Hobart, arriving on 5 March. After just two days to resupply, the ship headed for Macquarie Island, where he spent two days before returning to Melbourne on 18 March 1980. In that 14 months he would spend 58 days at sea and 27 days on land, absorbing experiences that he summed up as … "The closest thing to God, if you believe in such things!"

While the prime purpose of the those journeys was to produce accurate, compelling and convincing designs for stamps, in achieving that end he also produced a large number of atmospheric landscape studies, mostly on paper, and filled sketch books with drawings of his fellow travellers aboard ship.

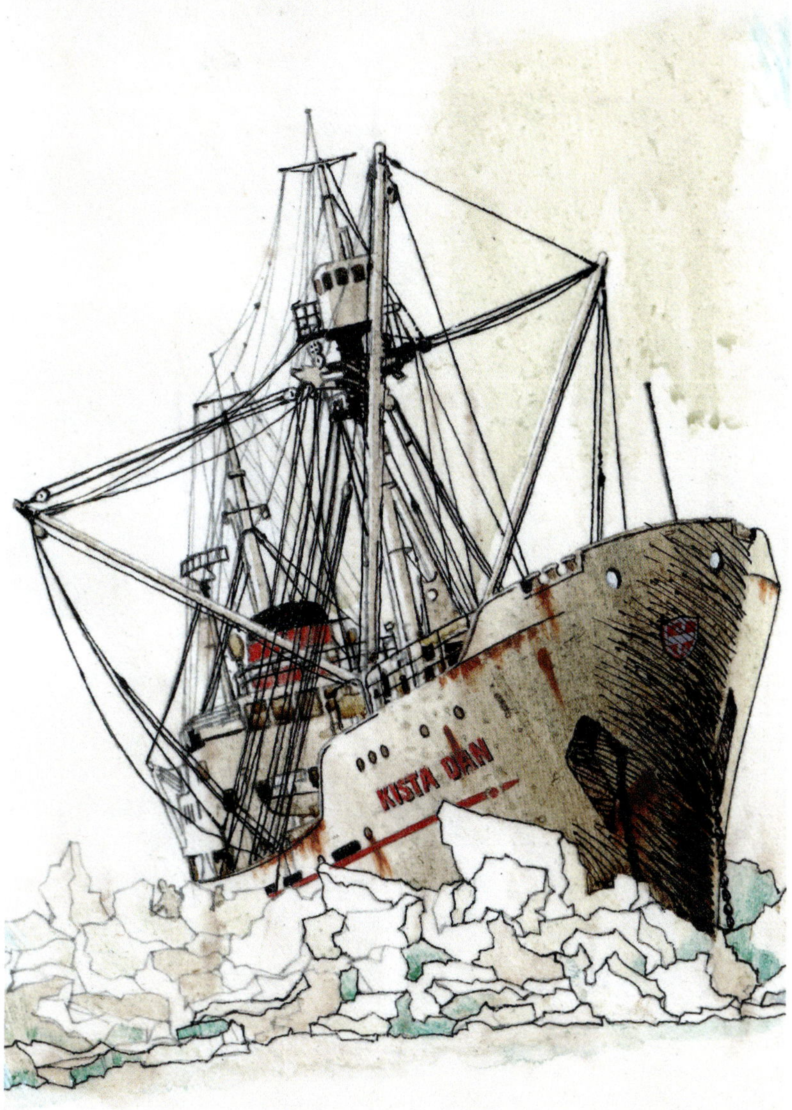

Above, Adopted artwork, ***Kista Dan****, watercolour and ink on paper*

Opposite, Adopted artwork. ***Thala Dan****, watercolour, ink and gouache on paper (Digitally repaired and cropped) This image is reproduced at approximately twice the size of the original artwork.*

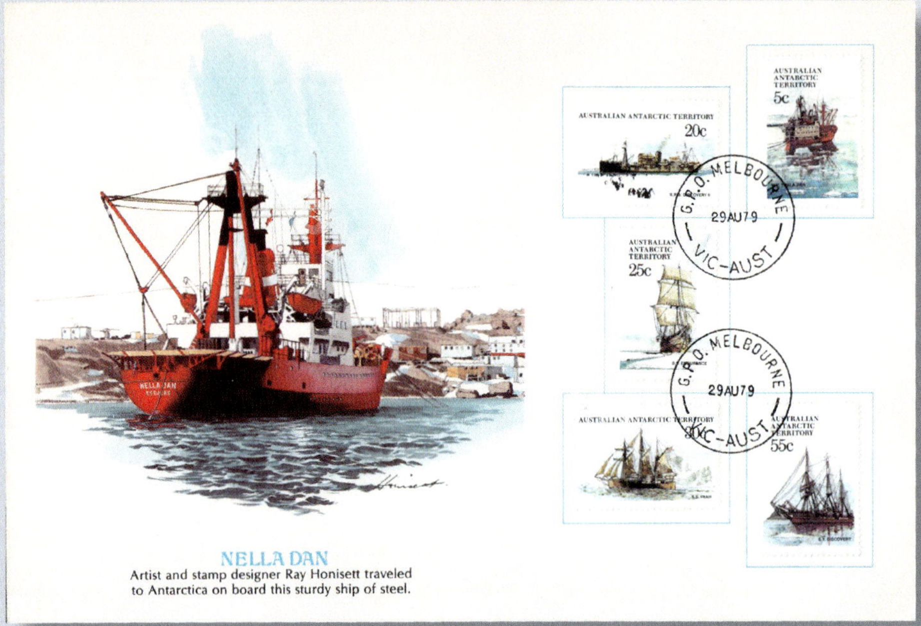

ANTARCTIC SHIPS SERIES I

29 August 1979

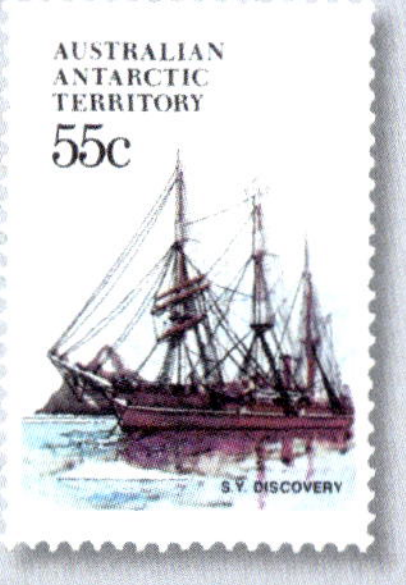

NOTE, The Antarctic ship stamps in this section have been reproduced at approximately twice the actual size of ± 37 x 25 mm, including perforations.

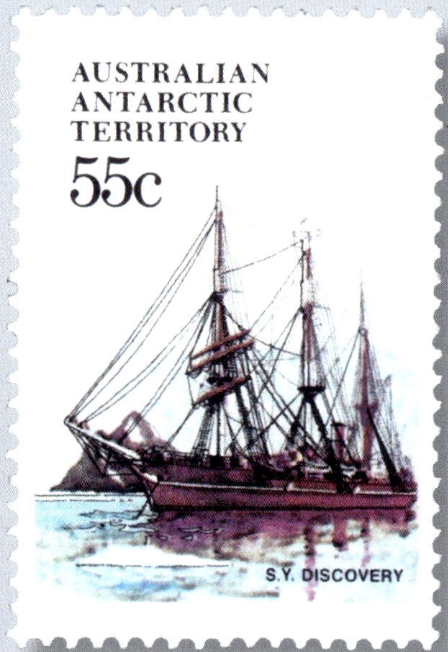

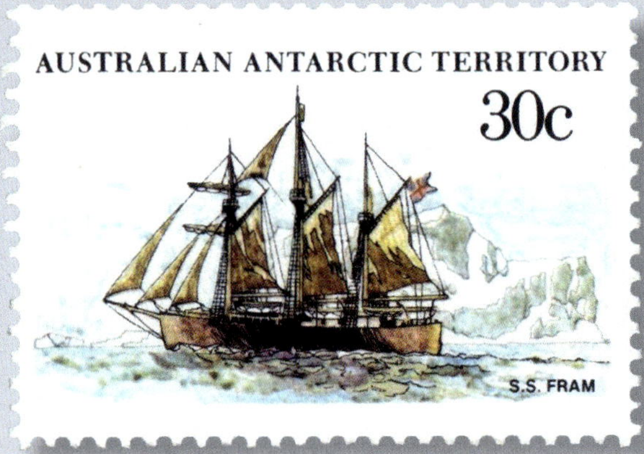

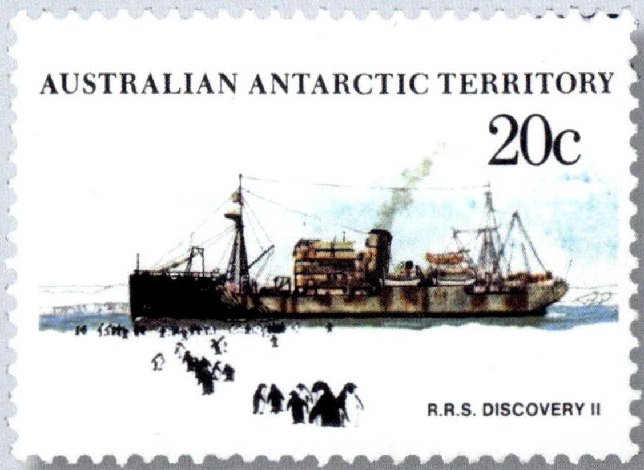

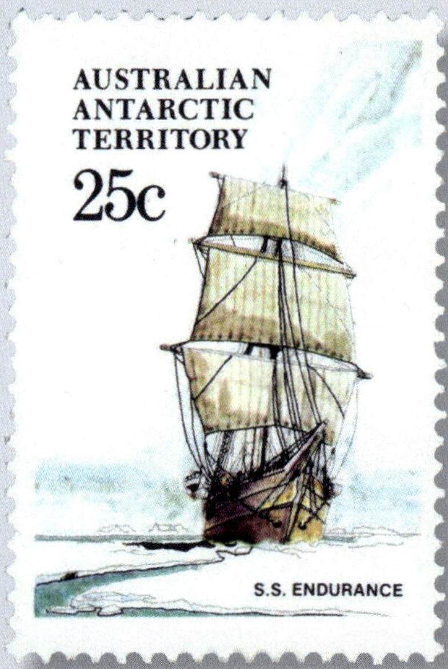

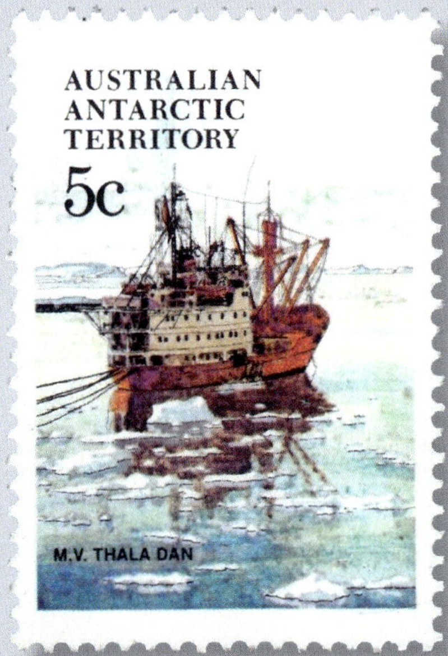

SERIES II
21 MAY 1980

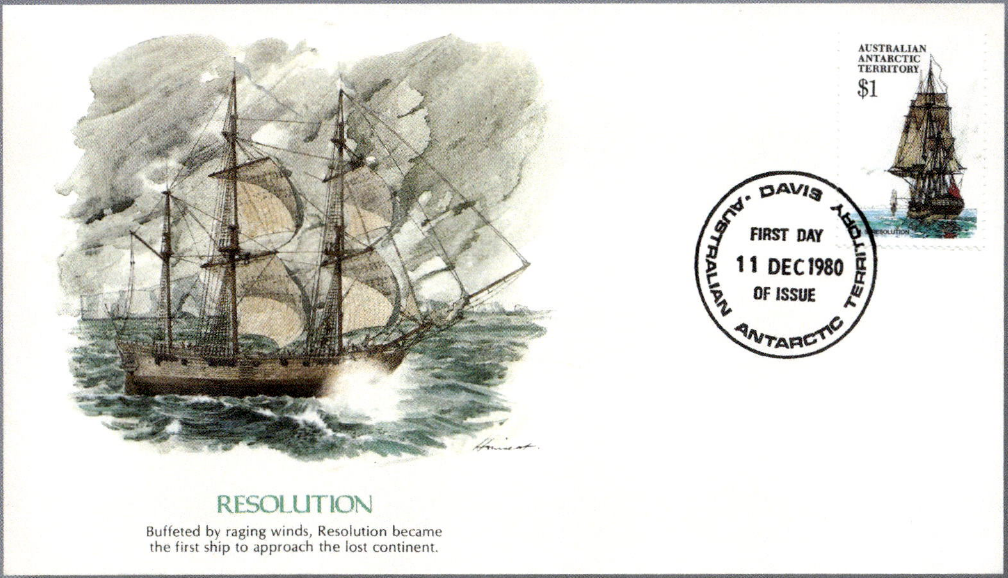

Wind River Studio's first day cover for ***HMS Resolution*** *- while the official day of release was in fact 21 May 1980, the American covers were postmarked 11 December 1980, the first day of issue at the Australian Antarctic base at Davis, providing the added kudos of a true Antarctic issue.*

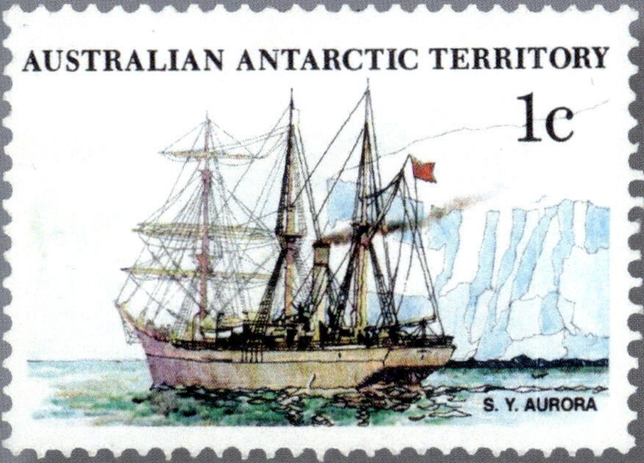

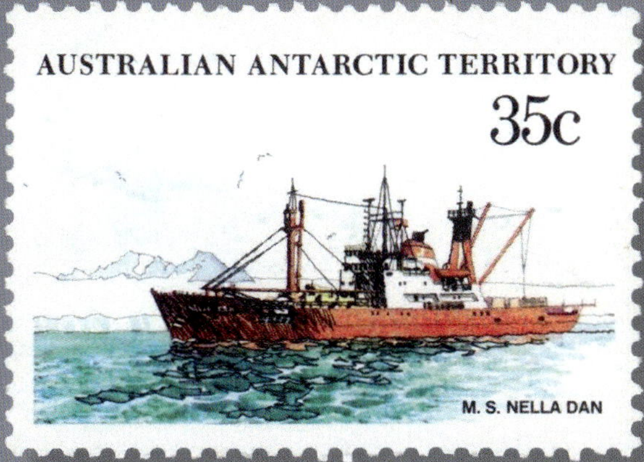

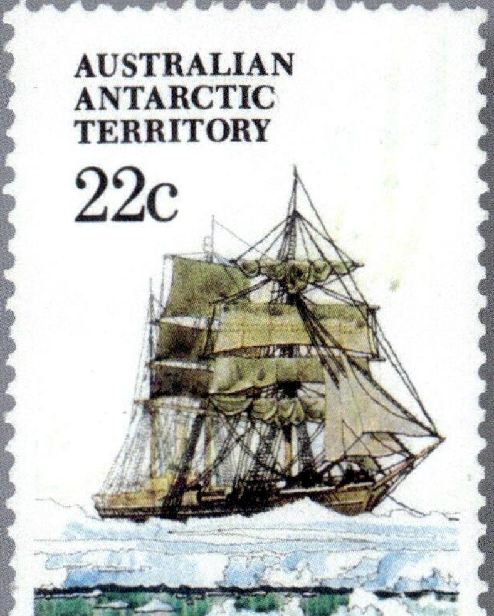

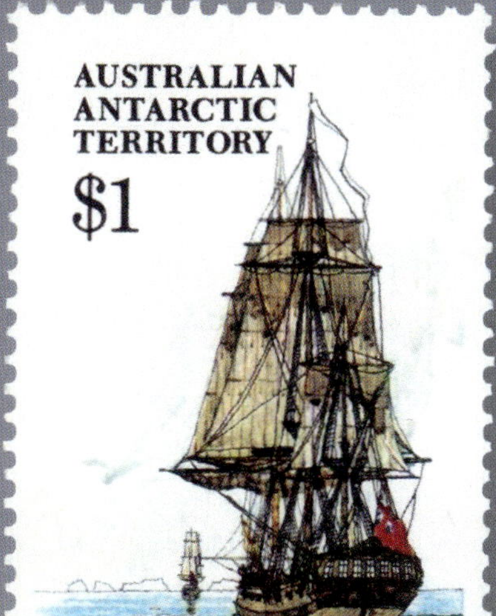

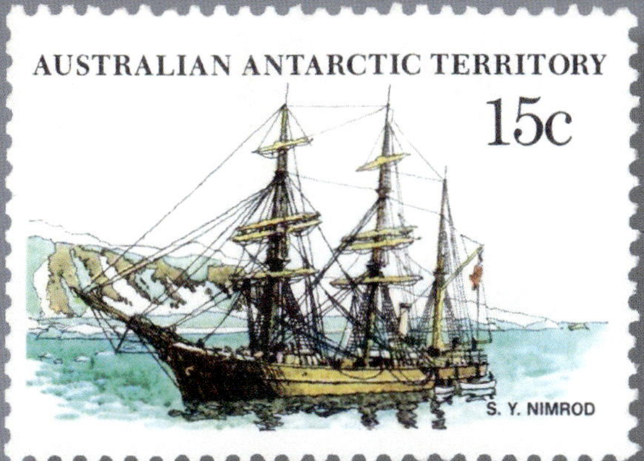

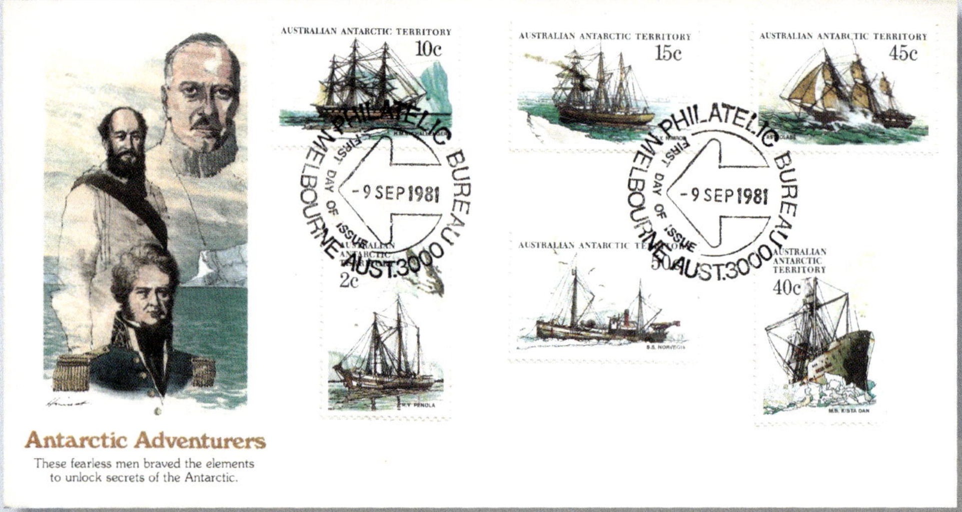

First Day Cover, Antarctic Ships of Exploration, Series III
9 September 1981

Below, Adopted artwork for HMS Challenger,
Watercolour, ink and gouache on card,
Australia Post Archives, Melbourne

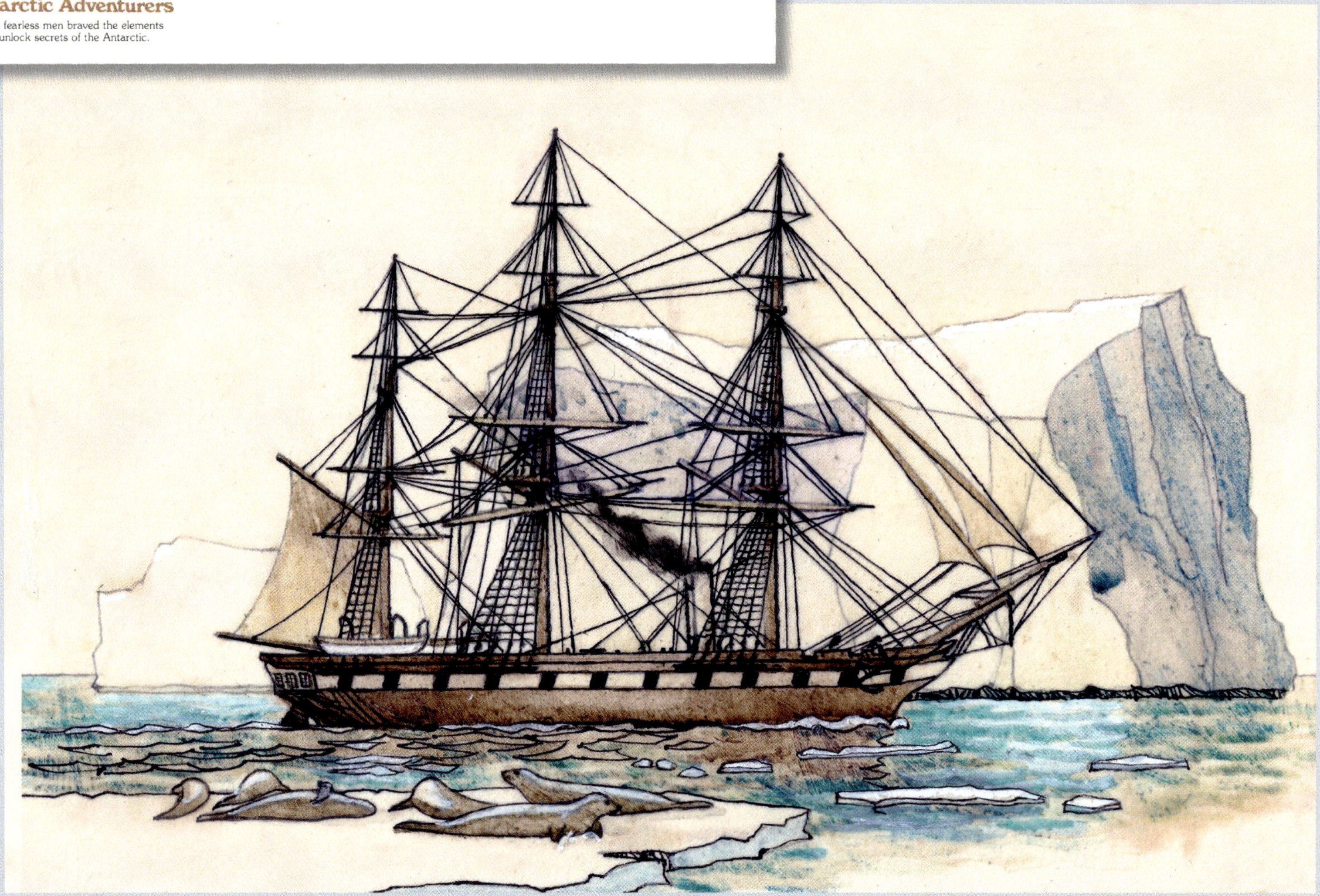

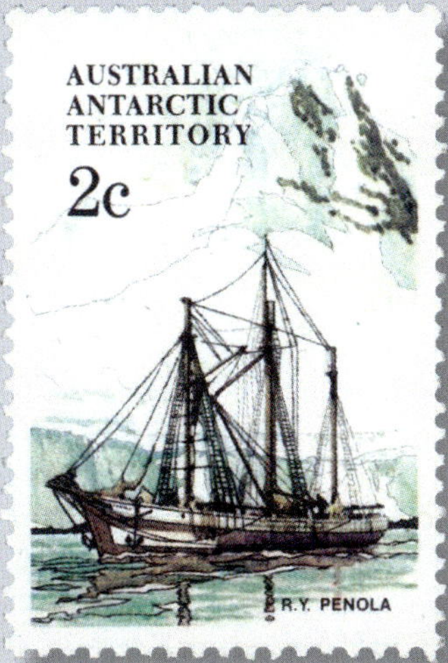

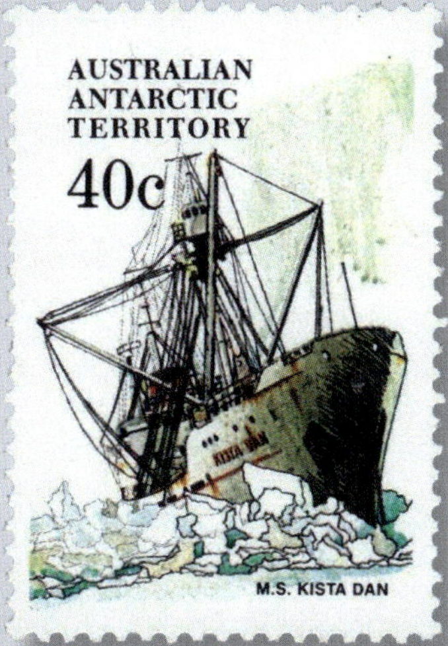

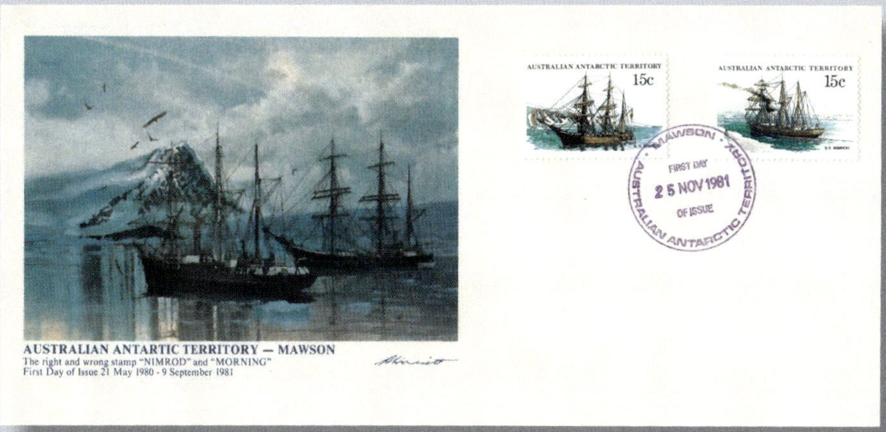

SERIES III
9 SEPTEMBER 1981

NOTE: Includes re-issue of Nimrod to replace incorrect ship in Series II

Above Right, With the re-issue of the Nimrod stamp, Wind River Studios made a virtue of necessity and produced a special first day cover featuring both stamps and a specially commissioned painting featuring the two ships together.

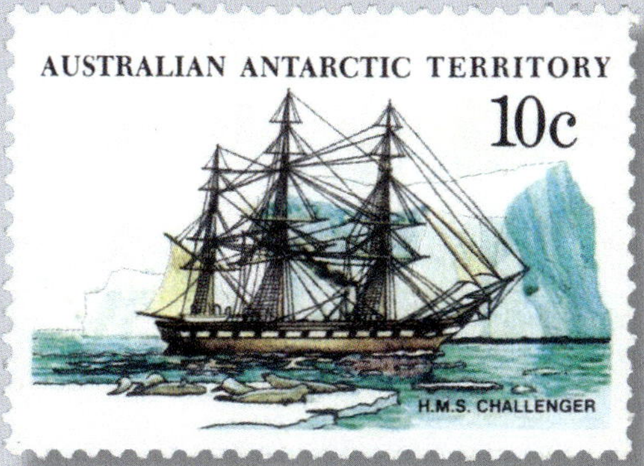

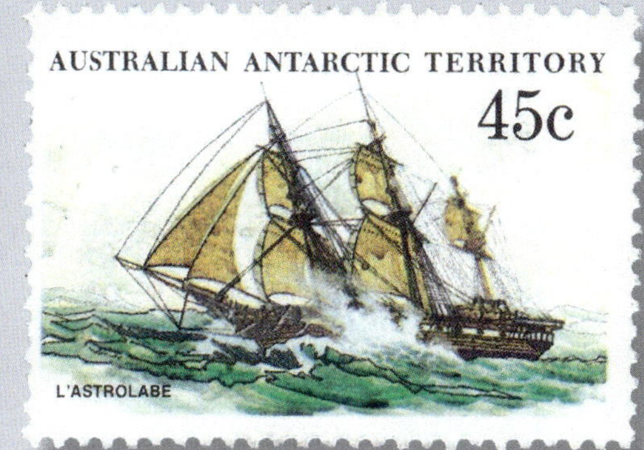

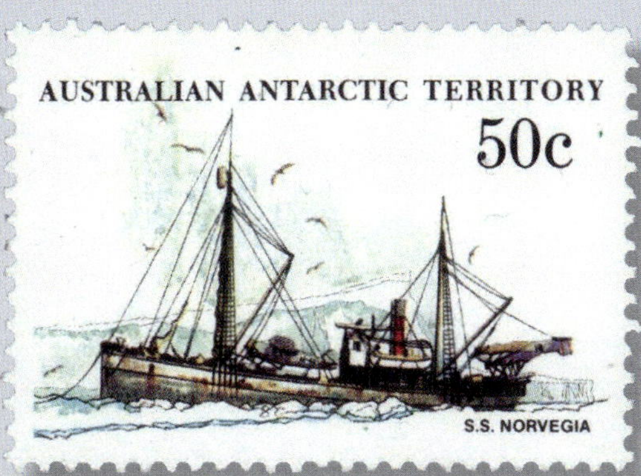

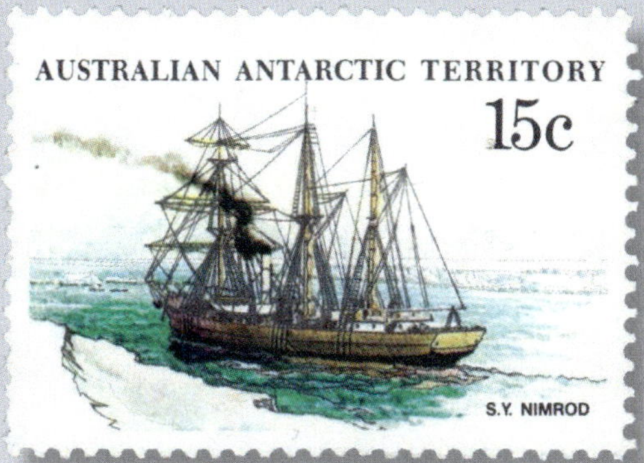

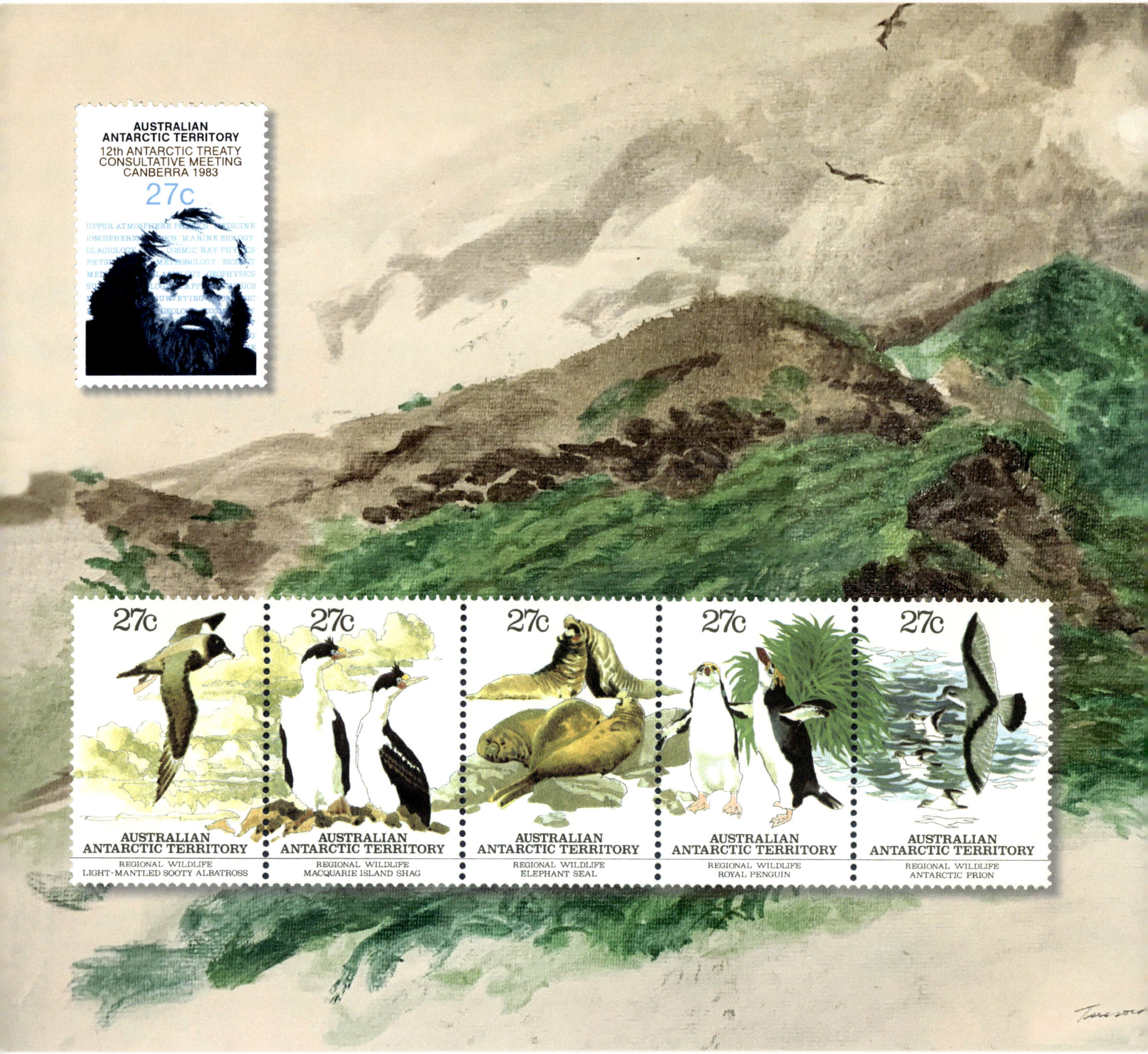
AUSTRALIAN
ANTARCTIC TERRITORY
12th ANTARCTIC TREATY
CONSULTATIVE MEETING
CANBERRA 1983
27c
27c
AUSTRALIAN
ANTARCTIC TERRITORY
REGIONAL WILDLIFE
LIGHT-MANTLED SOOTY ALBATROSS
27c
AUSTRALIAN
ANTARCTIC TERRITORY
REGIONAL WILDLIFE
MACQUARIE ISLAND SHAG
27c
AUSTRALIAN
ANTARCTIC TERRITORY
REGIONAL WILDLIFE
ELEPHANT SEAL
27c
AUSTRALIAN
ANTARCTIC TERRITORY
REGIONAL WILDLIFE
ROYAL PENGUIN
27c
AUSTRALIAN
ANTARCTIC TERRITORY
REGIONAL WILDLIFE
ANTARCTIC PRION

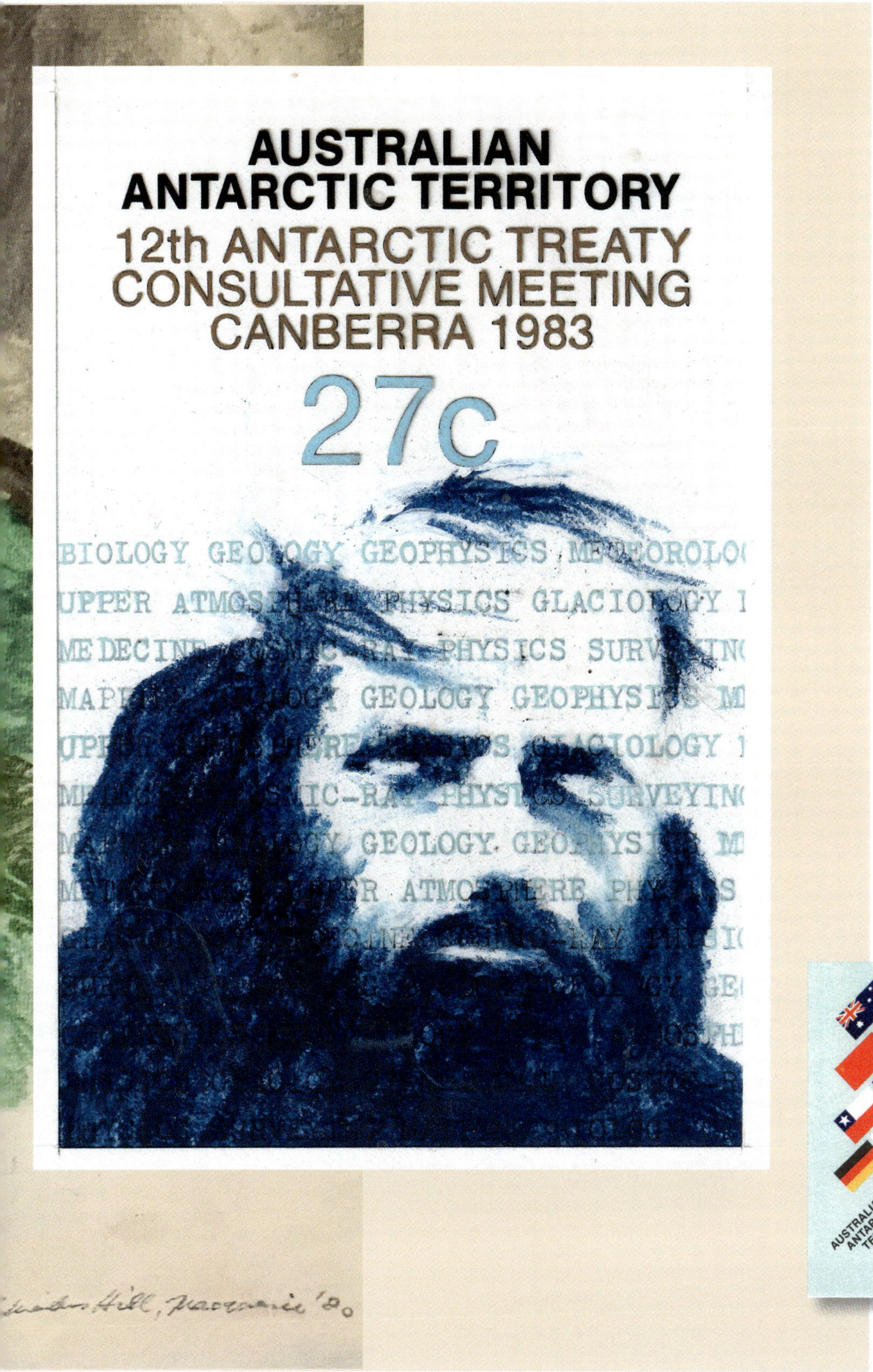

Ray Honisett's final two Antarctic stamp commissions came in 1983. The first was a *Se-tenant* arrangement of five stamps featuring the wildlife of Macquarie island, a format where images are printed from the same plate and adjoin one another, unsevered in a strip or block. Each image is discrete, while the background flows seamlessly from one to the next, creating one overall image when seen as a group. (*se-tenant* translates from French as "joined together" or "holding together".) His last Antarctic stamp came in September of that year, a commemorative issue for the 12th Antarctic Treaty Consultative Meeting being held in Canberra. The image was a graphic rendering of a windblown figure, a scientist working in the outdoors, seen behind a typed list of the many scientific disciplines at work in Antarctica. His first day cover featured the flags of the fourteen nations, the signatories to the treaty taking part in the meeting.

Left, adopted artwork, ***12th Antarctic Treaty Consultative Meeting****, acrylic and Letraset on card*

Opposite, foreground, ***Macquarie Island Wildlife*** *stamps, 1983*

Opposite background, ***Tussock grass, Wireless Hill, Macquarie Island****, acrylic on paper, 1980*

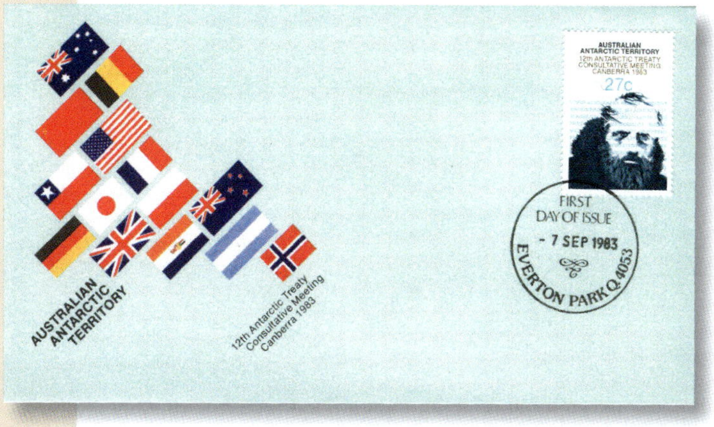

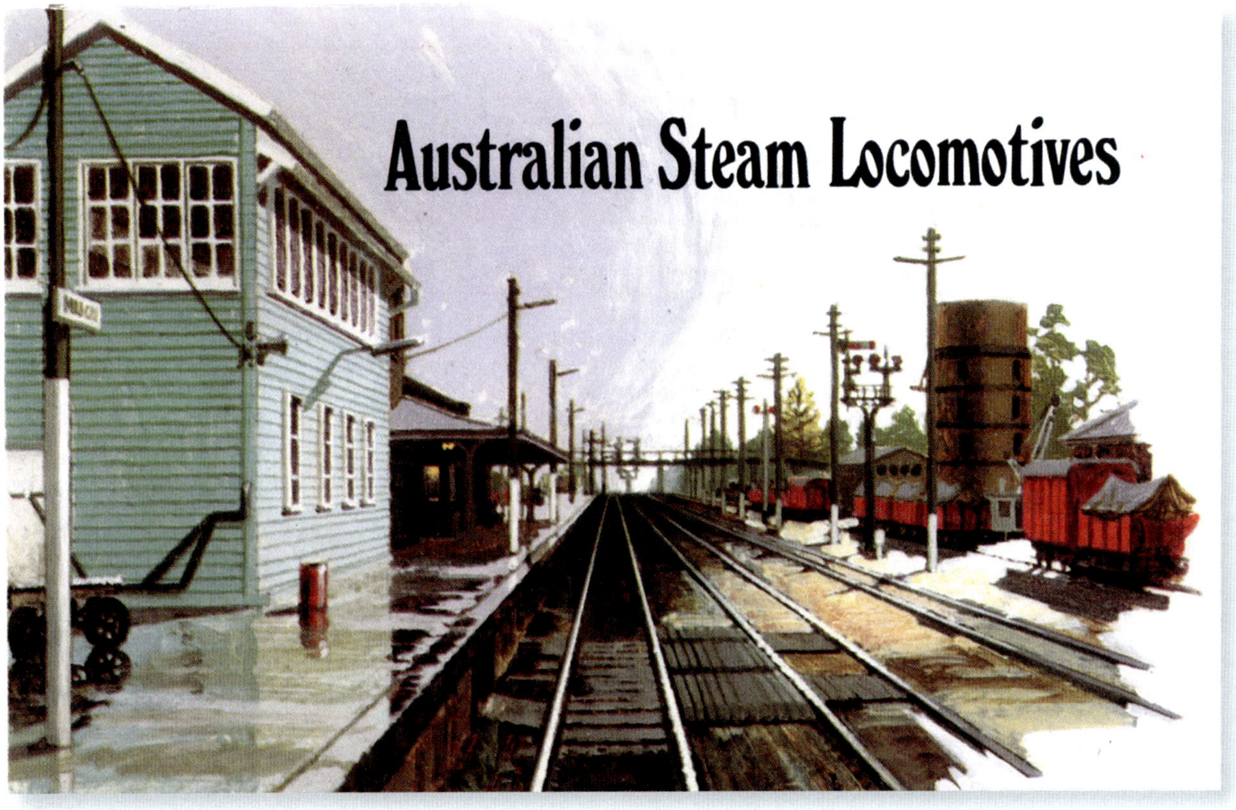

Above, card folder for Australia Post souvenir stamp issue, 16 May 1979

*Below, **Wangaratta Up Passenger S306**, 29 December 1977, Photograph by Weston Langford*

*Opposite, (**Wangaratta Station**), 1979*
Acrylic on card, 30 x 36 cms
Wind River Studios First Day Cover art
Private collection

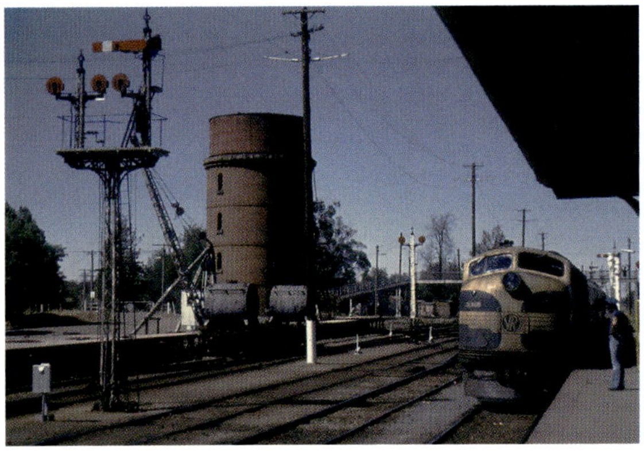

Just as the Macquarie Island wildlife series had shown that Ray Honisett wasn't completely type-cast as an aviation artist, a new challenge came his way with a commission to design four stamps featuring 19th century steam locomotives, three of which were still operating on tourist railways. The appeal of the stamp issue was broadened with a major promotion by the American philatelic company Wind River Studios, who would commission five paintings to be reproduced as first day covers. This featured the four locos shown on the stamps, as well a generic Australian railway station of the period, with Wangaratta being selected as the model. The stamp designs provided the opportunity for landscape content as a scene setter for each loco, to be further expanded upon with the cover paintings. The set was issued on 16 May 1979, one month before the Admiral Byrd commemorative stamps and three months before the first series of the Antarctic ships. It was an exceptionally busy time for the artist, so it was perhaps understandable that the odd error would might slip through. It happened there had been a mistake in the printing of the reference photograph for his depiction of Wangaratta Station and the subject was shown back to front. As the artwork had been sent to the USA for print production the error was not noted until the final first day covers were returned to Australia. A second version of the same scene, designed for the cover of the Australia Post souvenir stamp set, also suffered from the same error.

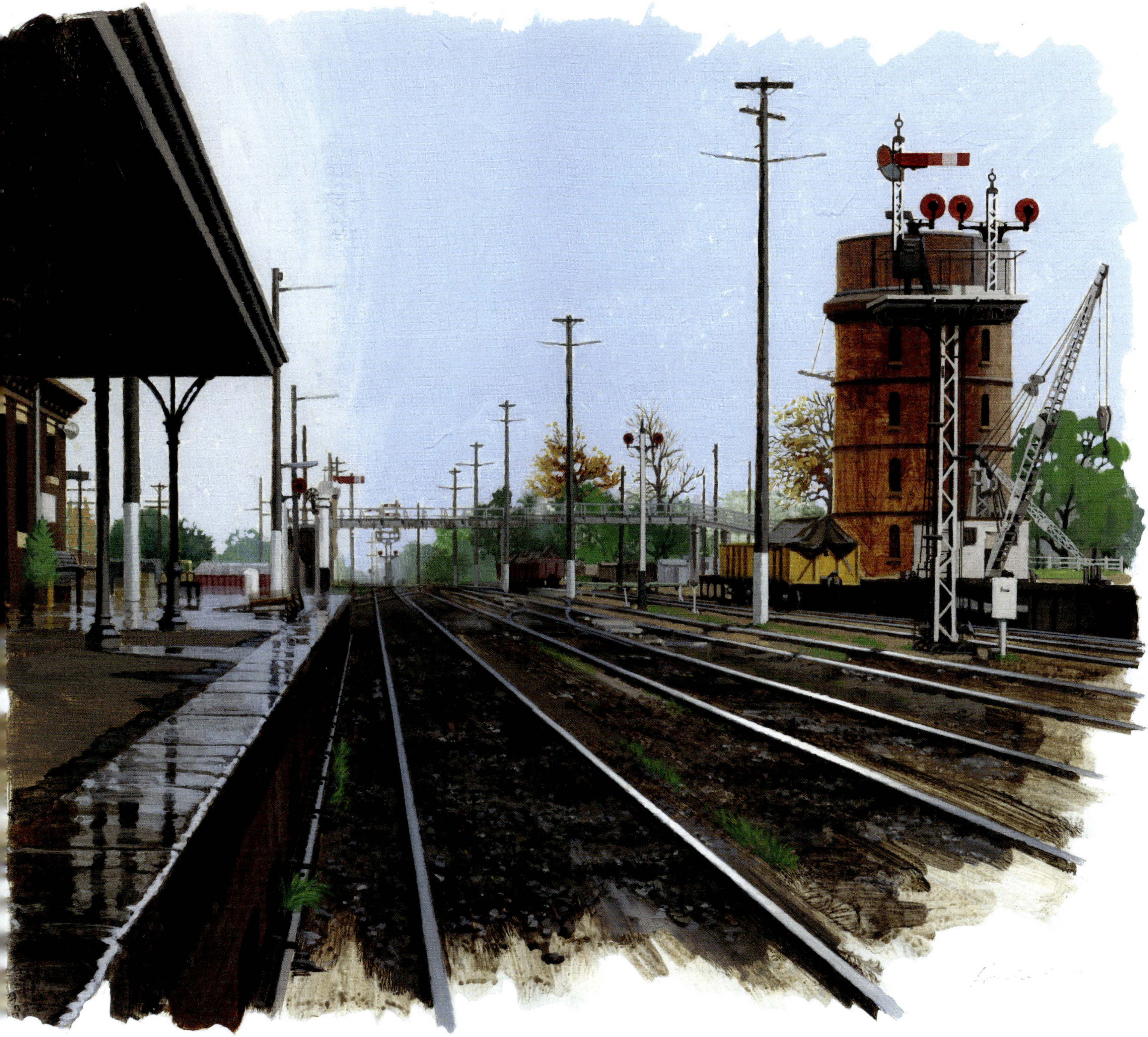

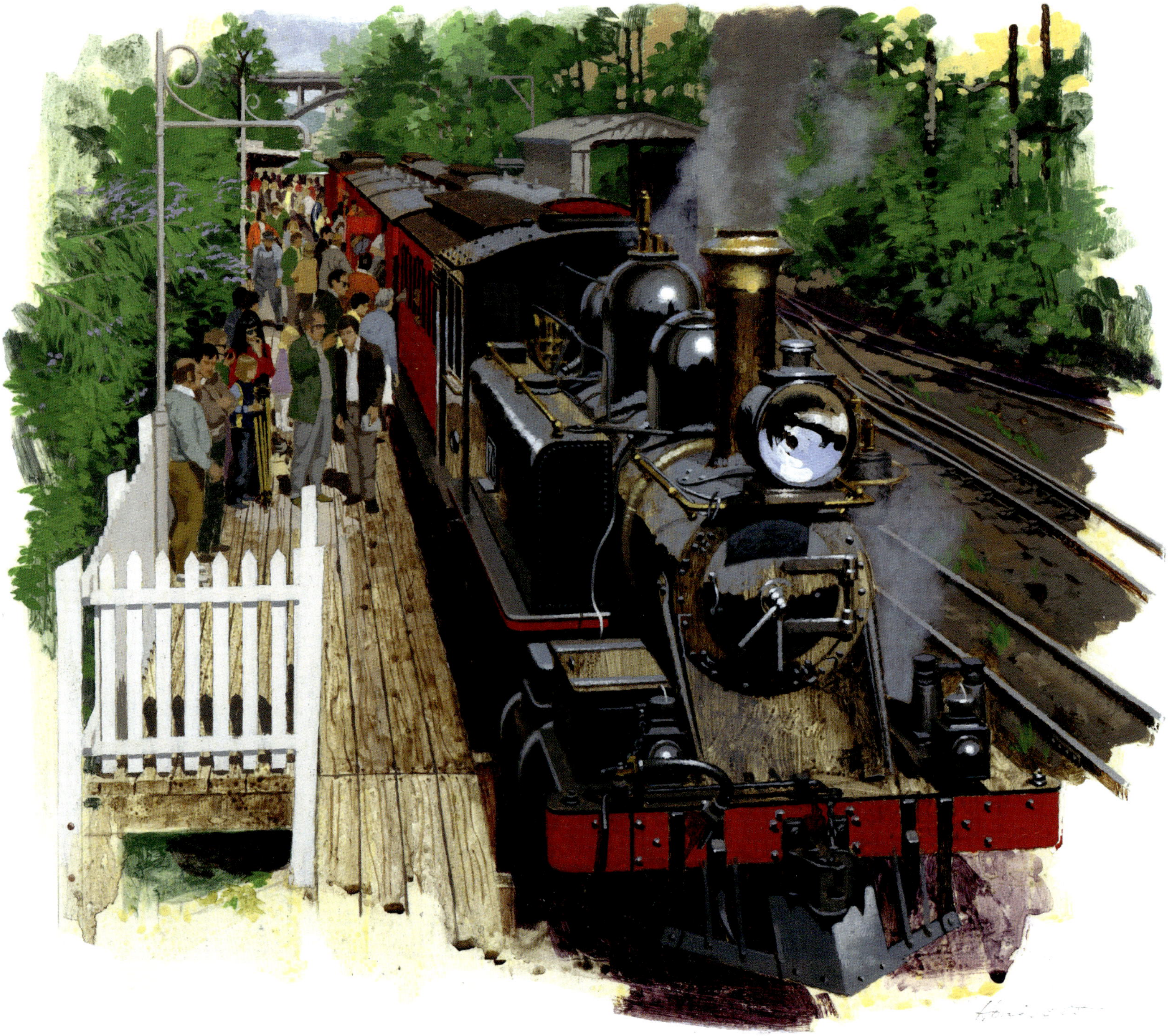

Puffing Billy Railway, *1979*
Acrylic on card, 30 x 38 cms (variable)
Wind River Studios, USA
Private collection

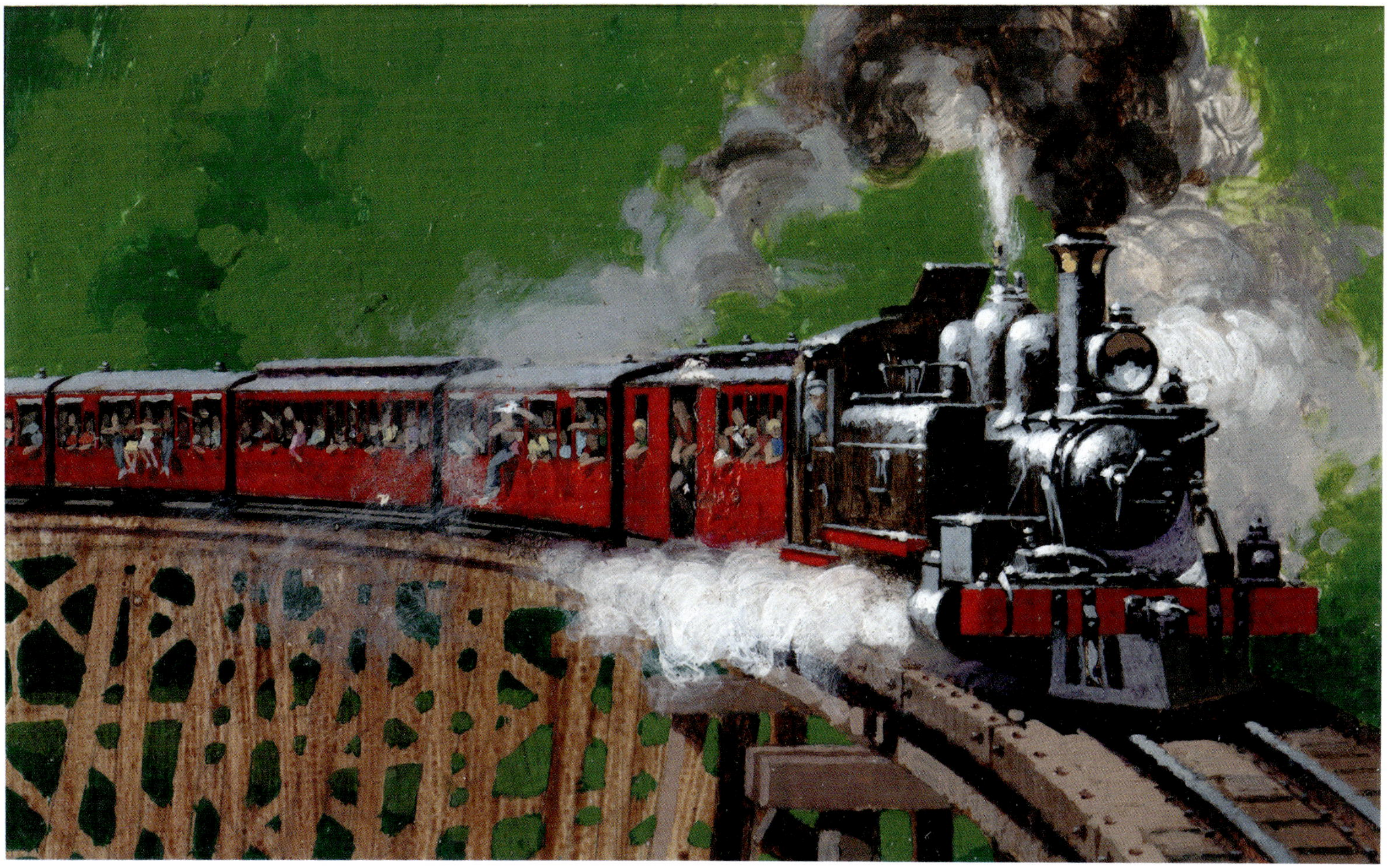

Adopted design, ***Puffing Billy Railway****, 1979*
Acrylic on card, 12.6 x 20.7 cms,
Australia Post Archives, Melbourne

PUFFING BILLY

Many things come into play when a piece of working transport infrastructure changes its status to become a heritage asset and a tourist attraction of worldwide standing. The narrow gauge Puffing Billy line was built to serve the rural hinterland of the Dandenong ranges, to the east of the city of Melbourne. Built at the time of Federation, the line served the small holdings and forest based industries along its route. A landslide blocked the line in 1952 and it was officially closed in 1954. Its value as a tourist attraction was quickly realised and local volunteers took over its operations just eight years later. Its route through the wooded valleys and hills of the Dandenongs and its start just an hour's drive from the city saw it become the most popular heritage railway in the country. Puffing Billy's life as a heritage railway has now outlasted its original working life by nearly a decade.

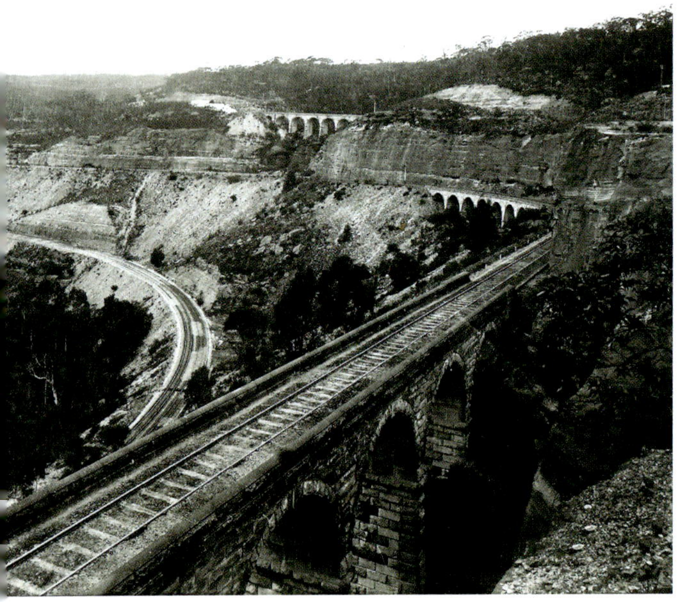

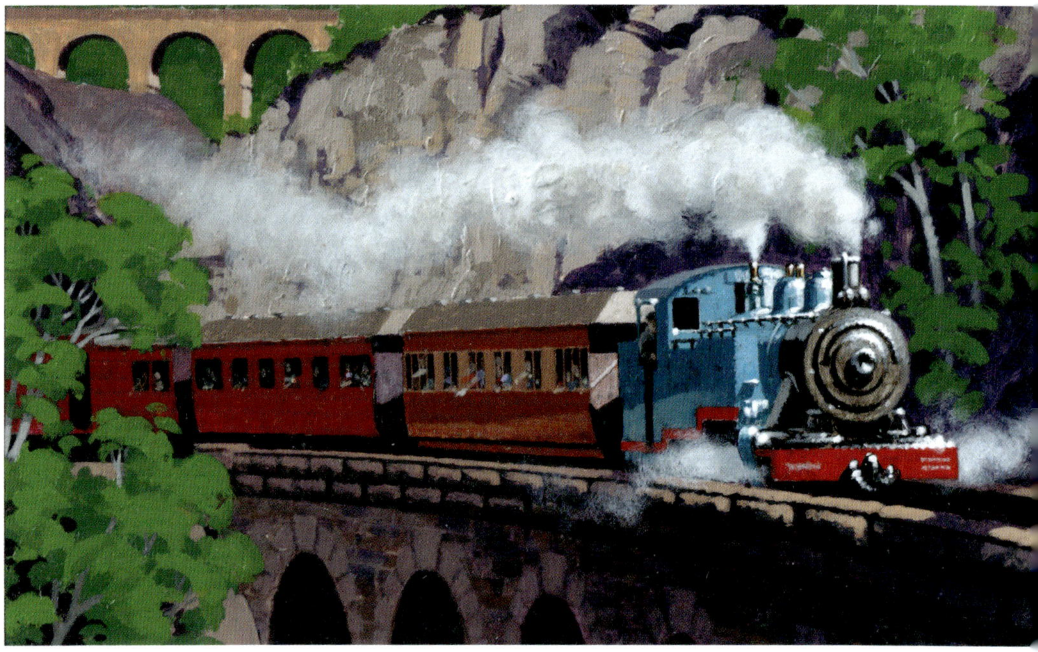

ZIG ZAG RAILWAY

The Great Zig Zag, built to allow the ascent and descent of the steep Western Escarpment of the Blue Mountains, was one of the engineering marvels of colonial Australia. Completed in 1869, the sweeping arrangement of viaducts seen against the mountain was a gift for artists and many made the journey to depict its majesty, a moment of classical splendour against the Australian bush. Conrad Martens famously painted it in 1873, just over a century before Ray Honisett accepted the challenge of showing the train on a piece of card just 126 by 207 millimetres. The locomotive was the subject rather than the engineering of the line, although glimpses appear in both the adopted stamp version and the larger painting done for the Wind River Studio first day cover of 16 May 1979.

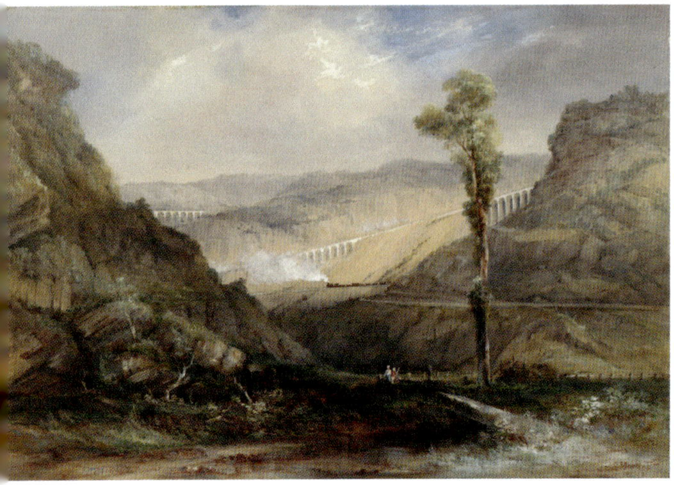

Above, Conrad Martens, ***(Viaducts on the descent to the Lithgow Valley)****, 1873*
Watercolour and gouache, 46.4 x 67.2 cm
National Gallery of Australia

Above right, adopted design, ***Zig Zag railway****, 1979*
Acrylic on card, 12.6 x 20.7 cms
Australia Post Archives, Melbourne

Above left, ***The Great Zig Zag****, Kerry photograph,*
Macleay Museum, Sydney University

Opposite, ***Zig Zag railway****, 1979*
Acrylic on card, 30 x 38 cms (variable)
Wind River Studios, USA
Private collection

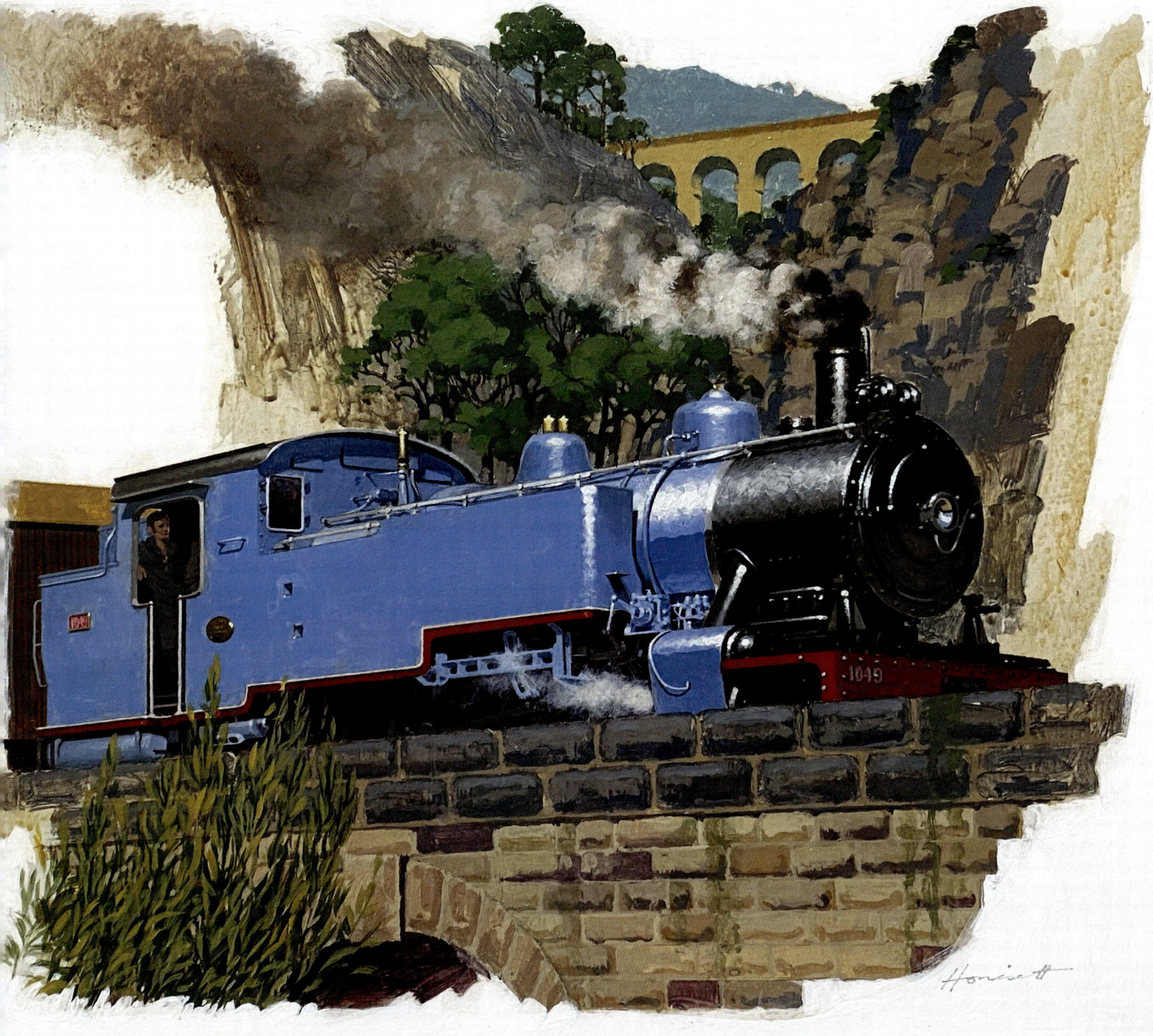
1049

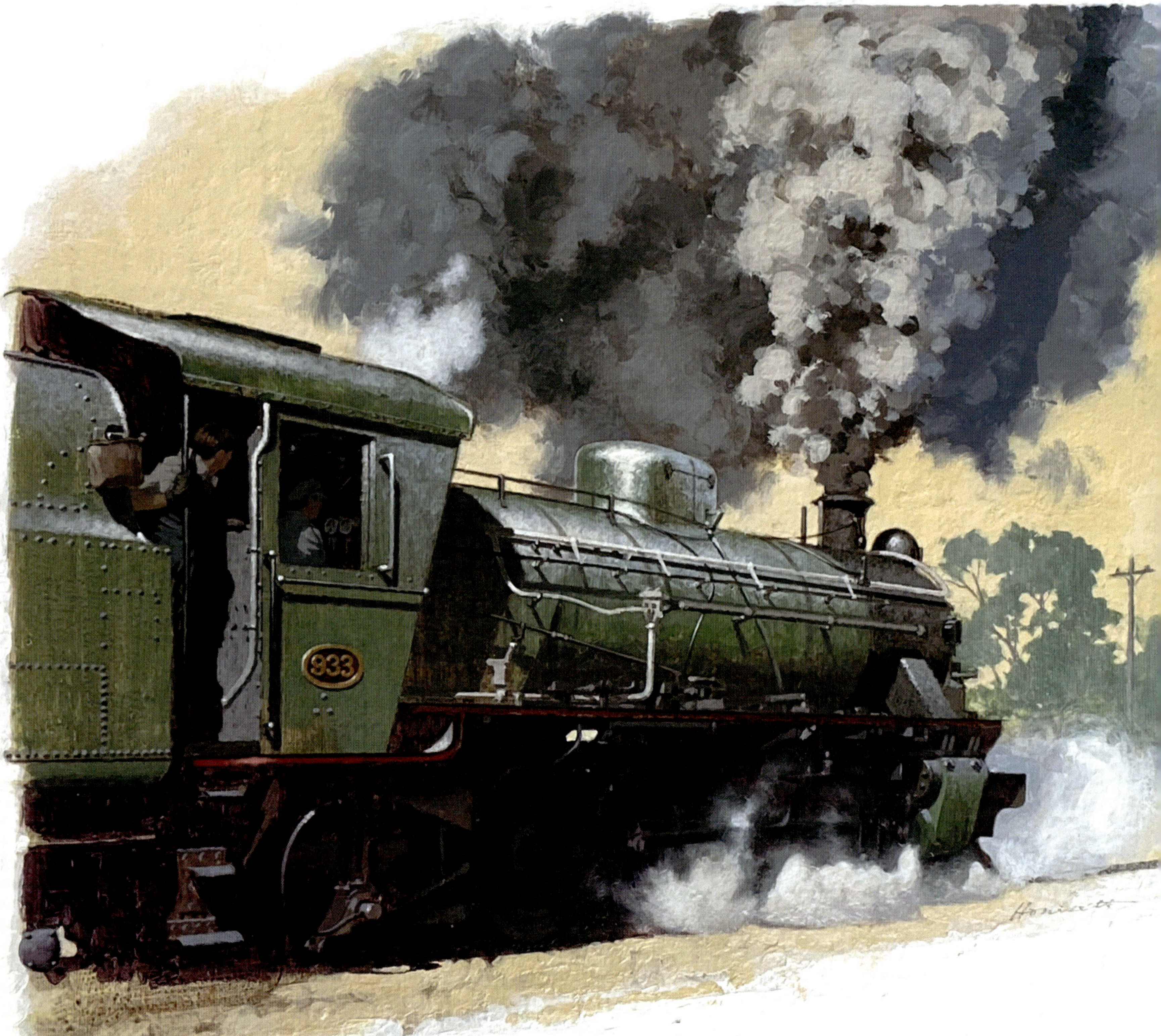
933

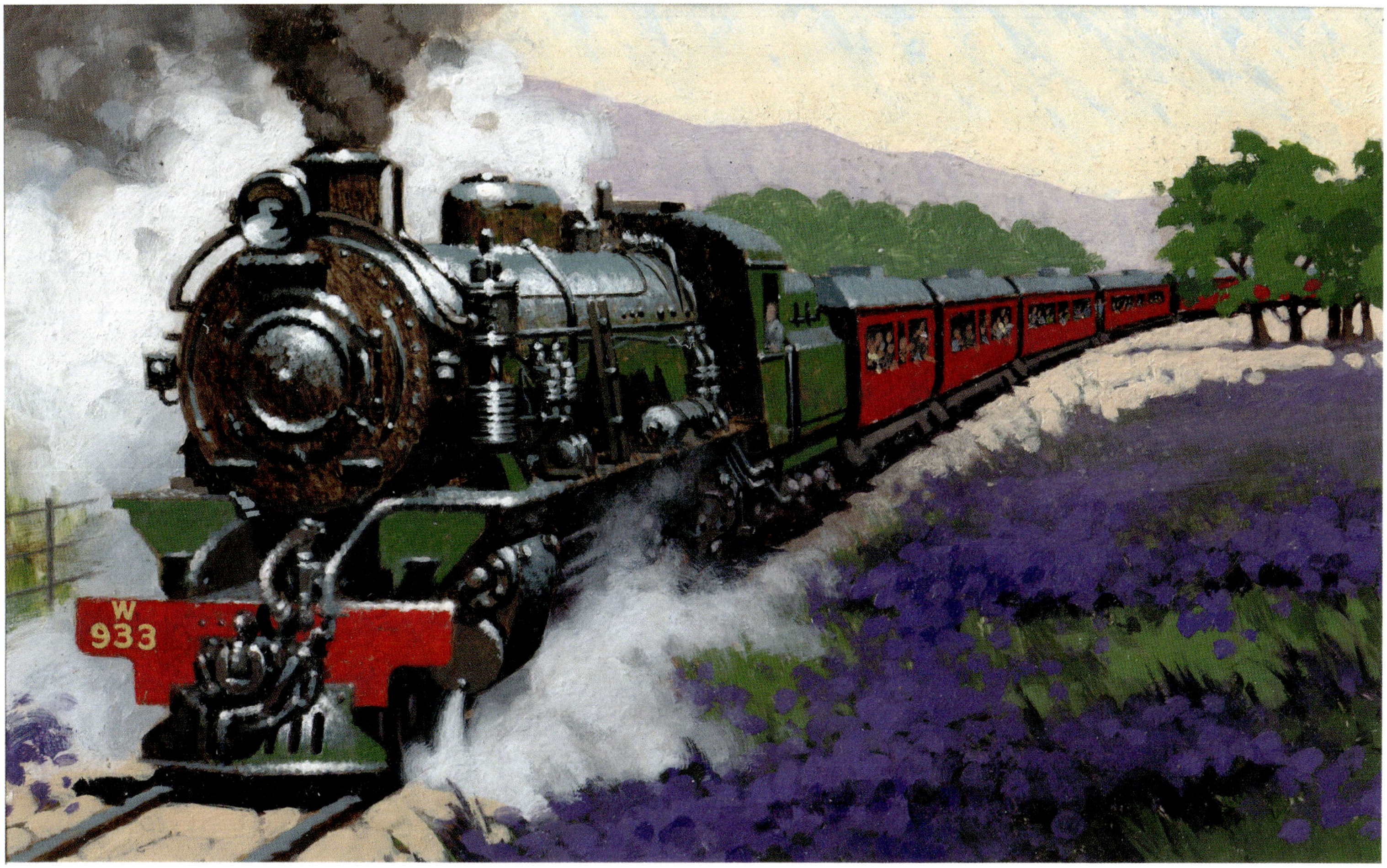

PICHI RICHI RAILWAY

For the 50 cent stamp Ray depicted the Pichi Richi Railway in rural South Australia, running from Port Augusta through the Pichi Richi Pass to Quorn, a mining and grazing community on the edge of the arid inland. Opened in 1879, it was part of the first stage of the Great Northern Railway that was intended to link Port Augusta with Darwin. That connection was not made for another century, with the completion of the Alice Springs - Darwin section in 2003. In a nod to local interest, and a chance to introduce some vibrant colour, he chose to include a brilliant field of purple flowers, the introduced species *Echium plantagineum*, a declared noxious weed throughout the country. Universally known in the eastern states as 'Paterson's Curse', the plant is welcomed in drought-prone South Australia as 'Salvation Jane', a good source of fodder for cattle in the dry months.

*Adopted design, **Pichi Richi Railway**, 1979*
Acrylic on card, 12.6 x 20.7 cms,
Australia Post Archives, Melbourne

***Pichi Richi Railway**, 1979*
Acrylic on card, 30 x 38 cms (variable)
Wind River Studios, USA
Private collection

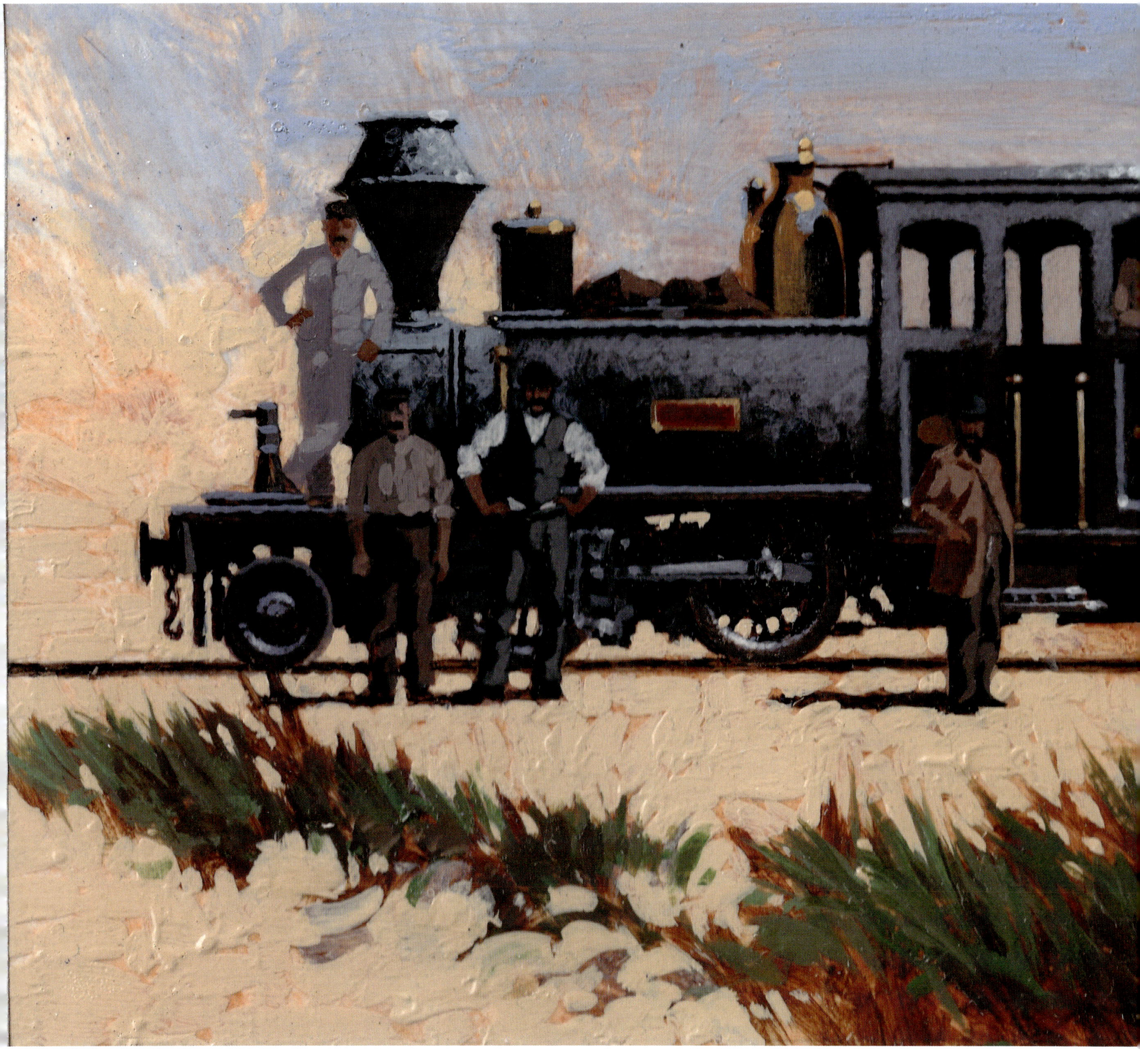

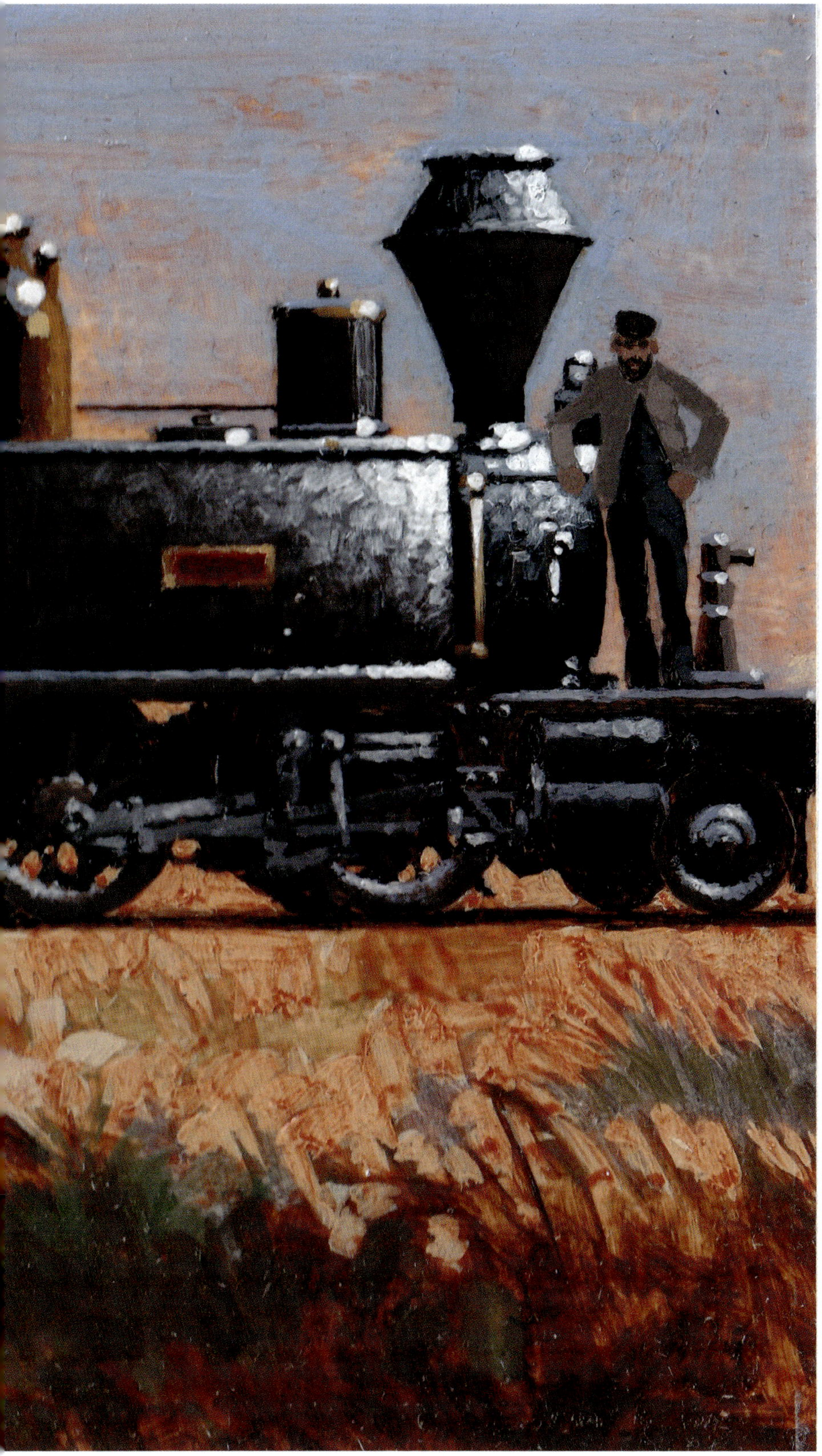

THE DOUBLE FAIRLIE 1879

Introduced by Nicholas Fairlie, a British designer, the Double Fairlie Engine was an early attempt to solve the problems of hauling freight and mineral traffic up steep grades and around tight curves. The Double Fairlie is of a type called an articulated engine - two engines attached together, sharing a firebox at the center. The two engines had pivots where they joined, and this allowed the engineer to apply power to the drive wheels even when the wheels were not parallel to each other. The first articulated engine was built in 1832, but the Fairlie design was not patented until 1863, and this locomotive, owned by the Western Australian Railway, was delivered in 1879, and was one of two shipped to Australia by the Avonside Engine Company, Ltd., of England. The Double Fairlie engines were a great boon to railroad men. Their power under difficult conditions allowed for rapid transit time in the journey to and from the mines, and their ability to adapt to tight curves without loss of speed cut down on the accident rate along the line. But by the 1880's, the Fairlie design was already twenty years old, and more modern locomotives using similar pivot principles were being introduced, solving the same problems more efficiently than did the Double Fairlie. And when mineral traffic declined with the exhaustion of the mines, the Double Fairlies were retired to yard work in Perth. These locomotives went out of service in 1893, but they live on in the hearts of railroad enthusiasts everywhere.

Descriptive text from the Wind River Studios First Day Cover Double Fairlie, 16 May 1979

Opposite, Adopted design, ***Double Fairlie****, 1979*
Acrylic on card, 12.6 x 20.7 cms
Australia Post Archives, Melbourne

This image is reproduced at almost double the original size of the painting, giving a clear idea of intricate technique required for philatelic illustration. The work has to be fine enough to fit in the tiny area of the stamp, yet boldly painted so that detail is not lost in the printing process.

ISLAND HOPPING
1976 - 1991

In simple numerical terms, the largest group of stamps designed by Ray Honisett were for the two smallest jurisdictions, Cocos (Keeling) Islands and Christmas Island, remote Australian outposts in the Indian Ocean. He undertook 29 designs for Cocos and 10 for Christmas Island. The Christmas Island stamps were in the main pictorial, depicting aspects of life of the Island, such as the set of four showing police work and a set of early exploration and naming of the island. The Cocos Island designs were almost exclusively maritime, celebrating the various historical events associated with the remote islands, 2750 kilometres from Perth. The novel 'barrel mail' tradition was celebrated in a set of stamps in 1984, the 75th anniversary of the unique delivery system whereby mail was sealed into a barrel and cast overboard from a passing ship, to be collected by the islanders in a small boat.

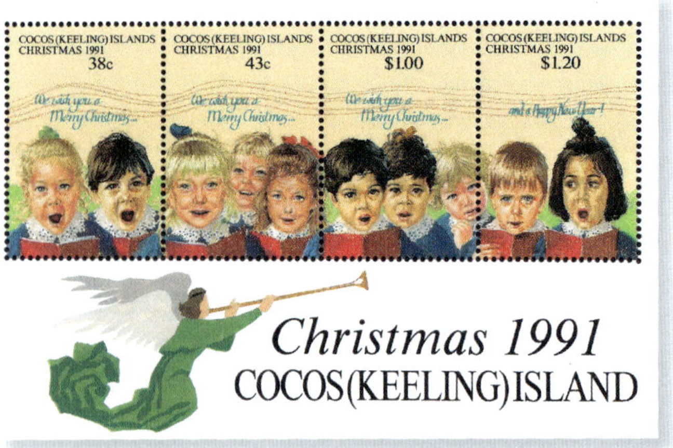

For his very last stamp commission, the 1991 Christmas stamps for Cocos Island, the artist allowed himself a small personal indulgence, using his young daughters Asha (left) and Anja (Right) as models for the 43 cent stamp.

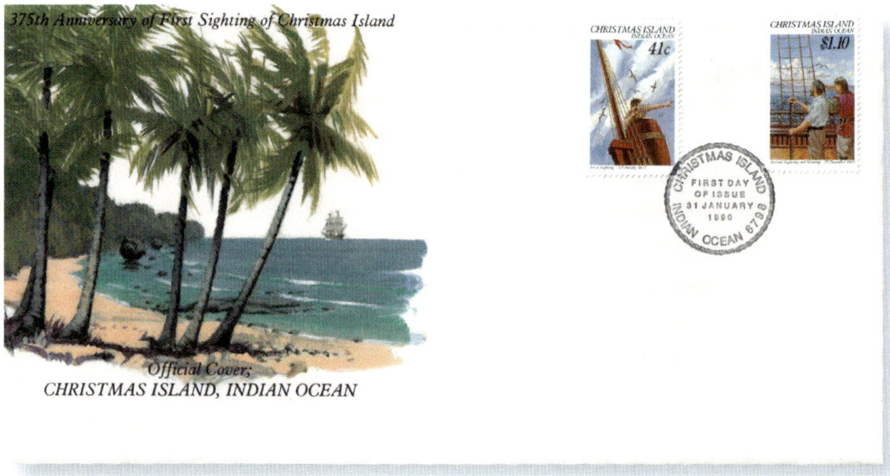

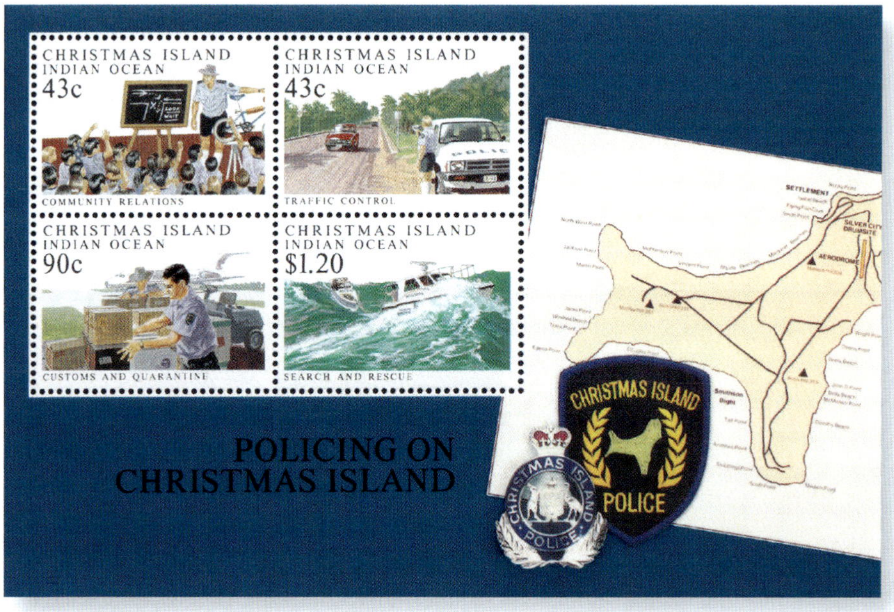

Between July 1970 and November 1991 Ray Honisett designed a total of 84 stamps for Australia and its territories.
The total was made up as follows:

Australia	9
Australian Antarctic Territory	32
Christmas Island	10
Cocos (Keeling) Islands	29
Samoa	4
TOTAL	**84**

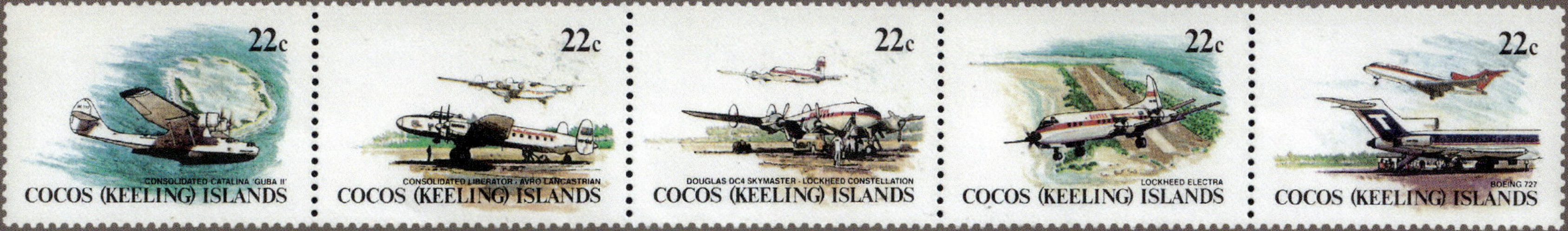
22c
CONSOLIDATED CATALINA 'GUBA II'
COCOS (KEELING) ISLANDS
22c
CONSOLIDATED LIBERATOR · AVRO LANCASTRIAN
COCOS (KEELING) ISLANDS
22c
DOUGLAS DC4 SKYMASTER · LOCKHEED CONSTELLATION
COCOS (KEELING) ISLANDS
22c
LOCKHEED ELECTRA
COCOS (KEELING) ISLANDS
22c
BOEING 727
COCOS (KEELING) ISLANDS

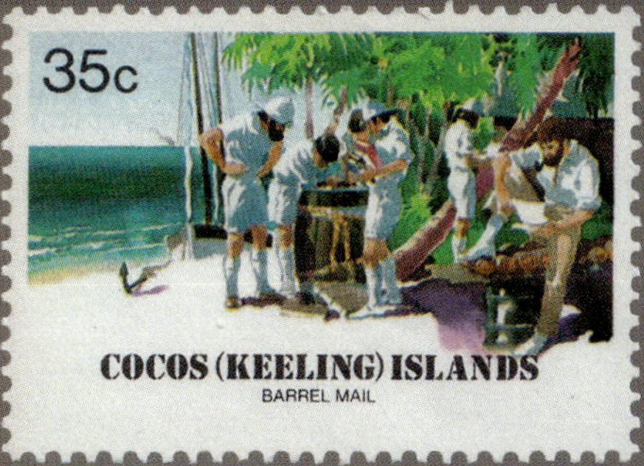
35c
COCOS (KEELING) ISLANDS
BARREL MAIL

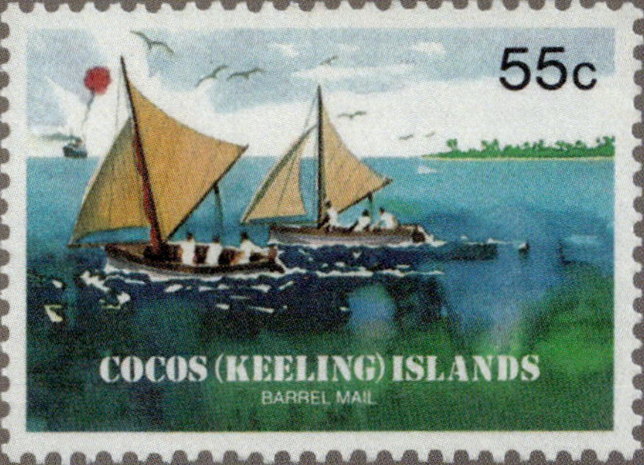
55c
COCOS (KEELING) ISLANDS
BARREL MAIL

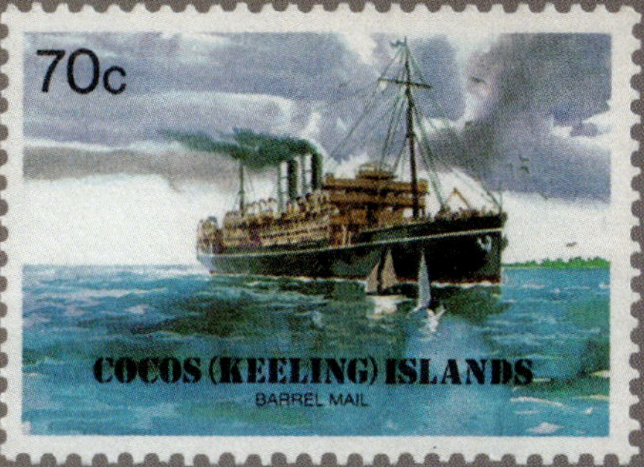
70c
COCOS (KEELING) ISLANDS
BARREL MAIL

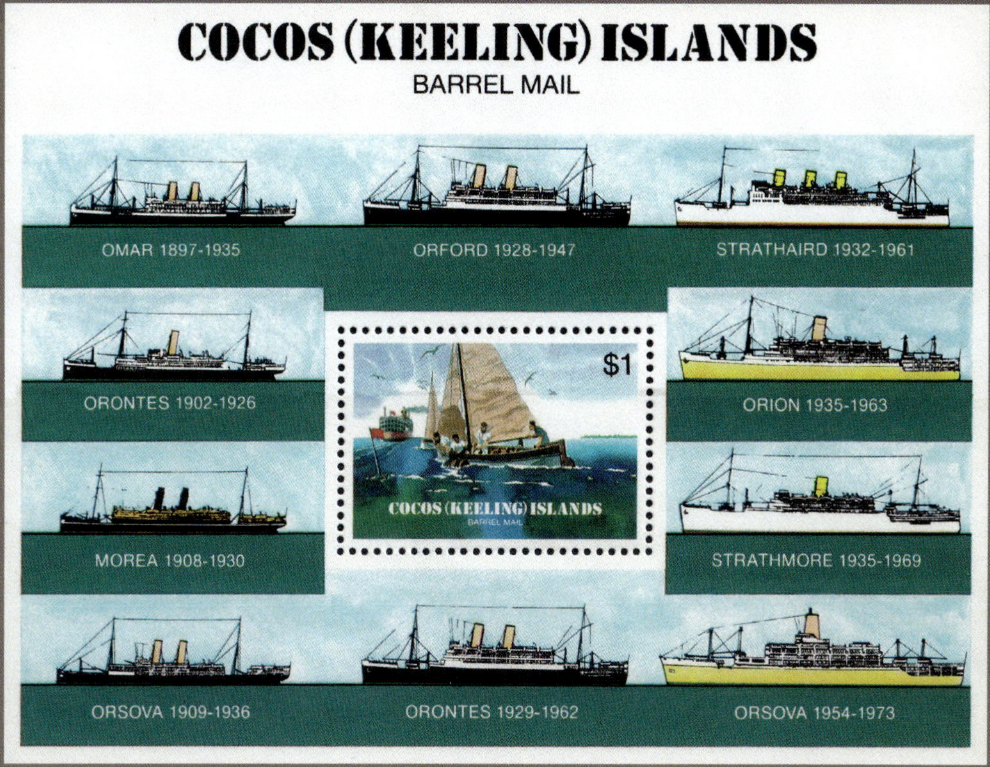
COCOS (KEELING) ISLANDS
BARREL MAIL
OMAR 1897-1935
ORFORD 1928-1947
STRATHAIRD 1932-1961
ORONTES 1902-1926
$1
COCOS (KEELING) ISLANDS
BARREL MAIL
ORION 1935-1963
MOREA 1908-1930
STRATHMORE 1935-1969
ORSOVA 1909-1936
ORONTES 1929-1962
ORSOVA 1954-1973

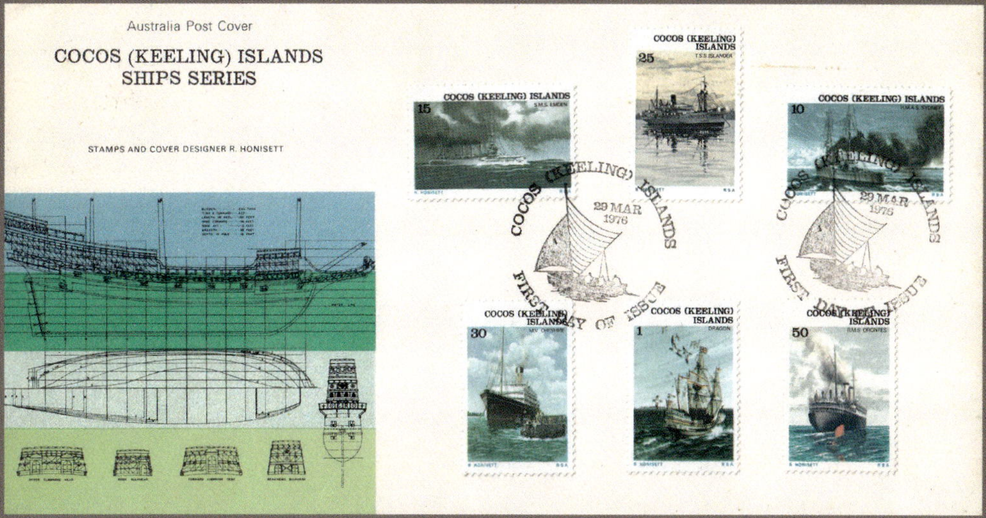
Australia Post Cover
COCOS (KEELING) ISLANDS
SHIPS SERIES
STAMPS AND COVER DESIGNER R. HONISETT
COCOS (KEELING) ISLANDS
29 MAR 1976
FIRST DAY OF ISSUE

THE McDONNELL DOUGLAS COLLECTION

A BICENTENNIAL GIFT TO AUSTRALIA

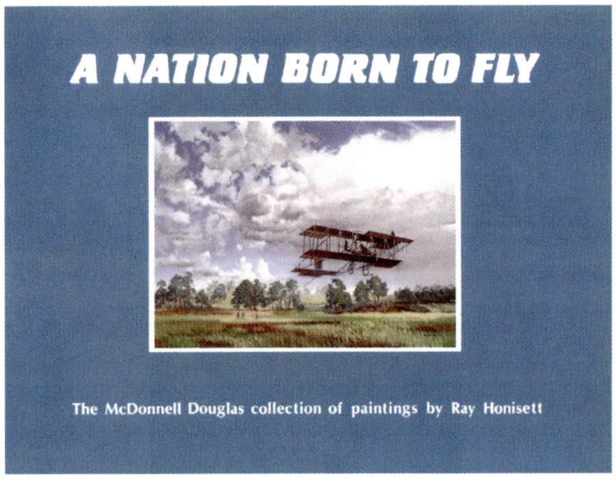

Above, ***A Nation Born to Fly****, McDonnell Douglas Limited, Canberra, 1988*
Documentary brochure accompanying published prints, text by Keith Isaacs AFC

Opposite
Ray Honisett, ***Sea Birds***
Oil on canvas, 75.5 x 98.5 cms
Powerhouse Museum Collection
Gift of McDonnell Douglas Limited, 1988

Ray Honisett's Australian War Memorial commissions in the mid 1980s had cemented his position as the nation's leading aviation artist, reinforced by a steady stream of stamp designs featuring aircraft subjects. A great deal of attention was focused on the Australian Bicentennial, an event heavily promoted by the Hawke government in Canberra, that had come to power in 1983 on a platform of Australian character and identity. The Bicentennial was billed as the 'Celebration of a Nation', taking many cues from the American Bicentennial celebrations of 18 years before. Governments and corporations were keen to show their appreciation of the event by making appropriate gifts to the nation through a number of public institutions. Because of the maritime origins of the event being celebrated, it was not surprising that new Australian National Maritime Museum in Darling harbour, Sydney, should be the recipient of a number of significant gifts, including the sailing yachts *Akarana* and *Kathleen Gillett*, as well as a $US 5 million donation from the Government of the United States.

The American aircraft builder McDonnell Douglas had a presence in this country for much of the 20th century, its aircraft being crucial to many aspects of Australia's defence and economic development. The company decided to recognise that valuable partnership by commissioning a series of paintings illustrating the importance of air transport in the development of the nation. Ray Honisett was the obvious artist for such a commission and he set to work to produce the sixteen paintings, to a standard format that would make them ideal for both publication and display. The works were relatively small in comparison to his Australian War Memorial paintings, just 75.5 x 98.5 cms, but the variety of subjects would give the artist all the challenges he could wish for. With subjects ranging from the Antarctic to the highlands of New Guinea and the outback to city airports, there was scope to display all his illustration skills. A number of the subjects involved detailed figure work, allowing a different side of his work to come to the fore.

The paintings were reproduced in print form for the public and the originals were given to the Powerhouse Museum in Sydney, another Bicentennial project that greatly expanded the exhibition possibilities for the old Museum of Applied Arts and Sciences, situated across the road in Ultimo.

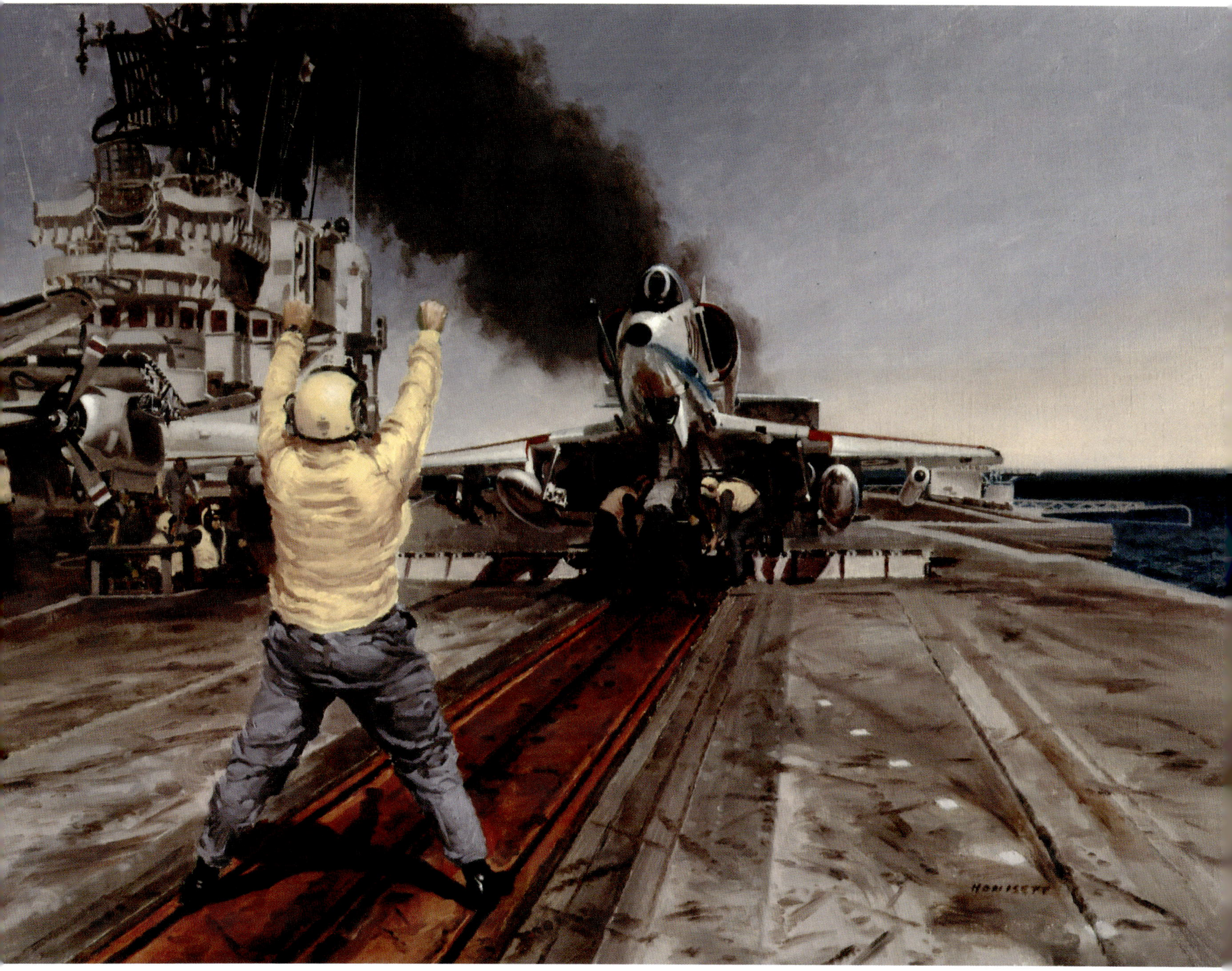

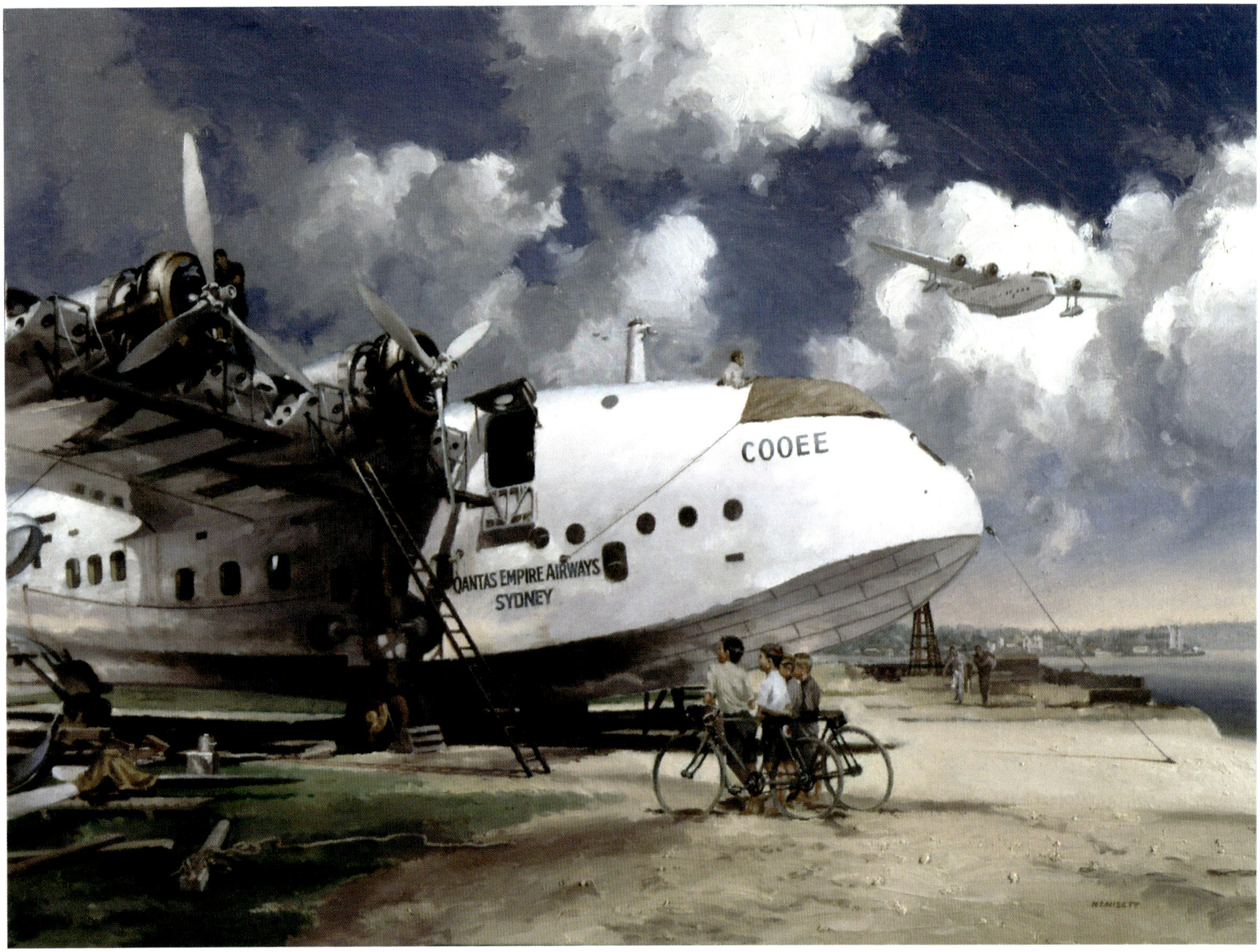

Ray Honisett, ***Great White***
Oil on canvas, 75.5 x 98.5 cms
Powerhouse Museum Collection
Gift of McDonnell Douglas Limited, 1988
Qantas Empire Airways' flying boat 'Coolangatta' (VH-ABB) in the air and 'Cooee' (VH-ABF) on the ground at the unfinished Rose Bay terminal, Sydney, in 1938

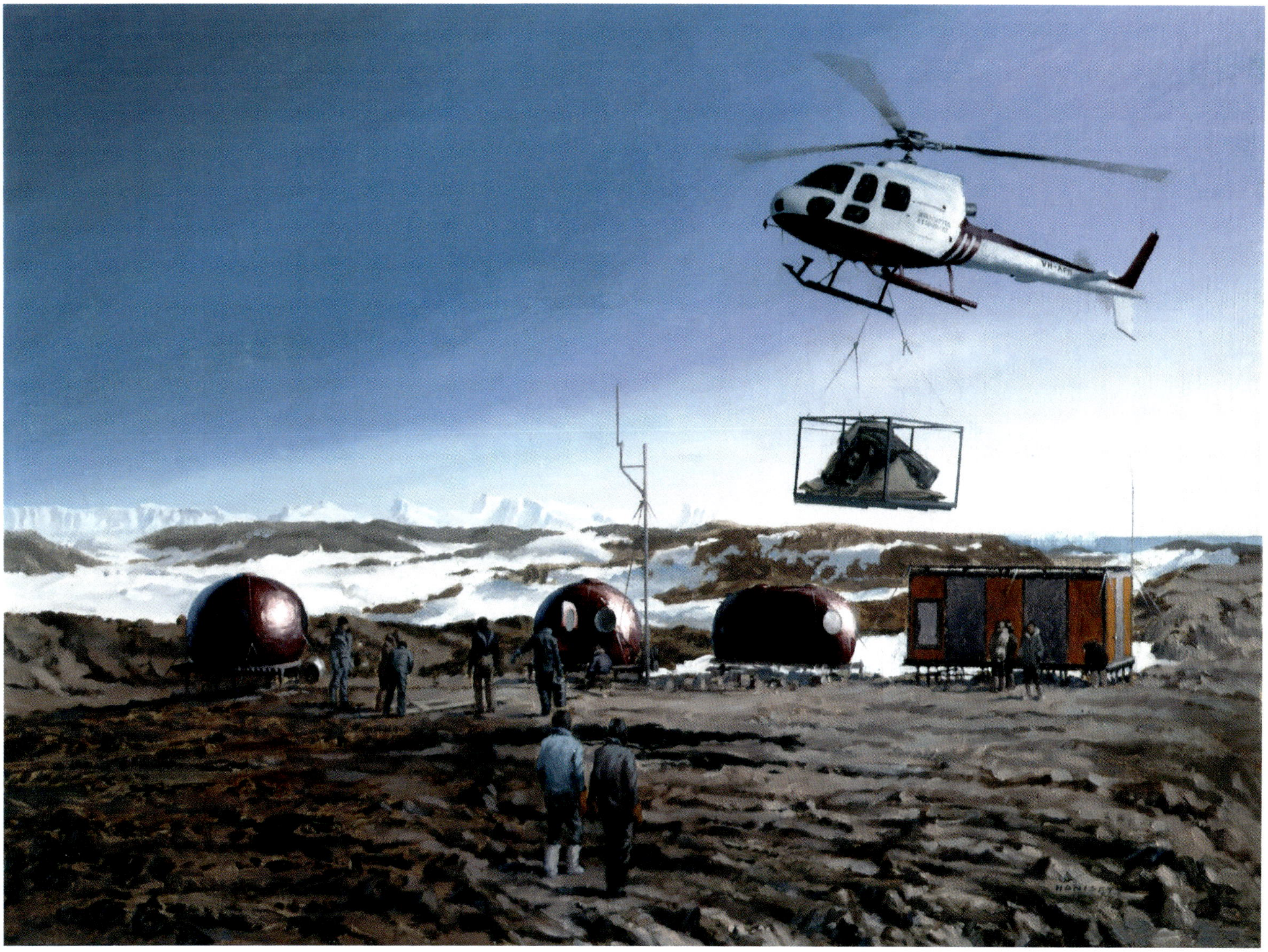

Ray Honisett, ***Antarctic Workhorse***
Oil on canvas, 75.5 x 98.5 cms
Powerhouse Museum Collection
Gift of McDonnell Douglas Limited, 1988

An Aerospatiale Squirrel helicopter transporting equipment at Larsemann Hills, a logistical hub in East Antarctica. The site is designated an Antarctic Specially Managed Area, the only one in which multiple countries have continuously occupied stations. The Aerospatiale Squirrel was first deployed during the 1986-87 ANARE expedition.

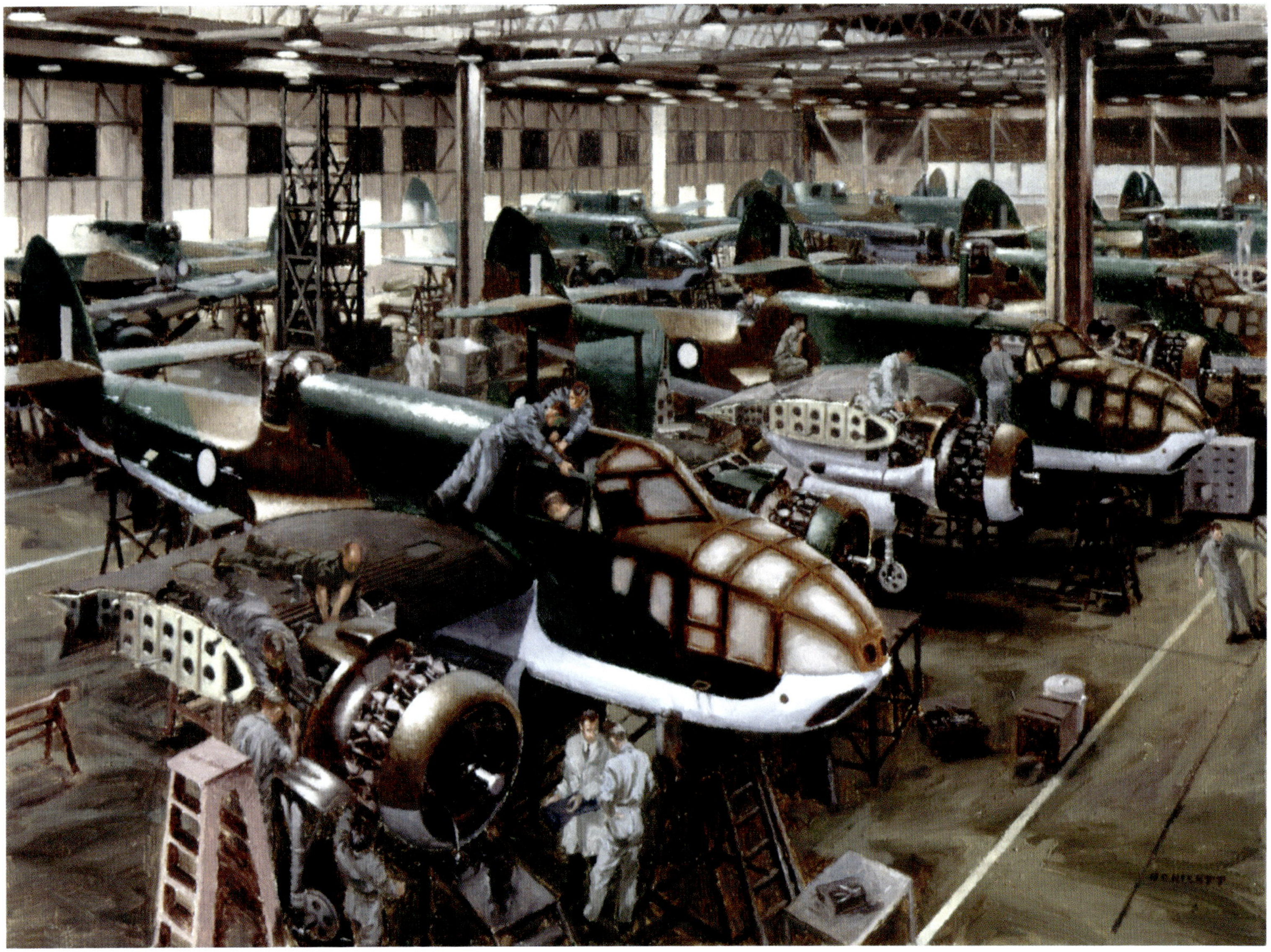

*Ray Honisett, **Beaufort Final Assembly***
Oil on canvas, 75.5 x 98.5 cms
Powerhouse Museum Collection
Gift of McDonnell Douglas Limited, 1988

Bristol Beaufort bombers being assembled in a Department of Aircraft Production (DAP) factory during World War II. The DAP factories were located at Fishermens Bend in Victoria and Mascot in New South Wales and produced 700 Bristol Beaufort torpedo bombers from 1941 to 1944 as well as 329 Beaufighters by August 1945.

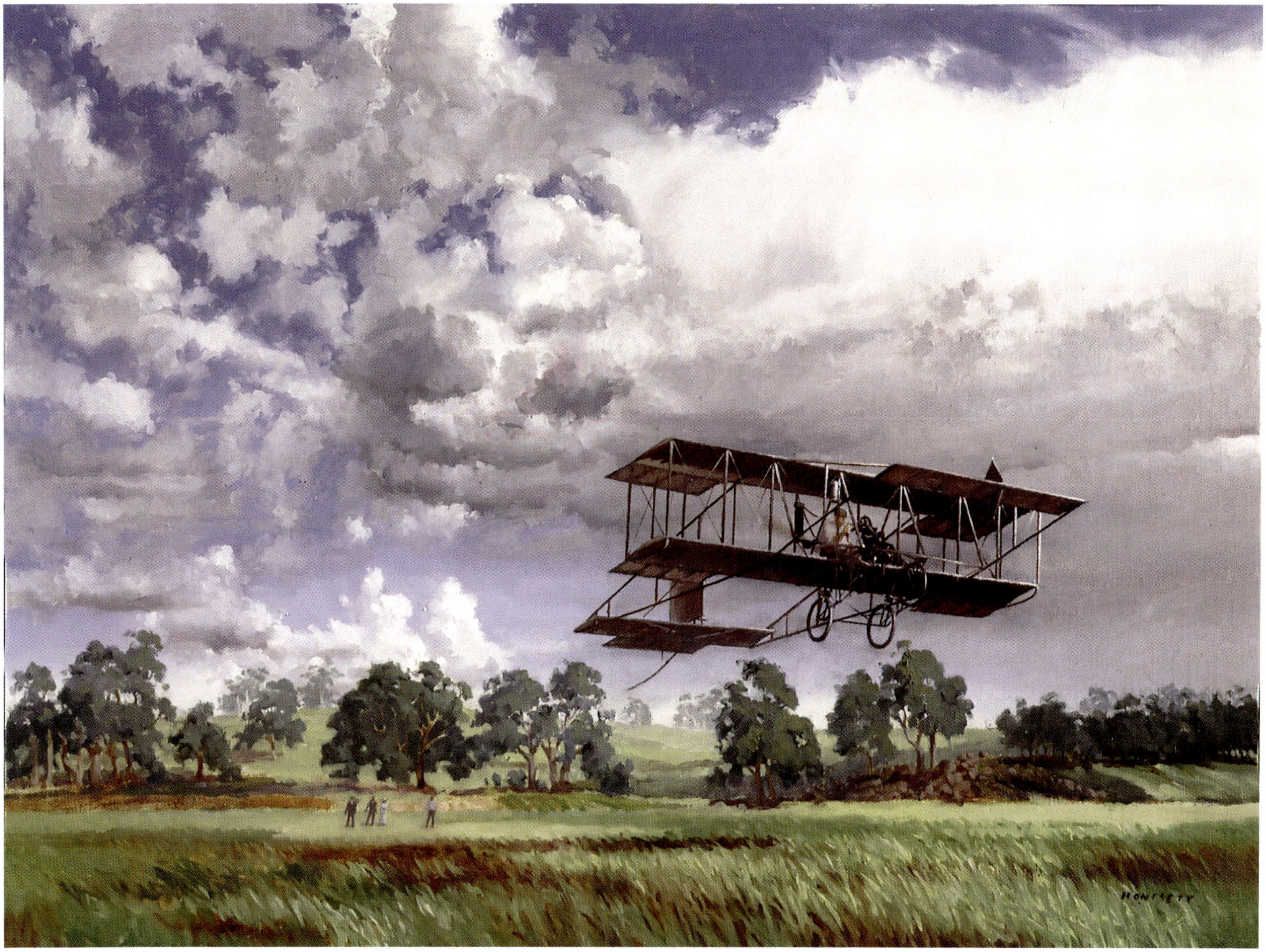

Ray Honisett, ***Airborne***
Oil on canvas, 75.5 x 98.5 cms
Powerhouse Museum Collection
Gift of McDonnell Douglas Limited, 1988

John Duigan's flight at Mia Mia, Victoria, Australia, in 1910, the first aircraft to be built in Australia and flown by an Australian

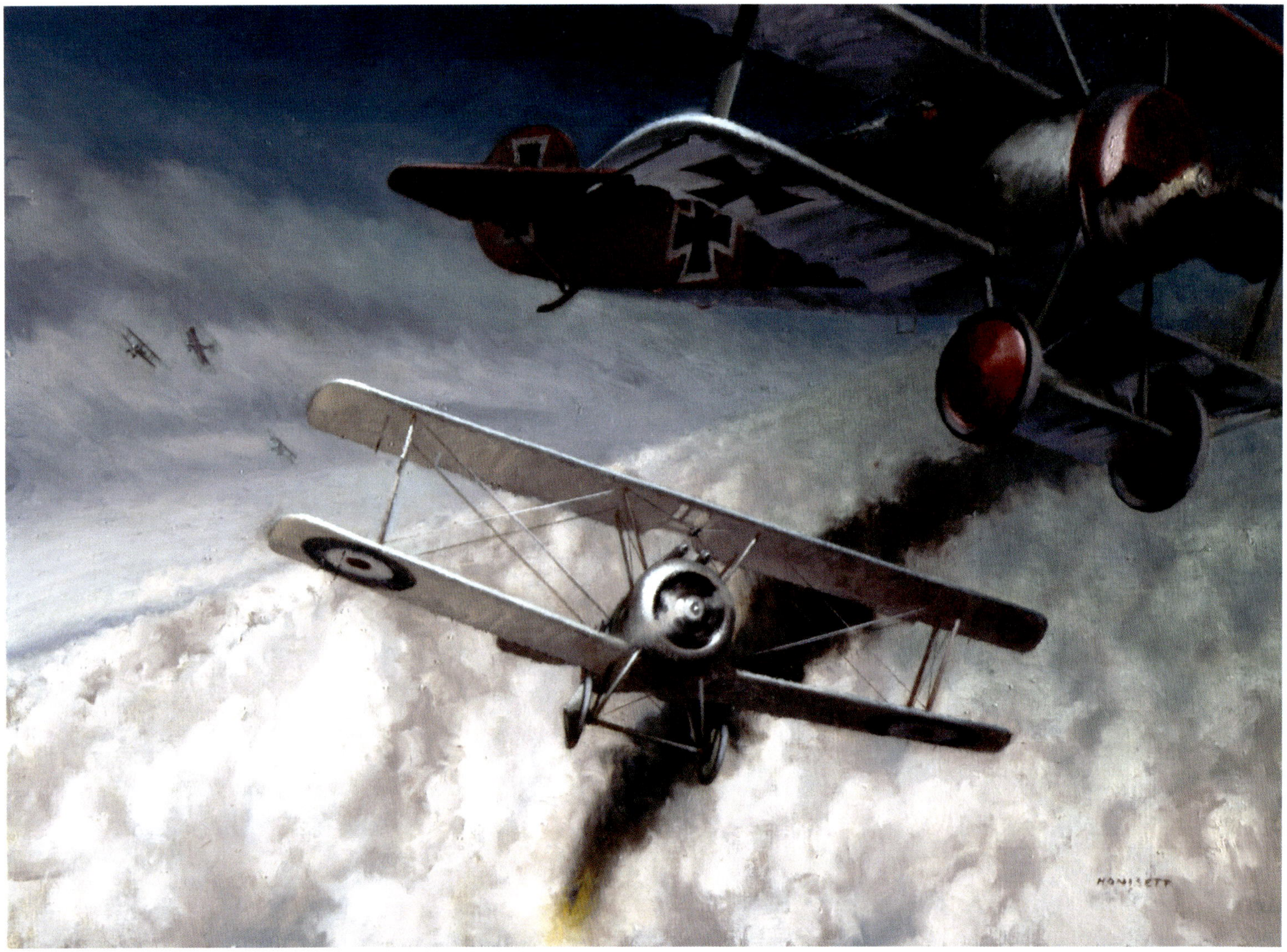

Ray Honisett, ***Morning Encounter***
Oil on canvas, 75.5 x 98.5 cms
Powerhouse Museum Collection
Gift of McDonnell Douglas Limited, 1988

Lieutenant Arthur Henry Cobby in a Sopwith Camel scout, No. 4 Squadron, AFC, attacking a Fokker Dr 1 of Captain Manfred von Richthofen's "Flying Circus" on 21 March 1918. Captain Cobby, DSO, DFC and two Bars, was the top Australian Flying Corps (AFC) ace with 29 aircraft and 13 balloons to his credit although he saw active service for less than a year during World War 1. The AFC formed four squadrons in 1916 for overseas service, with No. 4 Squadron in France destroying the largest number of enemy aircraft.

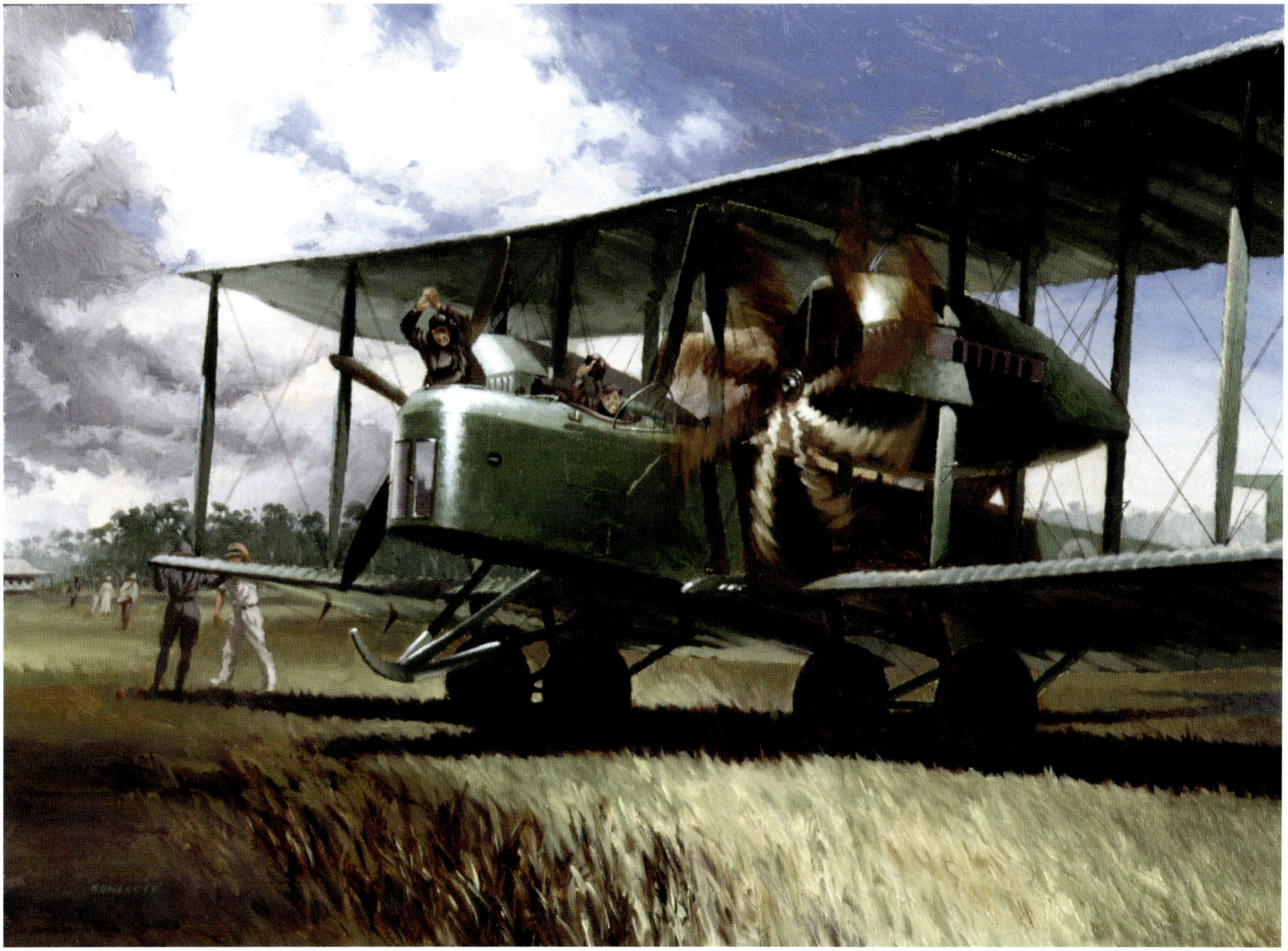

Ray Honisett, ***Winning Smiles***
Oil on canvas, 75.5 x 98.5 cms
Powerhouse Museum Collection
Gift of McDonnell Douglas Limited, 1988

Ross Smith, an Australian war hero, on 10 December 1919, having landed in Darwin to win the Great Air Race from England to Australia. Also in the crew were his brother Keith as assistant pilot, and mechanics James Bennett and Wally Shiers. The reworked WW1 Vickers Vimy bomber covered the dangerous flight of over 18,000 km in 28 days.

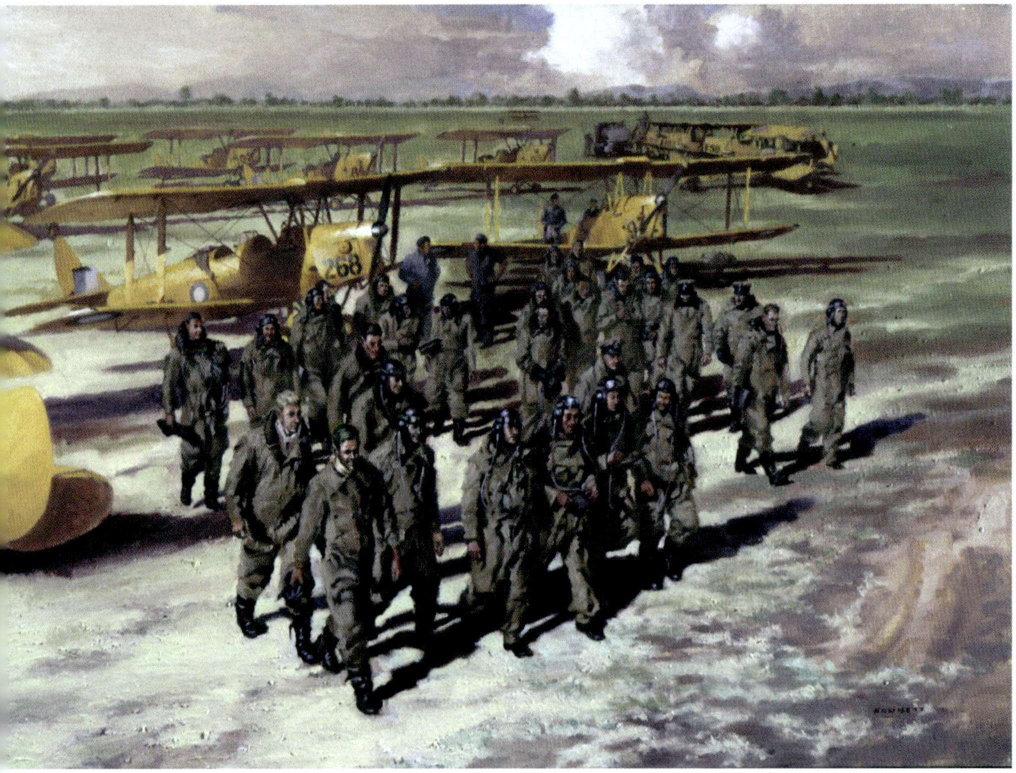

Ray Honisett, ***Fledglings***
Oil on canvas, 75.5 x 98.5 cms
Powerhouse Museum Collection
Gift of McDonnell Douglas Limited, 1988

Participants at one of the twelve Royal Australian Air Force (RAAF) Elementary Flying Training Schools in Australia. The group includes pilot instructors, trainees and maintenance personnel in front of a DH82A Tiger Moth. The RAAF Elementary Flying Training Schools were part of the Empire Training Scheme, developed during World War II. When the scheme ended in 1944 a total of 27,287 aircrew had been trained in Australia.

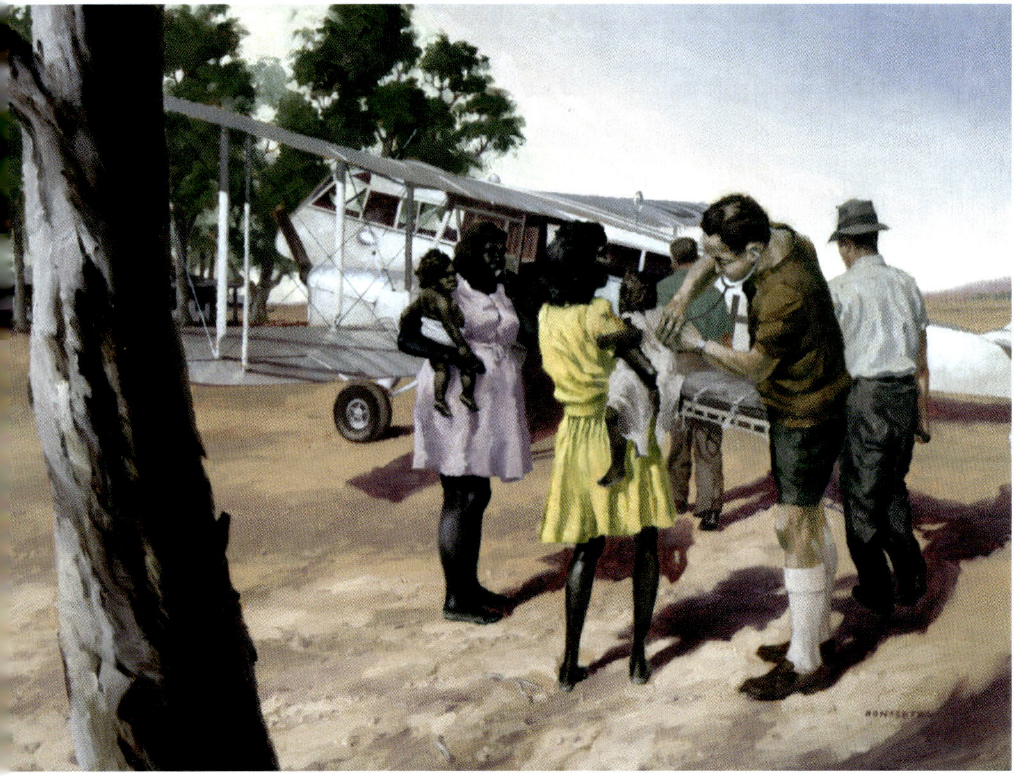

Ray Honisett, ***Flying Doctor***
Oil on canvas, 75.5 x 98.5 cms
Powerhouse Museum Collection
Gift of McDonnell Douglas Limited, 1988
Having just arrived in the DH84 Dragon aircraft VH-URE ‘L. M. Pattinson, a flying doctor attends to a child while a sick patient is stretchered aboard.

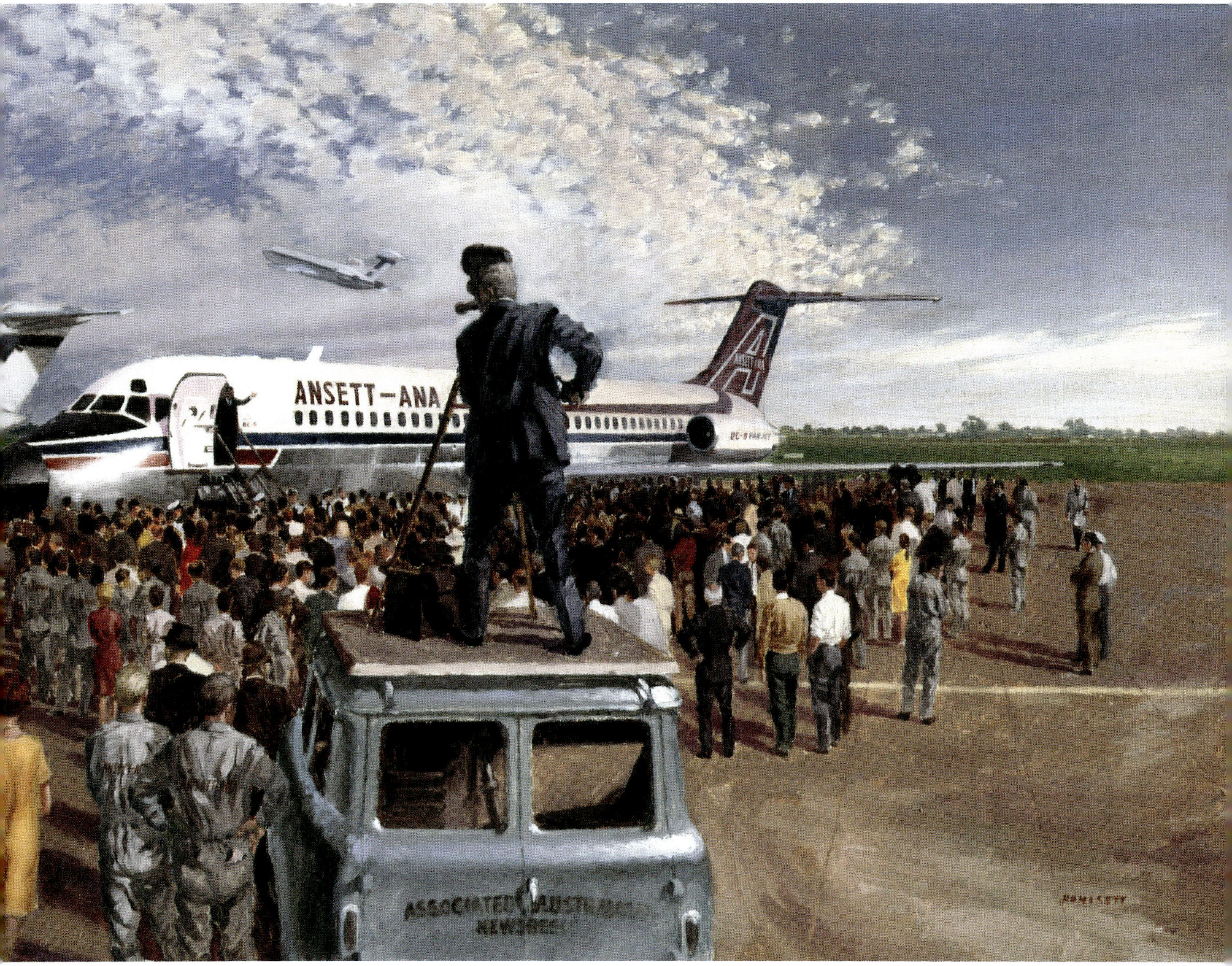

Ray Honisett, ***Crown Pleaser****,*
Oil on canvas, 75.5 x 98.5 cms
Powerhouse Museum Collection
Gift of McDonnell Douglas Limited, 1988

An excited crowd at Essendon airport welcomes VH-CZB, Ansett-ANA's first McDonnell Douglas DC9-31 airliner on 13 April 1967.

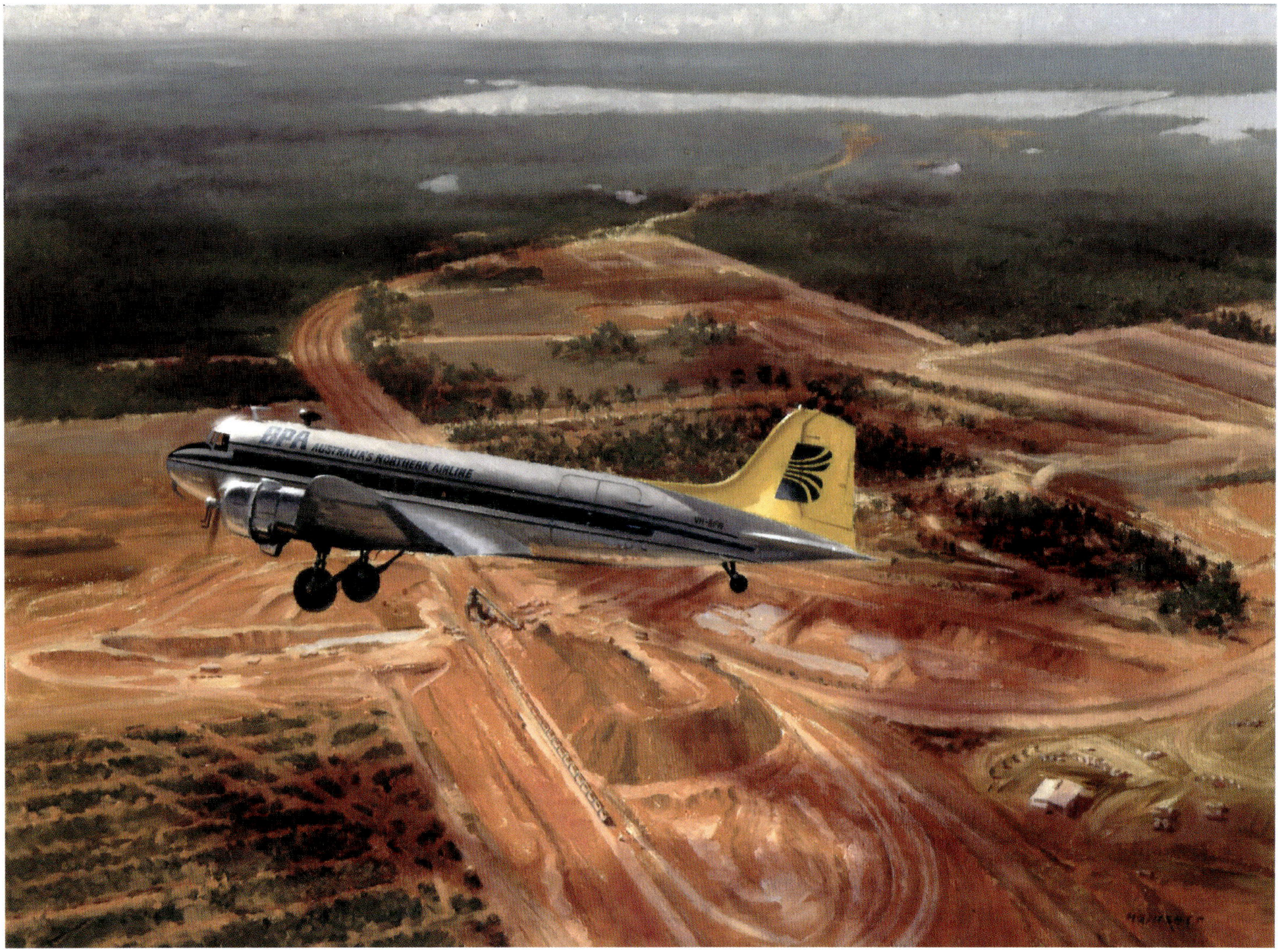

Ray Honisett, ***Classic***
Oil on canvas, 75.5 x 98.5 cms
Powerhouse Museum Collection
Gift of McDonnell Douglas Limited, 1988

A Douglas DC3 operated by Bush Pilots Airways flying over the Comalco bauxite mine at Weipa, Cape York, Queensland.

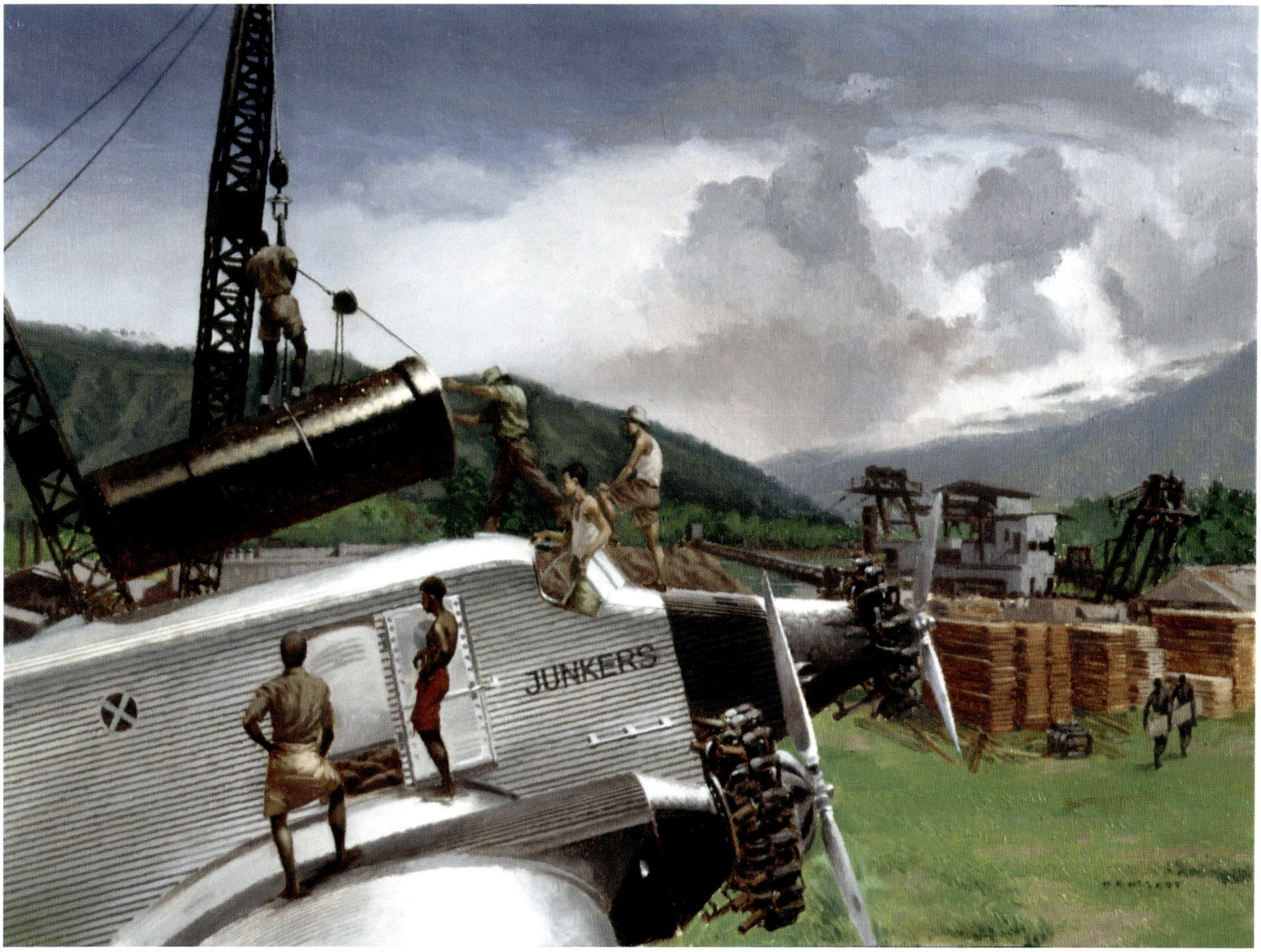

Ray Honisett, ***Piece by Piece***
Oil on canvas, 75.5 x 98.5 cms
Powerhouse Museum Collection
Gift of McDonnell Douglas Limited, 1988

Junkers G31, VH-UOU 'Bolulo Paul' was one of four Junkers Tri-motors purchased by the Bolulo Gold Dredging Company and operated by Guinea Airways, ferrying equipment from the base at Lae to Lau and Wae Airfields for the company, as there were no roads in the Highlands. The aircraft was destroyed on the ground at Bolulo by enemy fire on 21 January 1942.

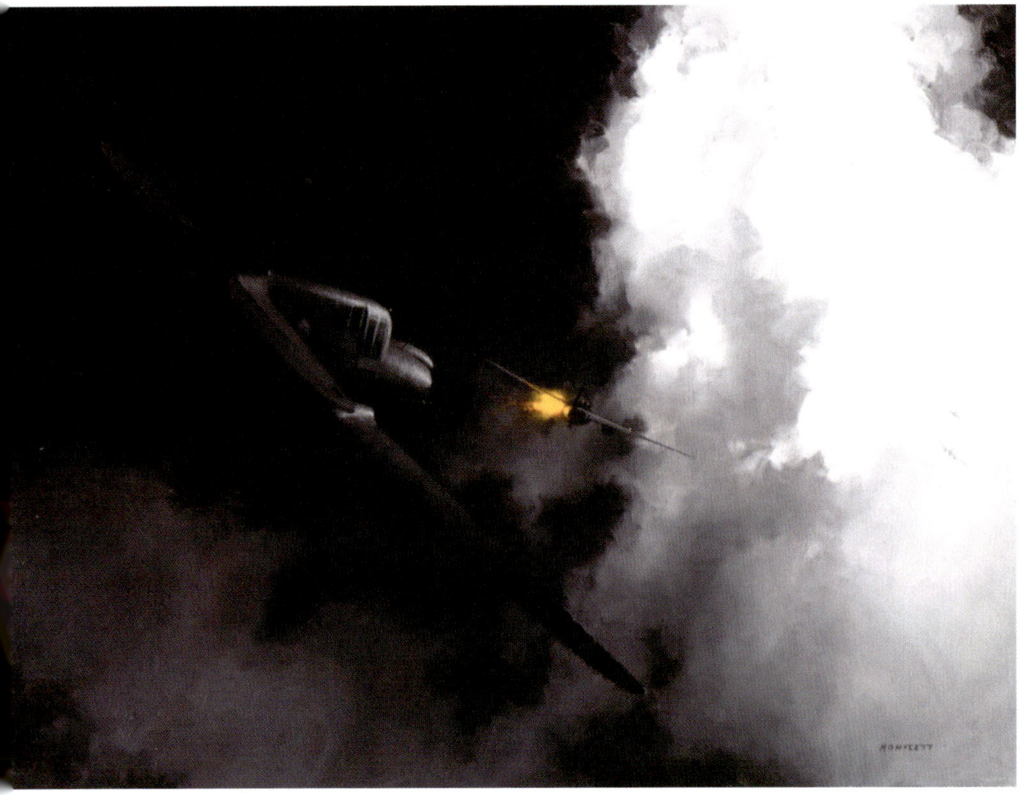

Ray Honisett, ***Predator***
Oil on canvas, 75.5 x 98.5 cms
Powerhouse Museum Collection
Gift of McDonnell Douglas Limited, 1988

Squadron Leader R.C. 'Dick' Creswell's WWII Kitty-hawk shooting down a Japanese bomber over Darwin on

23 November 1942

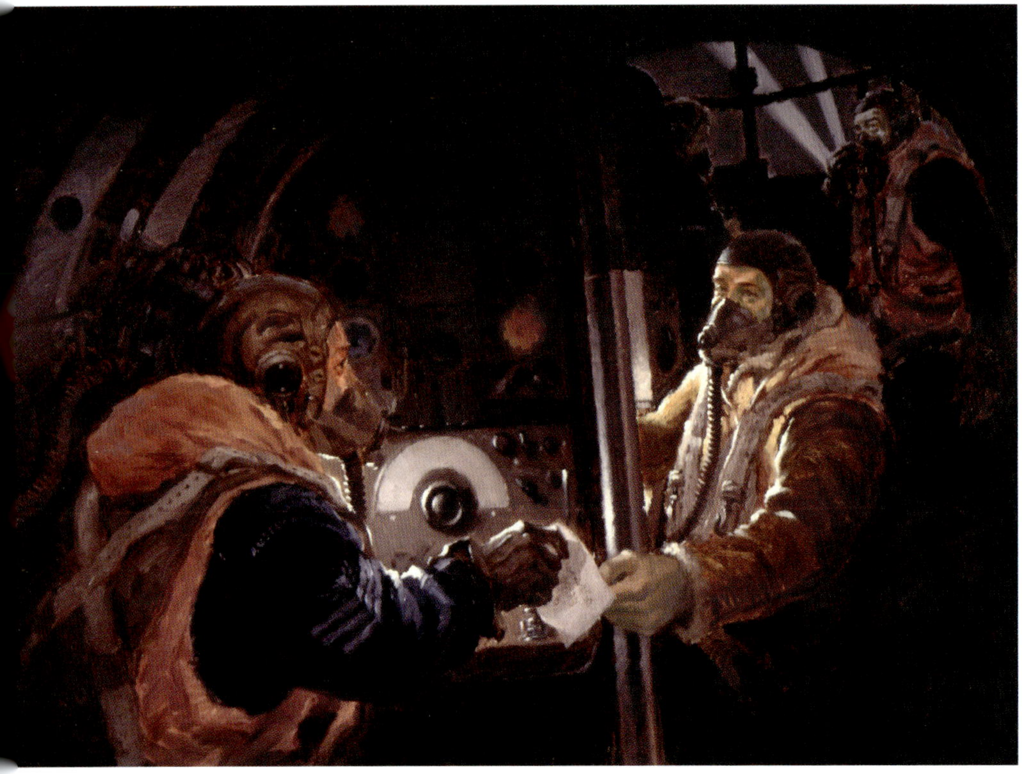

Ray Honisett, ***Doing the job***
Oil on canvas, 75.5 x 98.5 cms
Powerhouse Museum Collection
Gift of McDonnell Douglas Limited, 1988

RAAF crew members inside a WWII Avro Lancaster bomber approaching the target over Europe

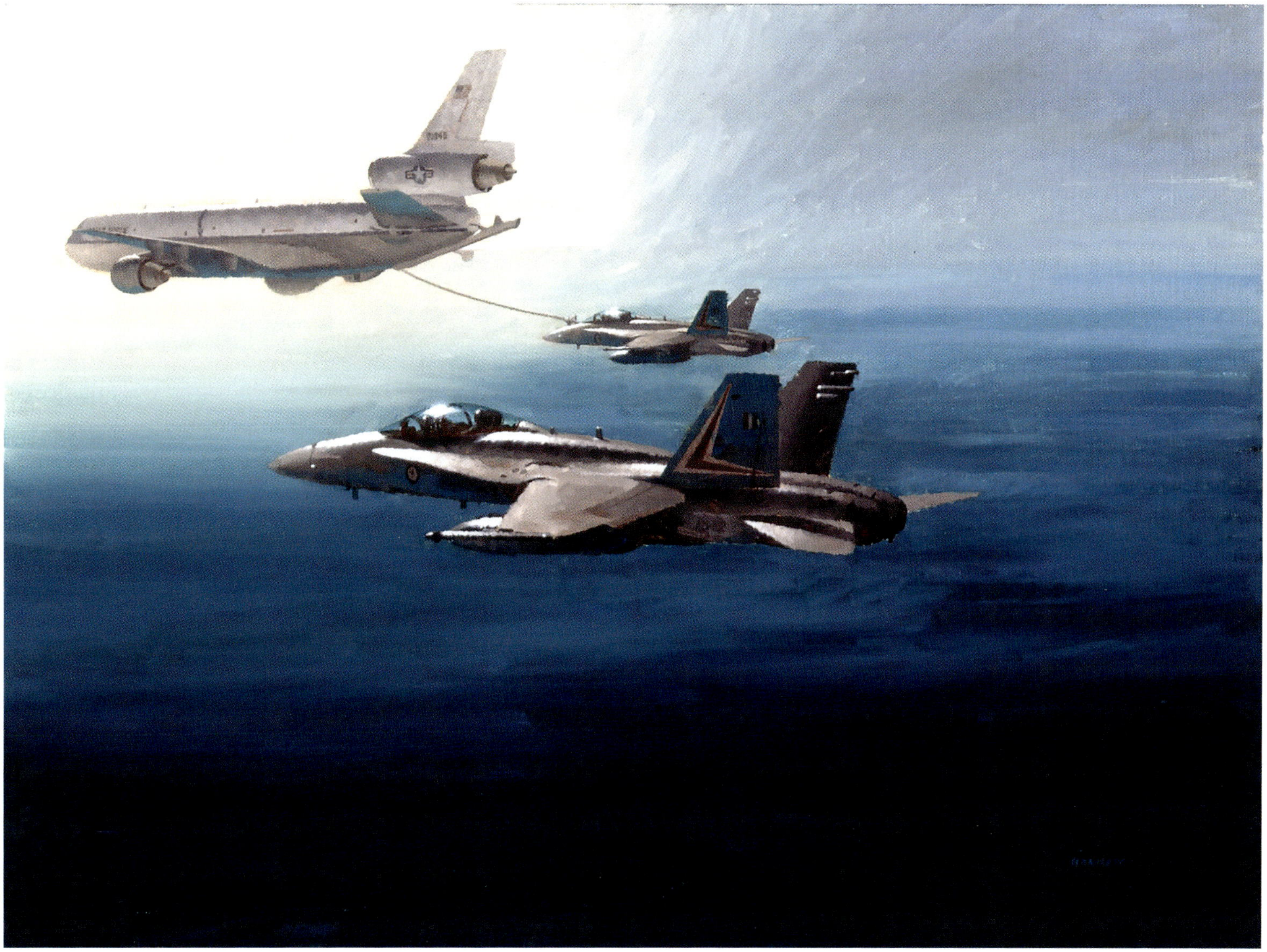

Ray Honisett, ***Express Delivery***
Oil on canvas, 75.5 x 98.5 cms
Powerhouse Museum Collection
Gift of McDonnell Douglas Limited, 1988
Two RAAF Hornets refuelling mid-air during their non-stop world record-breaking delivery flight from the USA to Australia on 17 May 1985

GUIDE AND MENTOR

AUSTRALIAN SOCIETY OF AVIATION ARTISTS

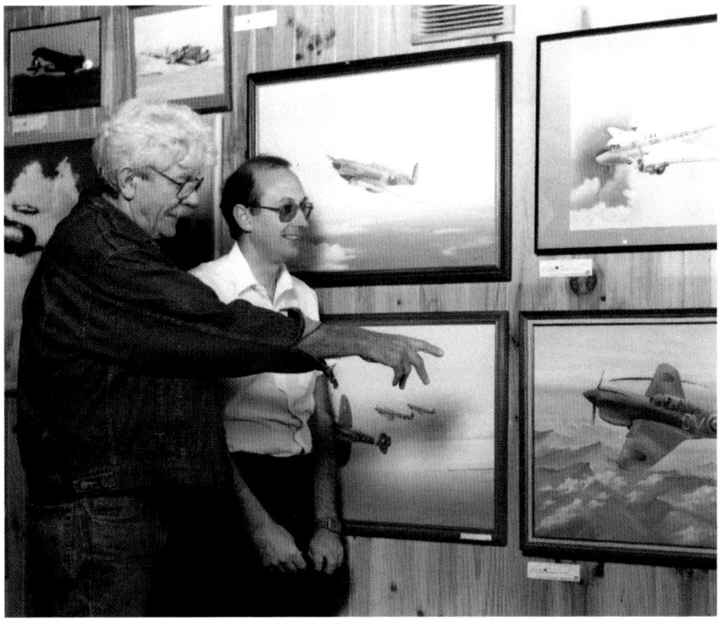

Ray Honisett and Alan Swann at 'Canvas Wings', the first exhibition of the Australian Society of Aviation Artists at Moorabin in 1989.

Ray Honisett, (RAAF Airspeed Oxfords), c.2000
Oil on canvas
Private collection, Melbourne

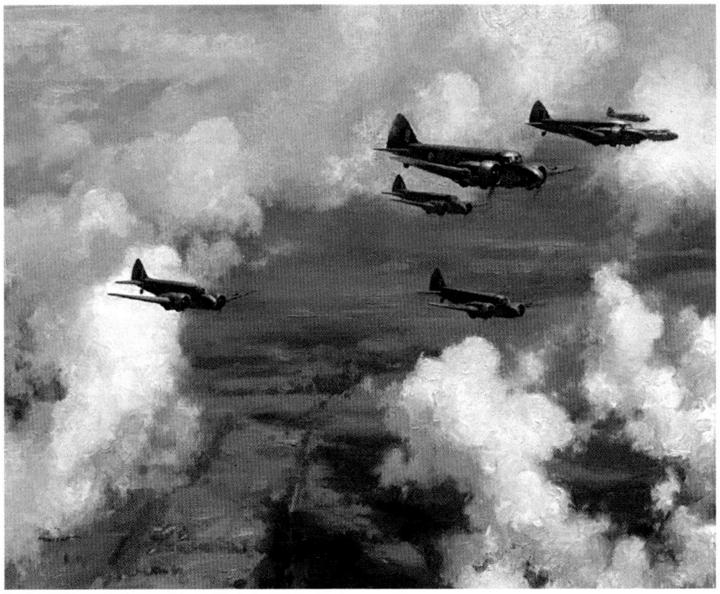

The art world is full of contradictions, a place where at one moment free expression and the outpourings of the soul are seen to be the central mantra, and at the same time there is the constant pressure to compartmentalise, define and constrain the outcomes of that expression. In painting it is often said that there are just three subjects or themes – the figure, landscape and still-life, and that anything else will be somehow a mix of those three elements – except of course pure abstraction, where the painting itself is in fact the subject. Within each form there will be subsets, each of which will have its own adherents and devotees. The major subject sub-set of landscape is probably the seascape, with or without ships, which might define it as 'maritime art'. Maritime subject matter has a long tradition, stemming in the main from Dutch painting of the 16th century and carrying through to the present day. In high art terms, art which is defined by its subject is often seen as something lesser, bordering on illustration, that no man's land that is the realm of the commercial artist. Until recent times art schools divided their classes between the 'trade' subjects of illustration, technical drawing, lettering and graphic design and those aimed at 'artists' — life drawing, painting, printmaking and sculpture. Ray Honisett was very much in the commercial end of the business, teaching illustration for more than 20 years at Melbourne's most important art school, the Royal Melbourne Institute of Technology.

The term aviation art is a relatively new, subject-based discipline, which has only come into use in the second half of the twentieth century. There were artists who specialised in aviation subjects, with Frank Wootton being the first to win international acclaim, but they were simply specialists in a broad field of endeavour. During the years of the Second World War there were artists who were employed to depict the air war, but they would not have been called 'aviation artists'. The Royal Australian Air Force operated its own program during WW2, employing three artists to work in Australia and the Pacific campaign. Harold Freedman, Eric Thake and Max Newton all depicted aircraft and airmen in action, but none was a 'specialist aviation artist' as the term would now be understood. Similarly other Australian artists, such as Ivor Hele, Frank Norton and Dennis Adams painted RAAF subjects, but usually as an aside to their figure and landscape work or, in the latter case, maritime subjects.

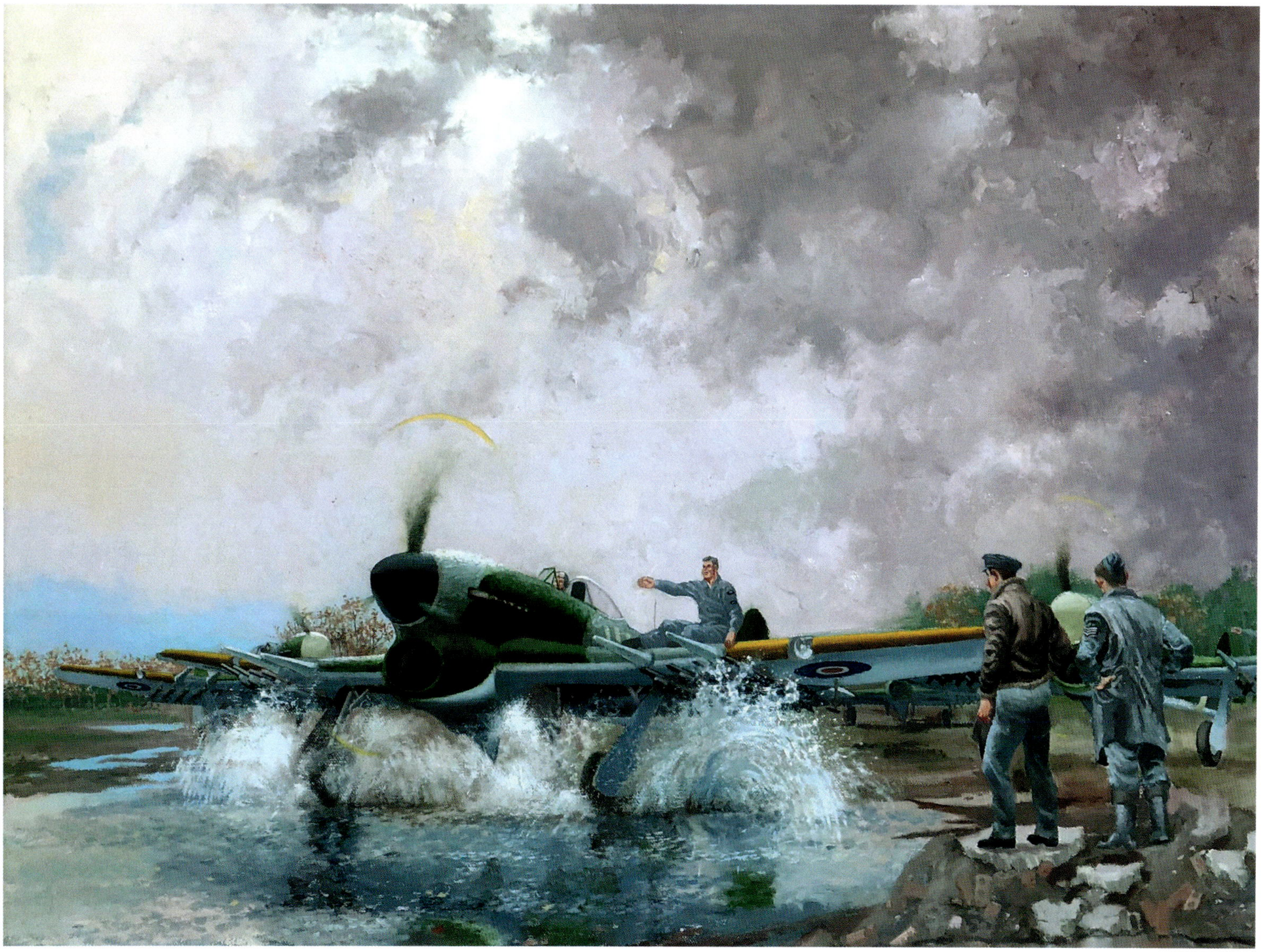

Working in aged care I had the privilege of getting to know Tom Hall, a WW2 D-Day veteran. With his memories adorning his walls, Tom and I shared many conversations as he reminisced of his early years as an air force pilot and the challenges of the ageing process. With multiple operational missions flying the Hawker Typhoon fighter aircraft, Tom in his later years was having to come to terms with the limitations an ageing body presented on the basic daily tasks we often take for granted. Knowing Ray's interest in aviation history, it was a pleasure to be able to connect the two of them. In what was initially planned as a brief meet and greet, minutes turned into hours with the shared interest of the era reflected in the atmosphere in the room, feeling like two old cobbers catching up. ***Path of the Storm*** *reflects Ray's genuine connection with Tom and his passion for portraying key aviation stories in his art.*

Murray Pitman, 2020

Ray Honisett, ***Path of the Storm***
Oil on canvas
Private collection, Melbourne

In Britain a new Society of Aviation Artists was formed in 1954, under the leadership of Norman Wilkinson and Frank Wootton, to be succeeded in 1971 by the larger, nationally based Guild of Aviation artists. An American society was founded in 1986, like its British counterpart, initially focused on one special individual, Keith Ferris, who in the USA held a similar status to Frank Wootton. It was the British Guild that attracted the attention of a Trans Australian Airlines engineer, Alan Swann, who was in 1988 passing through the UK on a training exercise for the company. Alan was a keen amateur painter, attracted to aircraft subjects because of his professional work with the airline and, as such, already aware of what the British was doing in the field. *"As it happened it was my responsibility to manage the training requirements of the Jetstream JS35 aircraft into the fleet. The training conference requirements required me to travel to Prestwick in Scotland (the home of the Twin Pioneer) and to Phoenix Arizona to make training arrangements. To my delight I was able to arrange a weekend stopover in London on the weekend of the GAvA art show. At the show I met with Charles Thompson who advised that one Harry Margulies had made inquiries about membership. On my return I contacted Harry and arranged a meeting; we compared notes and together contacted Kevin Limon of the Sport Aviation Association of Australia who we found ran an aviation art show at their regular Mangalore Air Show. We contacted past participants and arranged a meeting in the February of 1988 at the studio of Norm Clifford. Ten artists attended that meeting and unanimously agreed to become involved in the society. The first activities involved electing office bearers and becoming incorporated. Ray Honisett accepted the responsibility as inaugural President and retained that position until 1995 when Brian Evans took the reins".*

Having a professional with the skills and experience that Ray Honisett could offer was a coup for the new society. By the time the Society came together Ray had been teaching illustration for more than two decades and was happy to impart all he could to the enthusiastic group of artists. The group organised exhibitions, painting tours, workshops and classes for the expanding group of members, which at its peak numbered 67 members, although two decades later its number had dwindled to just 24. Like so many clubs and associations founded on the enthusiasm of a small number of dedicated members, it has not survived, with its communication role now being taken over by social media platforms that can bring together like-minded individuals world wide.

Ray Honisett, ***Pacific Morning*** *(Martin M-130 Clipper)*, ***1990***
Oil on canvas, 76 x 101 cm
Private Collection

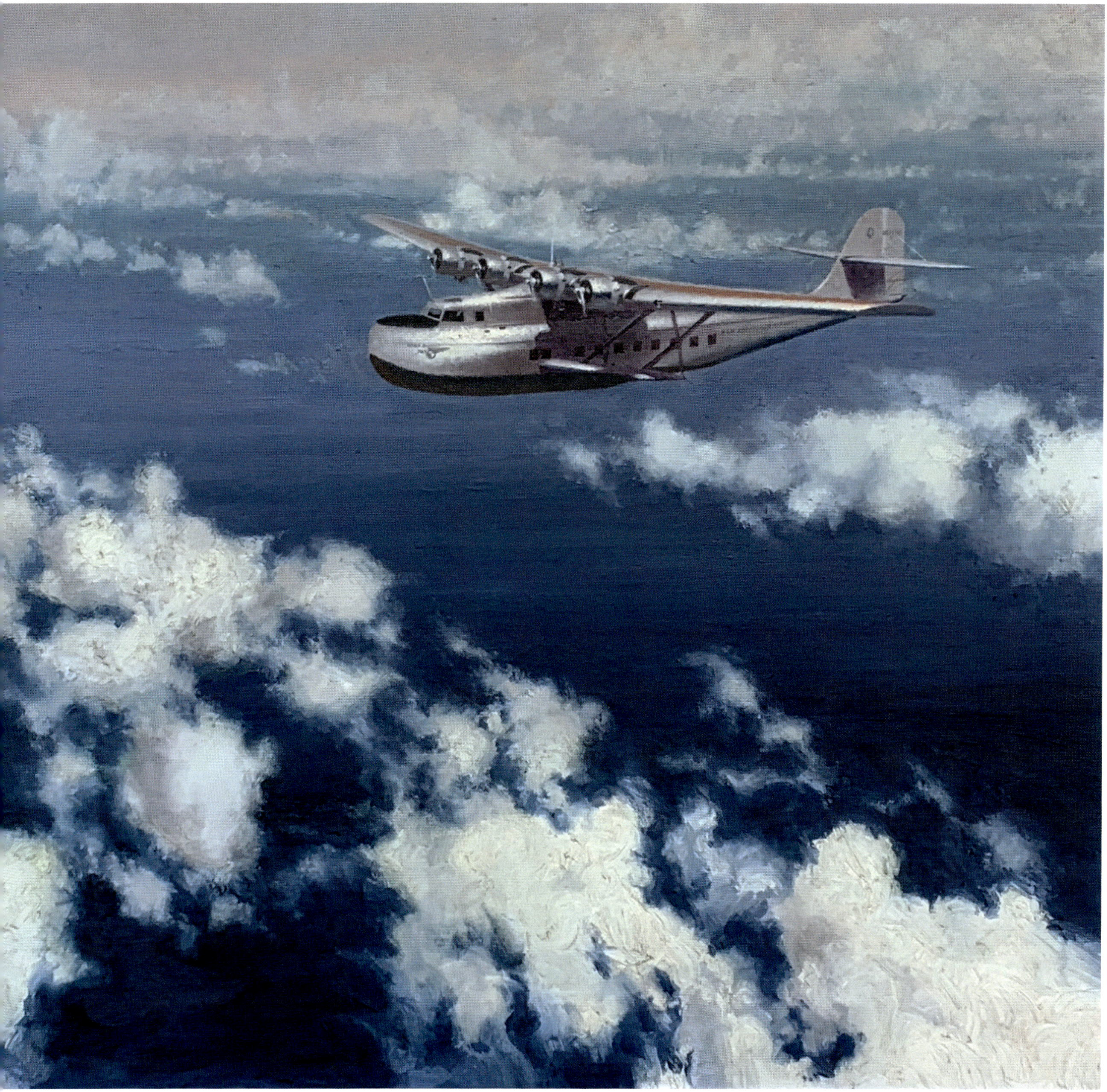

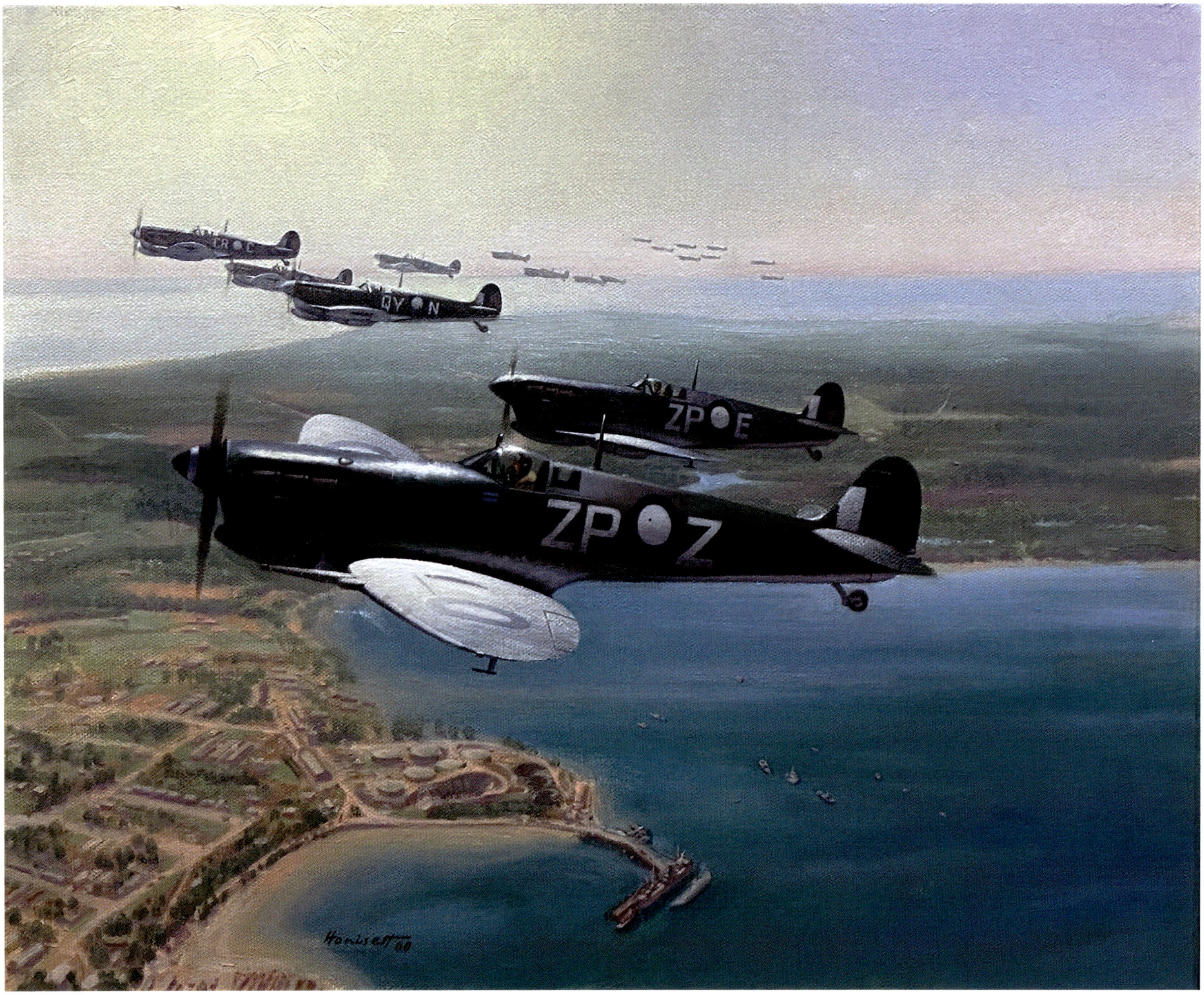

Ray Honisett (RAAF Squadron 457 Supermarine Spitfire Vcs (Tropical) over the port of Darwin, 1943), 2000
Oil on canvas
Private collection, Melbourne

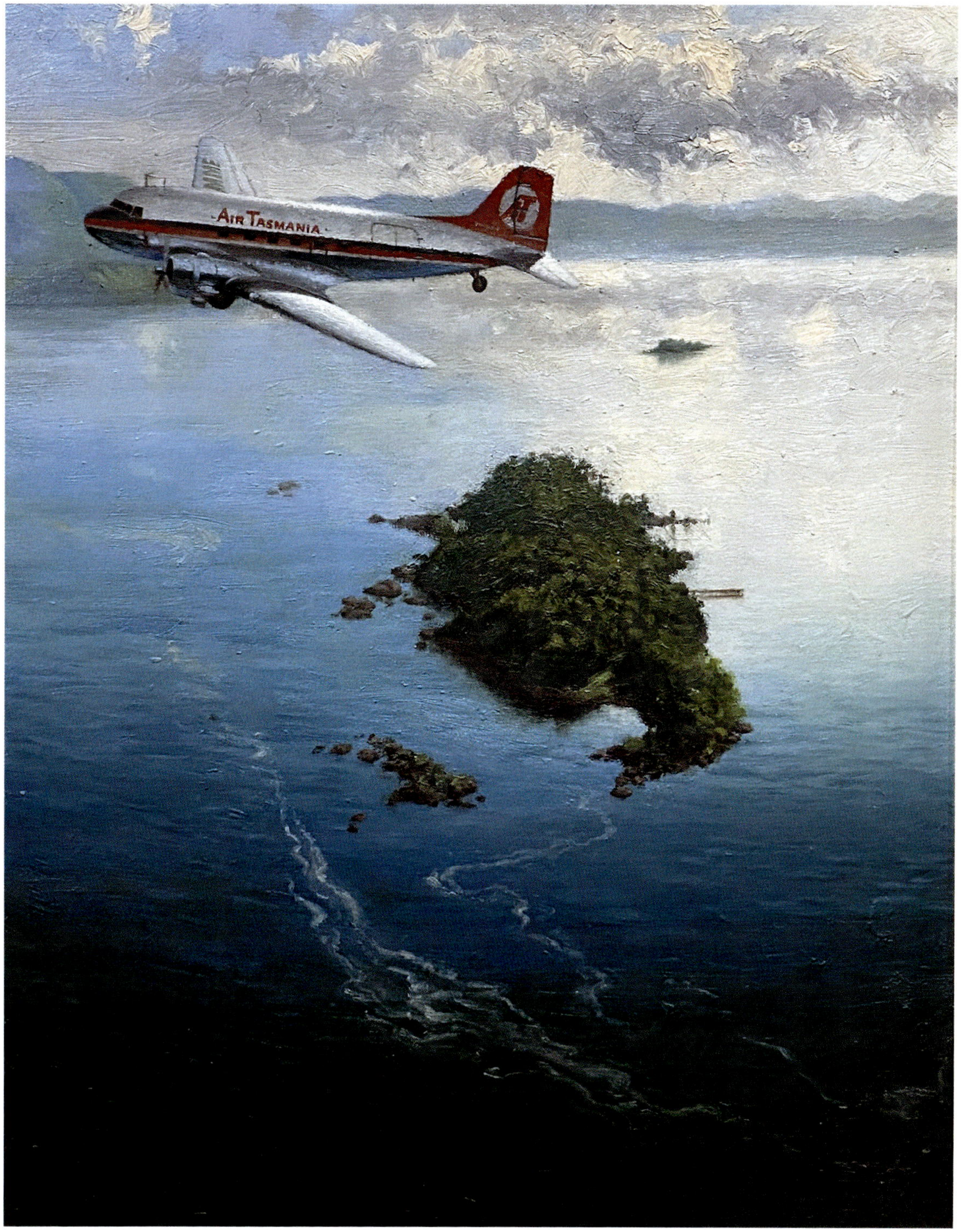

Ray Honisett (Douglas DC-3 over Sarah Island, Macquarie Harbour, Tasmania), c.2000
Oil on canvas
Private collection, Melbourne

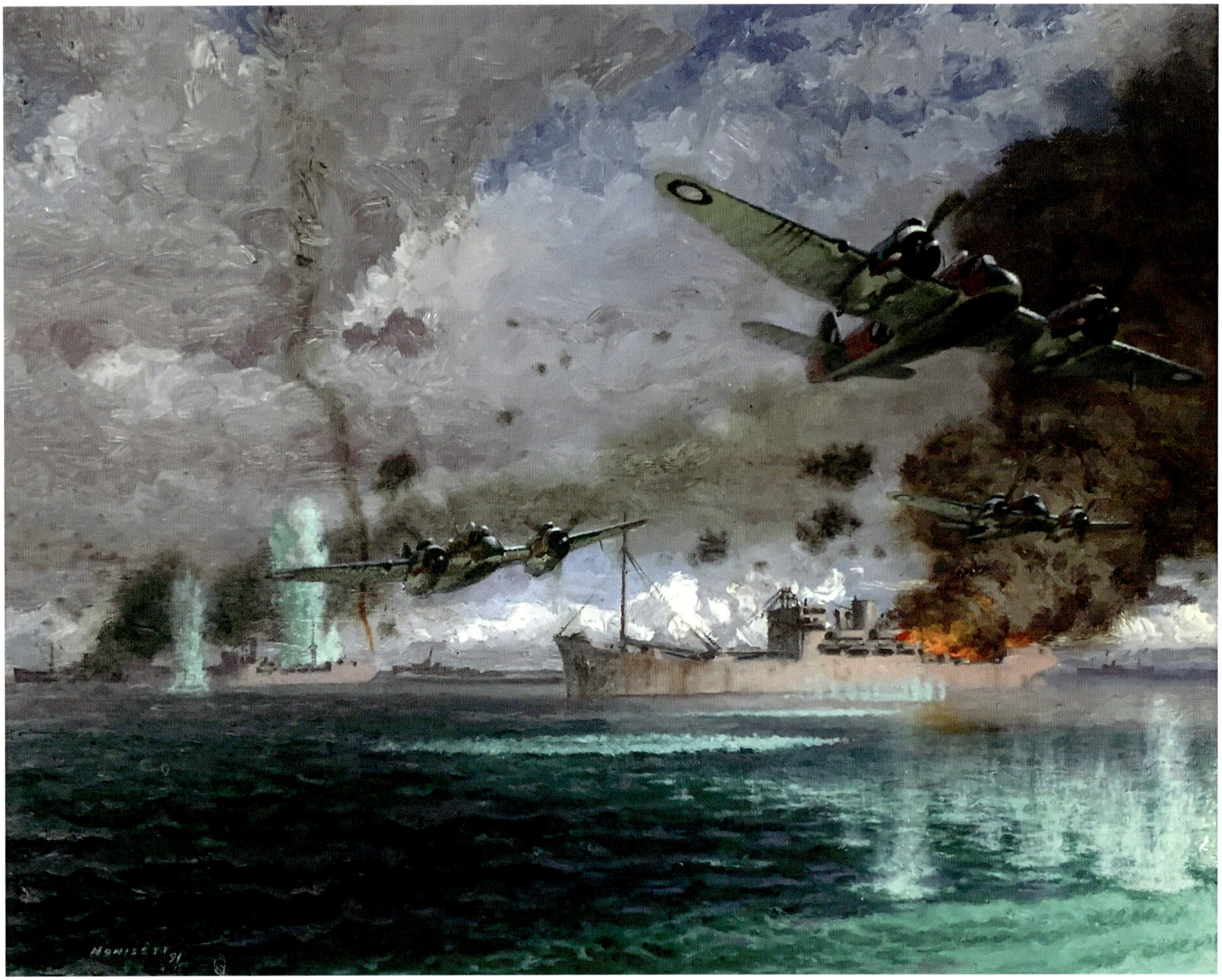

Ray Honisett, ***Sweeping the Decks****, 1991*
Oil on board, 47.5 x 57.5 cm
RAAF Museum, Point Cook
Winner, 1991 RAAF Heritage Award

Ray Honisett's winning entry for the RAAF Heritage Award of 1991 reprised his most successful painting, the 455 Squadron Beaufighters of 1969. In revisiting the subject he has brought the viewpoint down from a high level, as though seen from another aircraft, to deck level, imagining what crew members might have witnessed as the Beaufighters swept low.
In Ray's repertoire this was a relatively unusual viewpoint, for as much as possible he utilised the 'air-to-air' depiction, as in the Beaufort composition opposite. There was a simple message in this view, encouraging the viewer that they too were feeling the thrill of flight..

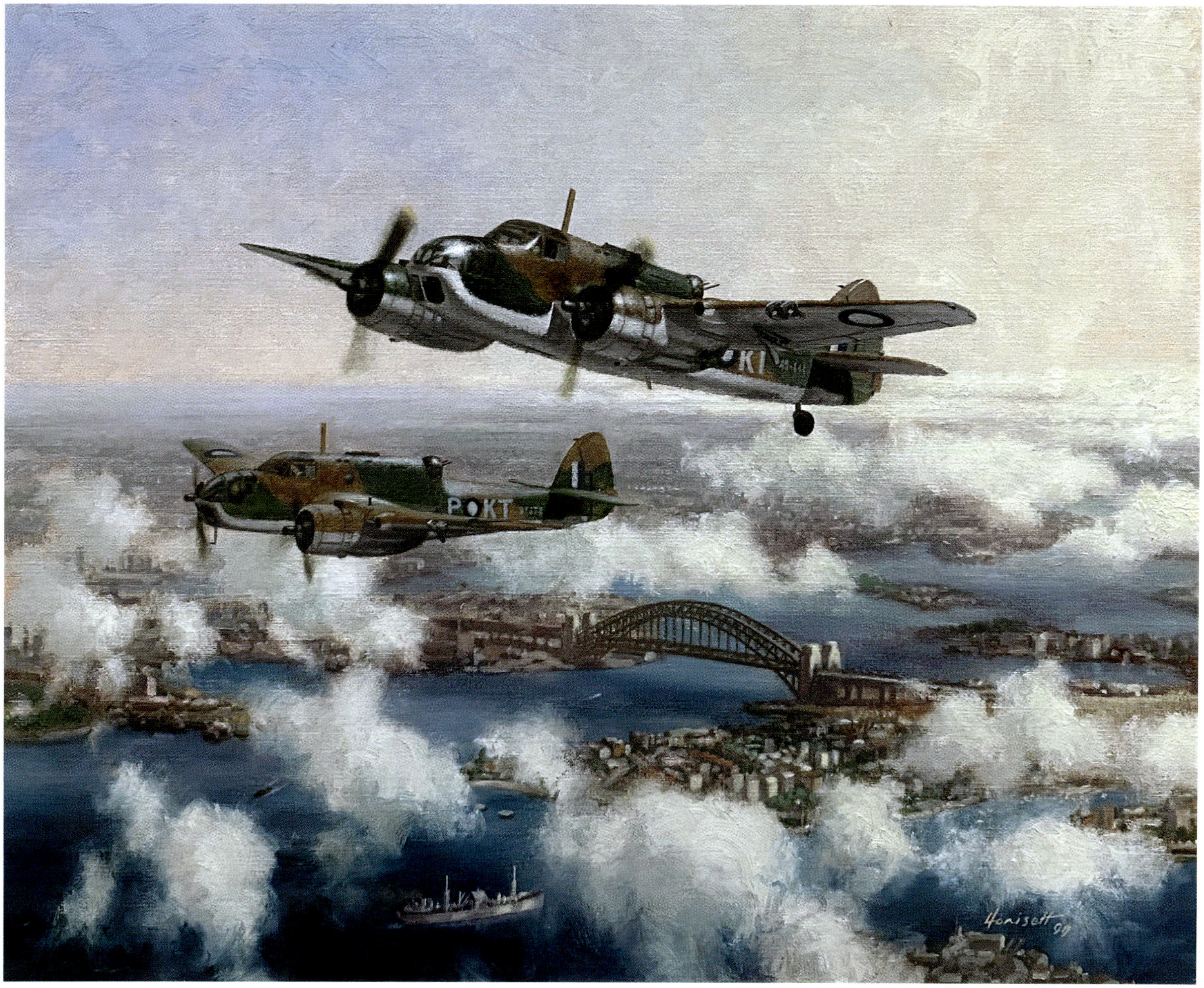

Ray Honisett (A pair of Australian built Bristol Beauforts over Sydney during World War Two), 1999
Oil on Canvas
Private collection, Melbourne

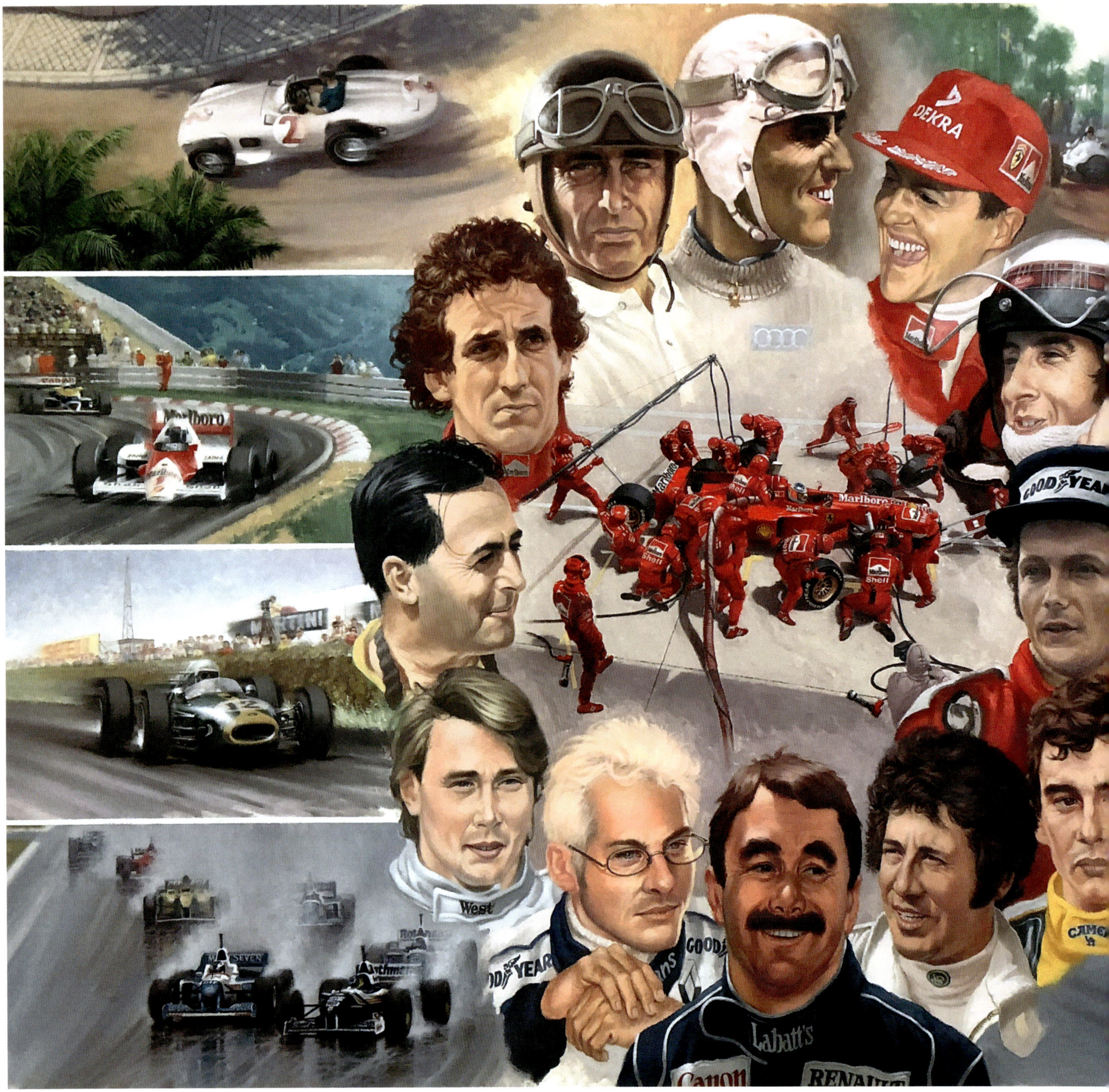
DEKRA
Marlboro
MARTINI
West
Labatt's
Canon

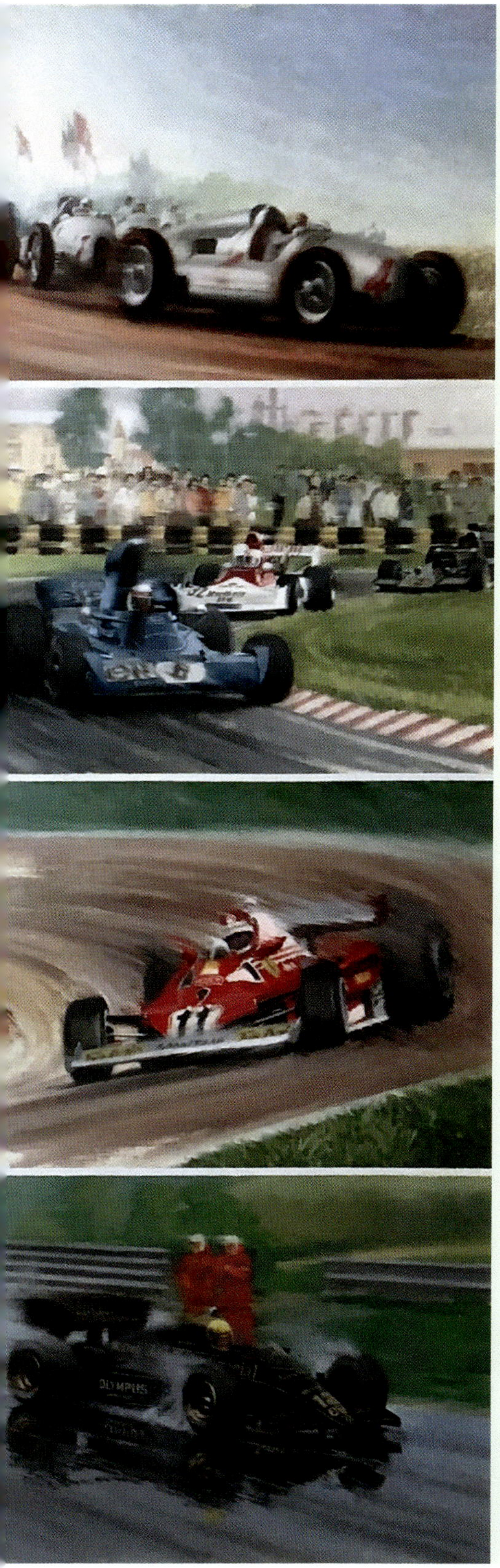

THE SPORTING FIELD

I met Ray in 1977 while working with the National Bank in Bulleen and, over the ensuing years, we kept in touch and I took an interest in his art as our friendship developed. In the late 1980s Ray experienced some financial difficulties and I offered him the use of a large shed on my property as a studio. Later on Ray used an unoccupied floor of the Bank to continue his works, as it was close to RMIT where he was teaching. I am a passionate sports fan, so in the late 1990s I convinced Ray to embark on a series of paintings based on my favourite Australian sporting champions. The first was of Australian world motor sport champions, which depicted Jack Brabham, Alan Jones, Wayne Gardiner and Mick Doohan and was completed in 1999. This was quickly followed by cricket champions Bradman, Lillee, Border and Warne. Being a longstanding hoarder and collector of all sorts of items, I decided that all Ray's Australian subjects needed to have a plaque signed by all of those depicted. This was a challenge in itself, tracking down the numerous sportsmen and women that Ray had painted. In all he completed 40 works, the last being of Cadel Evans, which he finished in 2018. A final, unfinished painting was to have been the 2019 Richmond Premiership team.

In the process of obtaining signatures from our Aussie Champions I met a great range of personalities and enjoyed listening to their comments on Ray's pictures, which were universally complimentary. Many of these champions requested either a reproduction or a photograph of the painting they were depicted in. This prompted me to have limited edition prints made of both the Holden and Ford Legends, which were individually signed by all the drivers depicted in the respective artworks. Unfortunately we only had them released on the market after Peter Brock was killed in a Motor Rally in Western Australia. Allan Moffat was most impressed, as he commented that most artworks by other artists had correct and accurate depictions of his race cars, but somehow his portraits were gorilla-like or with unrecognizable facial features. Holden Champion Jim Richards was so impressed that his canvas reproduction hangs proudly in the entrance way to his family home.

Aussie Champions painted by Ray covered a great range of sports from Motor Racing, AFL, Horse Racing, Swimming, Tennis, Boxing, Rugby League and Union and Golf. On one occasion it was great to see the reaction of both Ray and Johnny Famechon when I introduced them. After picking up Johnny and receiving driving tips regularly from him on the journey down to Ray's house at Rye, it was like a scene out of the TV Show 'This is your Life'. Ray was not aware that I was bringing a famous visitor to meet him and Johnny could not believe that Ray's artwork was so great. Each of these 90 x 60 cm works took Ray a minimum of 15 weeks to complete.

*Ray Honisett, **World Motor Racing Champions**, 2003*
Oil on canvas, 120 x 180 cm
Lachie Pitman Collection
Clockwise from top, Tazio Nuvolari, Michael Schumacher, Jackie Stewart, Nicki Lauda, Aryton Senna, Mario Andretti, Nigel Mansell, Jacques Villeneuve, Mika Häkkinen, Jack Brabham, Alain Prost, Juan Manuel Fangio

Part way through the Aussie Champions series, I convinced Ray to embark on something grander, as I have a passion for both Golf and Grand Prix racing. After some deliberation, Ray decided to accept the challenge. Ann Bradshaw, Press Secretary of the Williams Formula One team, who was introduced to me by my great friend Win Percy, selected the drivers to be included in the painting. Fangio and Nuvolari were considered the grandfathers of the sport and sit at the top of the composition. When Ray commenced the painting, Michael Schumacher was reigning supreme, so his Ferrari pitstop was made the centrepiece of the work. There is one driver from each country and, as Stirling Moss was never a world champion, Nigel Mansell was the English representative. Ray went to great lengths to consider the layout, balance and design of this work, as he did not want the viewer to focus on the same spot each time and, knowing this, it is evident he has achieved what he set out to do. This painting took 12 months to complete in 2003.

The next large painting would be the World Golfing Champions. My favourite at the time, Tiger Woods, was sitting at the top of the tree and it looked like he would easily surpass the number of majors held by Jack Nicklaus, so he was made the main focus. I chose all of the golfers depicted and gave it an Australian flavour by including Greg Norman, Peter Thompson and Karrie Webb. This painting sits proudly in my bar along with another 35 or so of Ray's works. Along our journey Ray painted three larger artworks — Craig Lowndes, Indigenous Australian Rules Champions and the Collingwood Football Club 2010 Premiership, which included President Eddie McGuire, Coach Mick Malthouse, along with six players. We reproduced 25 individually signed canvas prints and one of each was given to all depicted, with the remainder sold by the Club as a fund raiser. In addition to the Holden and Ford Champions, we produced a limited edition print of The Grand Prix Legends artwork signed by Sir Jack Brabham. My son Murray and I drove to meet Jack on the Gold Coast in 2009 to have him individually sign each of the prints. He was extremely pleased and complimentary about Ray's painting and we gave him one to hang in his home. The Australian Grand Prix Corporation has a 1200 x 1800 cm Canvas reproduction hanging in their office in Melbourne.

Over our Sporting Champions series Ray and I developed a great bond and mateship. In addition to meeting monthly, we looked forward to the Sports Gallery annual Christmas gathering, which we held for 20 years straight with 100% attendance - Ray and Me!

Lachie Pitman
2020

*Ray Honisett, **World Golf Champions**, 2004*
Oil on canvas, 120 x 180 cm
Lachie Pitman Collection
Left to right, top, Greg Norman, Karrie Webb, Peter Thomson, Nick Faldo Arnold Palmer, Gary Player, Tom Watson, Lee Trevino, Seve Ballesteros, (front) Ben Hogan, Tiger Woods, Jack Nicklaus

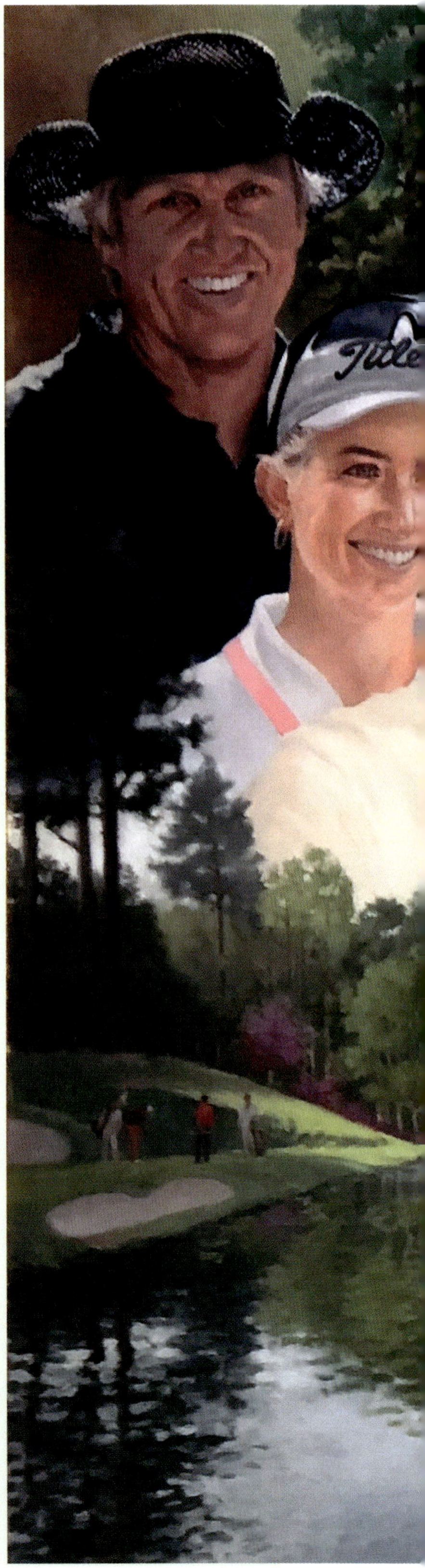

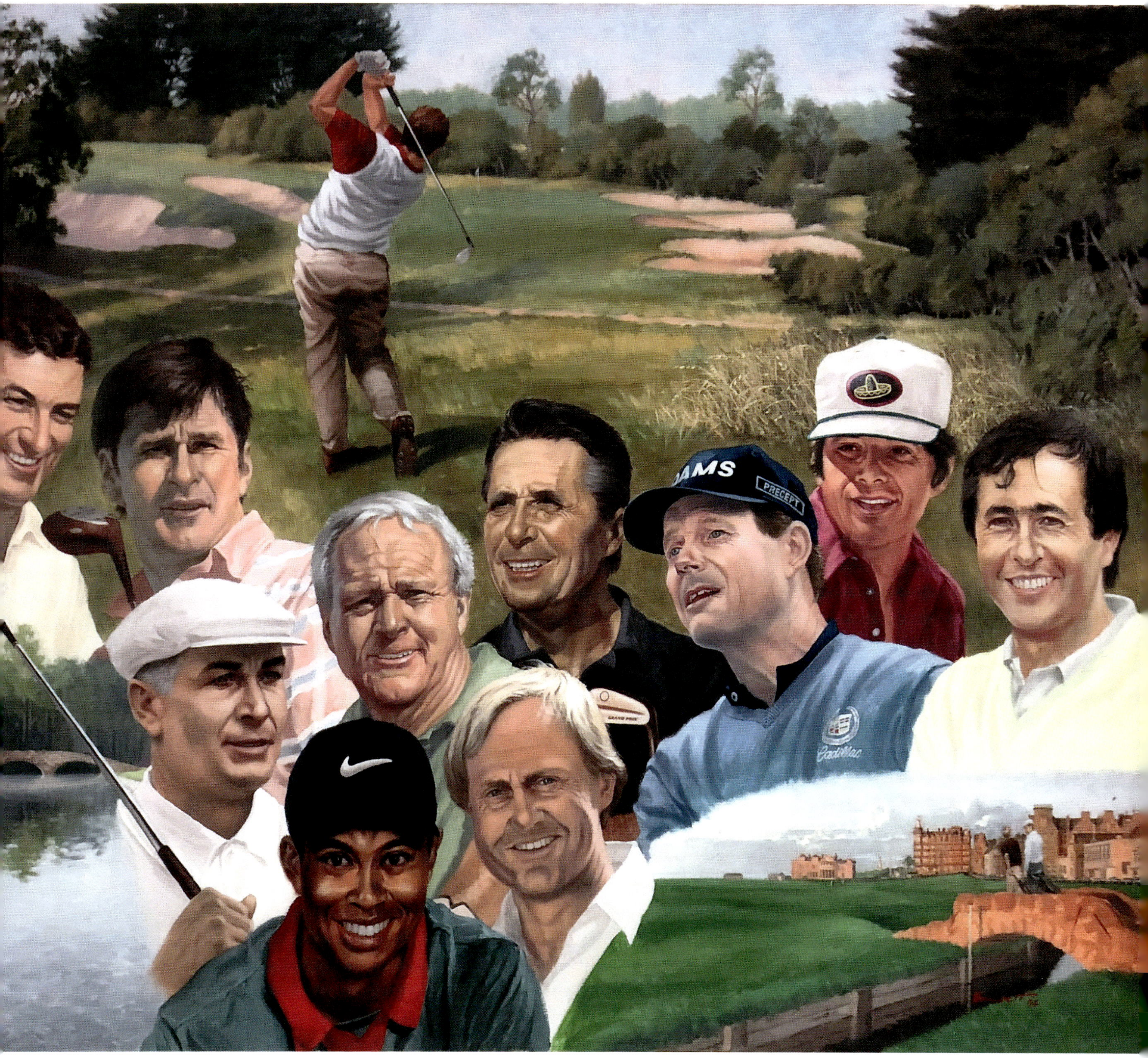
PRECEPT
Cadillac

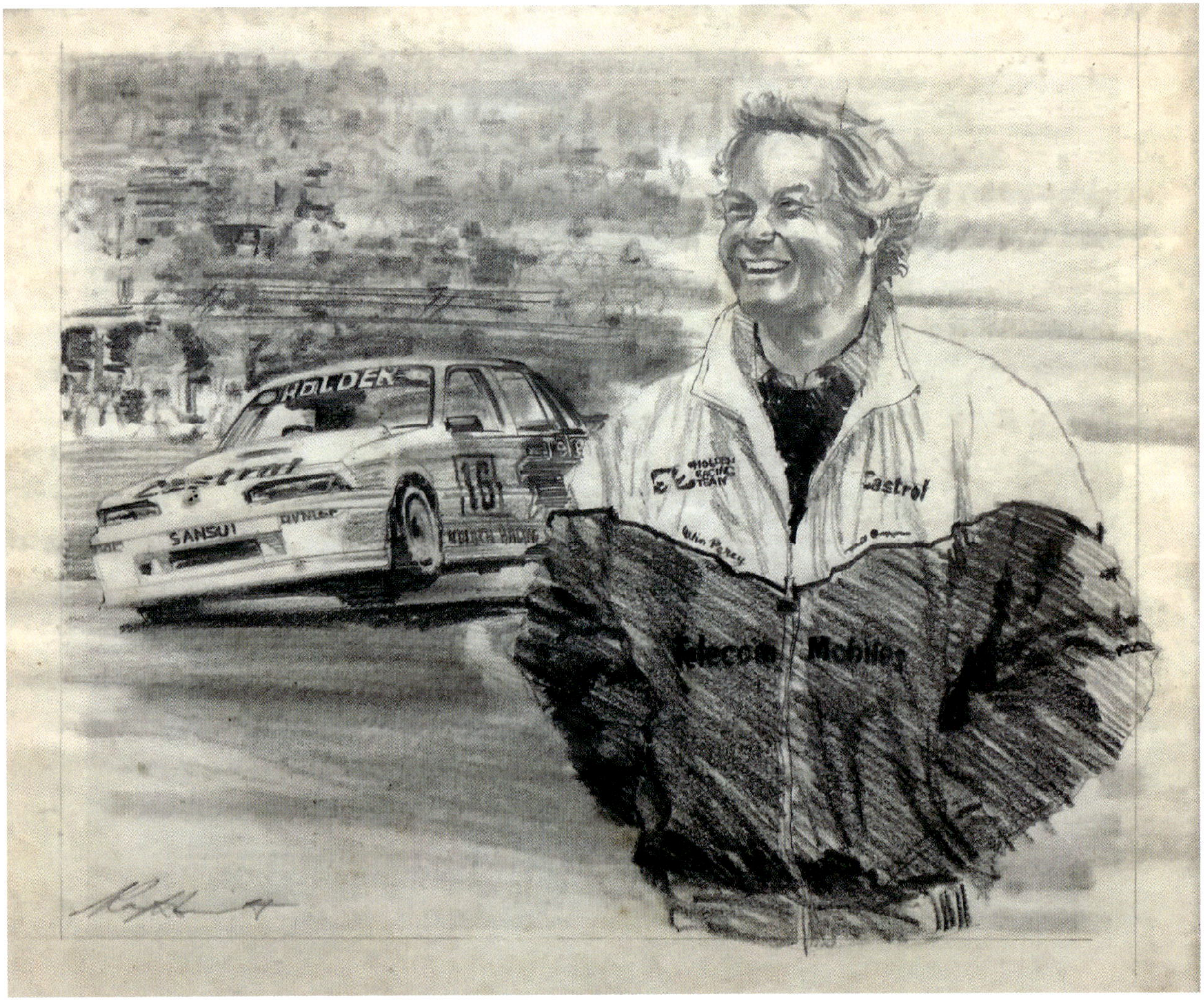

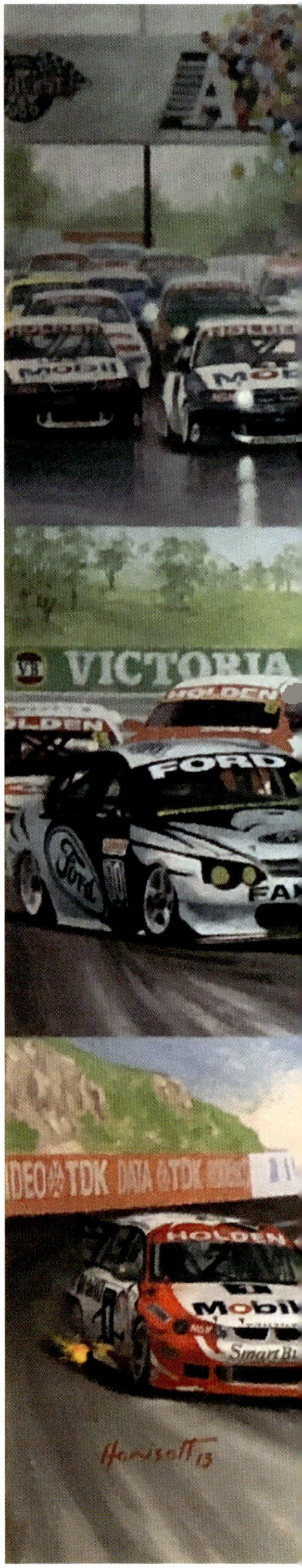

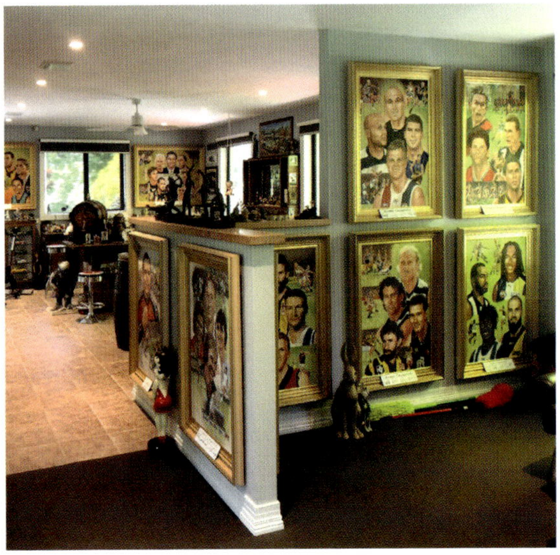

Above, Ray Honisett, ***Win Percy****,*
Pencil on paper
Lachie Pitman Collection

Right, Ray Honisett, ***Craig Lowndes****, 2013*
Oil on canvas, 90 x 120 cm
Lachie Pitman Collection

Left, the Lachie Pitman Collection — When too much Ray Honisett is barely enough …

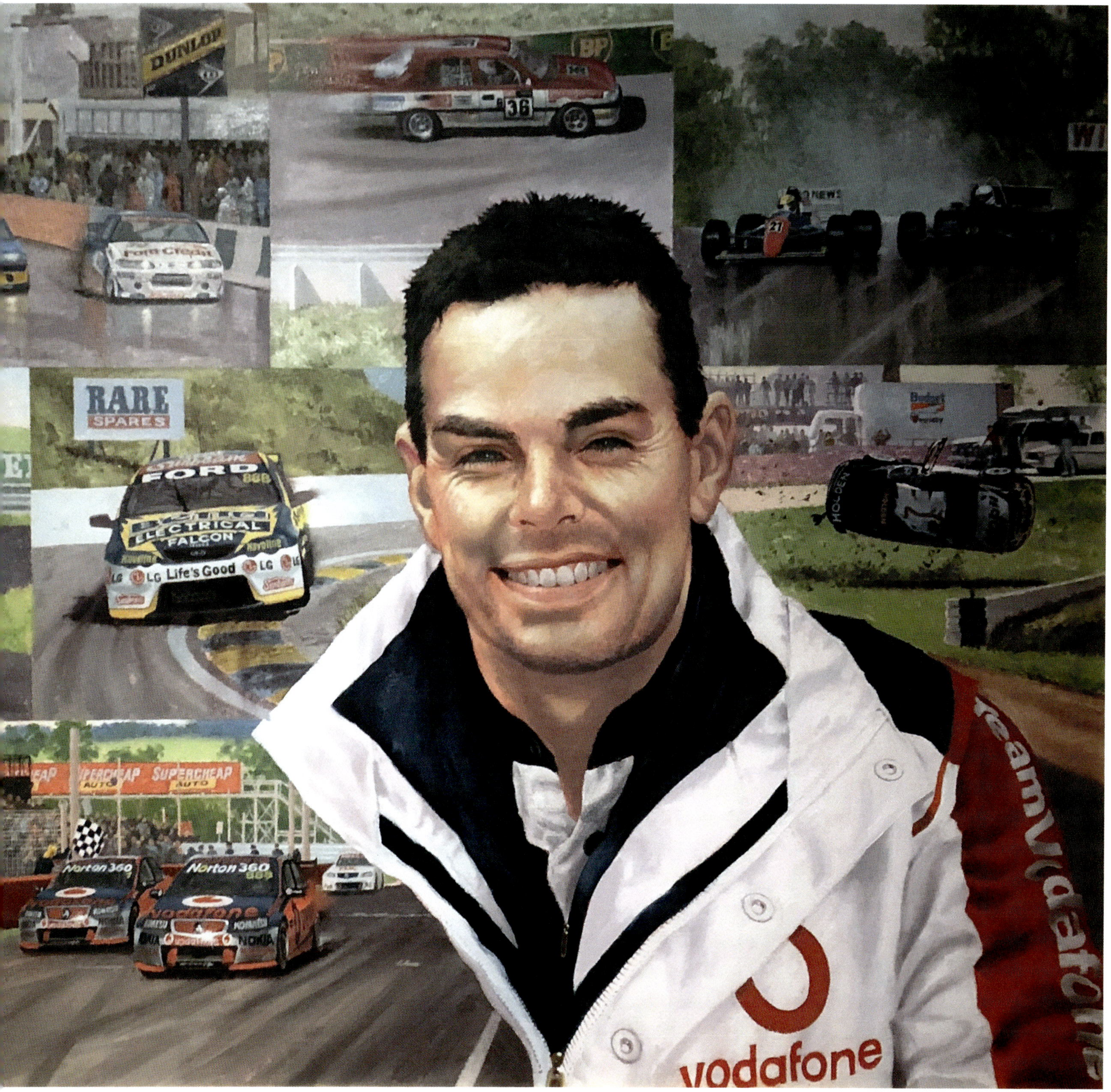
DUNLOP
BP
36
21
RARE
SPARES
FORD
ELECTRICAL
FALCON
LG Life's Good
HOLDEN
SUPERCHEAP
AUTO
Norton 360
vodafone
TeamVodafone

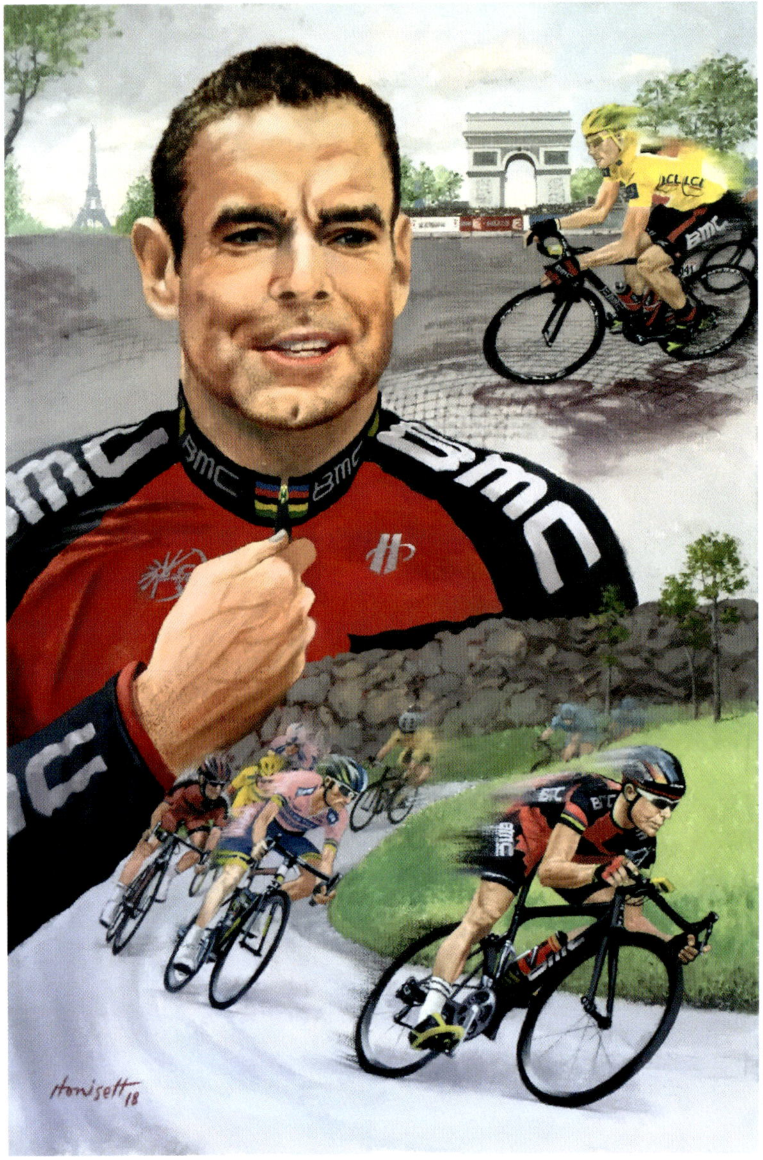

*Ray Honisett, **Cadel Evans**, 2018*
Oil on canvas, 90 x 60 cm
Lachie Pitman Collection

The last work in the Australian Sports Stars series, a remarkable achievement for an artist in his 87th year.

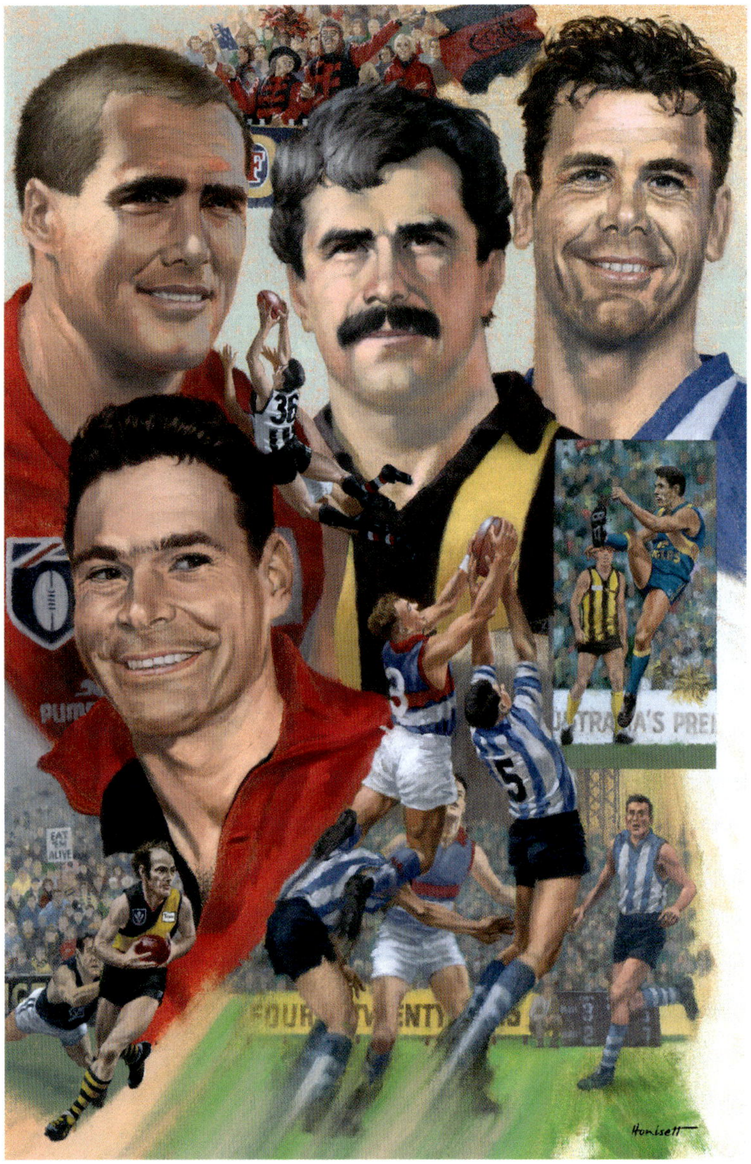

*Ray Honisett, **AFL Champions**, 2004*
Oil on canvas, 90 x 60 cm
Lachie Pitman Collection
Top, left to right, Tony Lockett, Leigh Mathews, Wayne Carey,
Front, Ron Barrassi

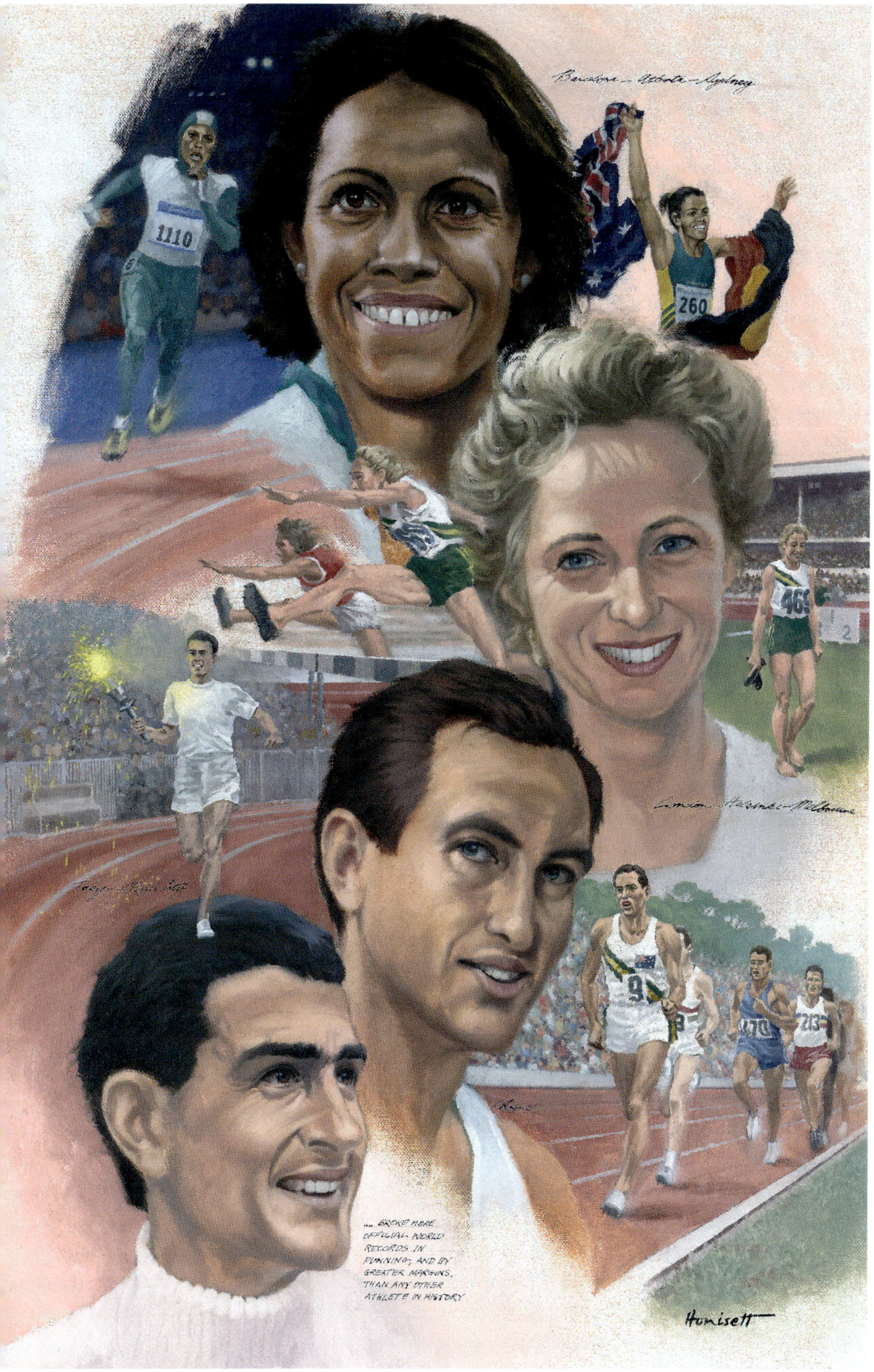

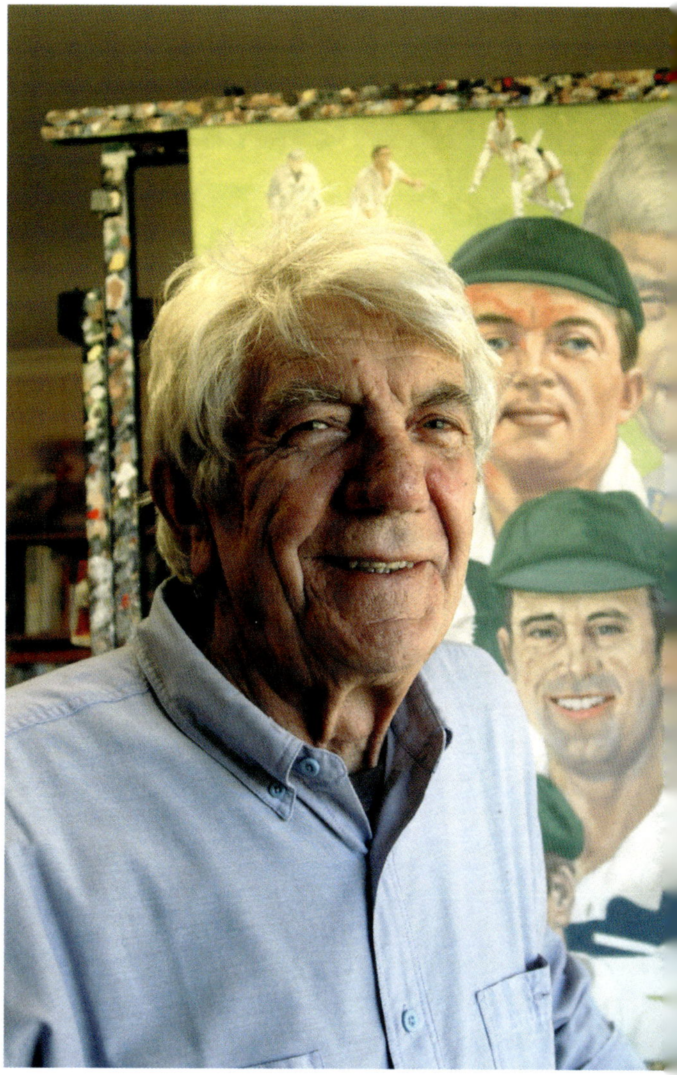

Above, Ray Honisett in the studio, painting Richie Benaud and Australian cricket captains.

Ray Honisett, ***Athletics Champions****, 2004*
Oil on canvas, 90 x 60 cm
Lachie Pitman Collection

Top to bottom, Cathy Freeman, Shirley Strickland, Herb Elliott, Ron Clarke

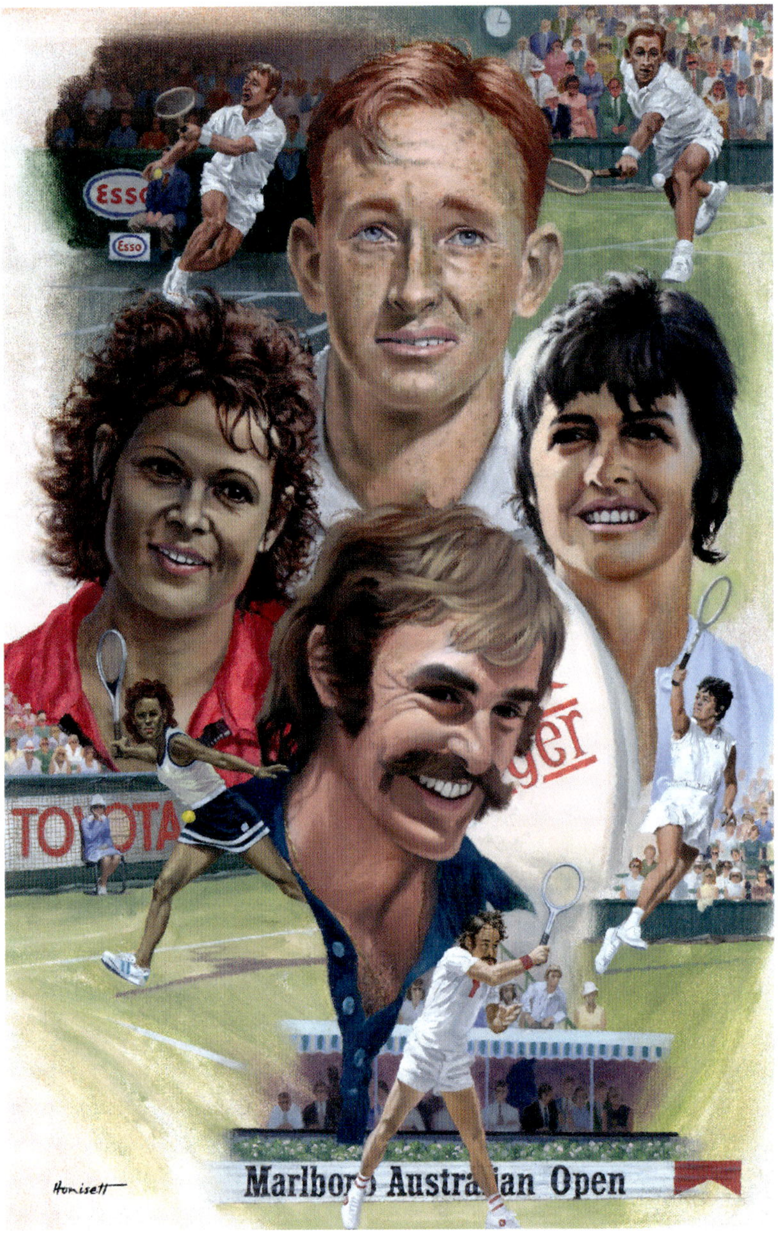

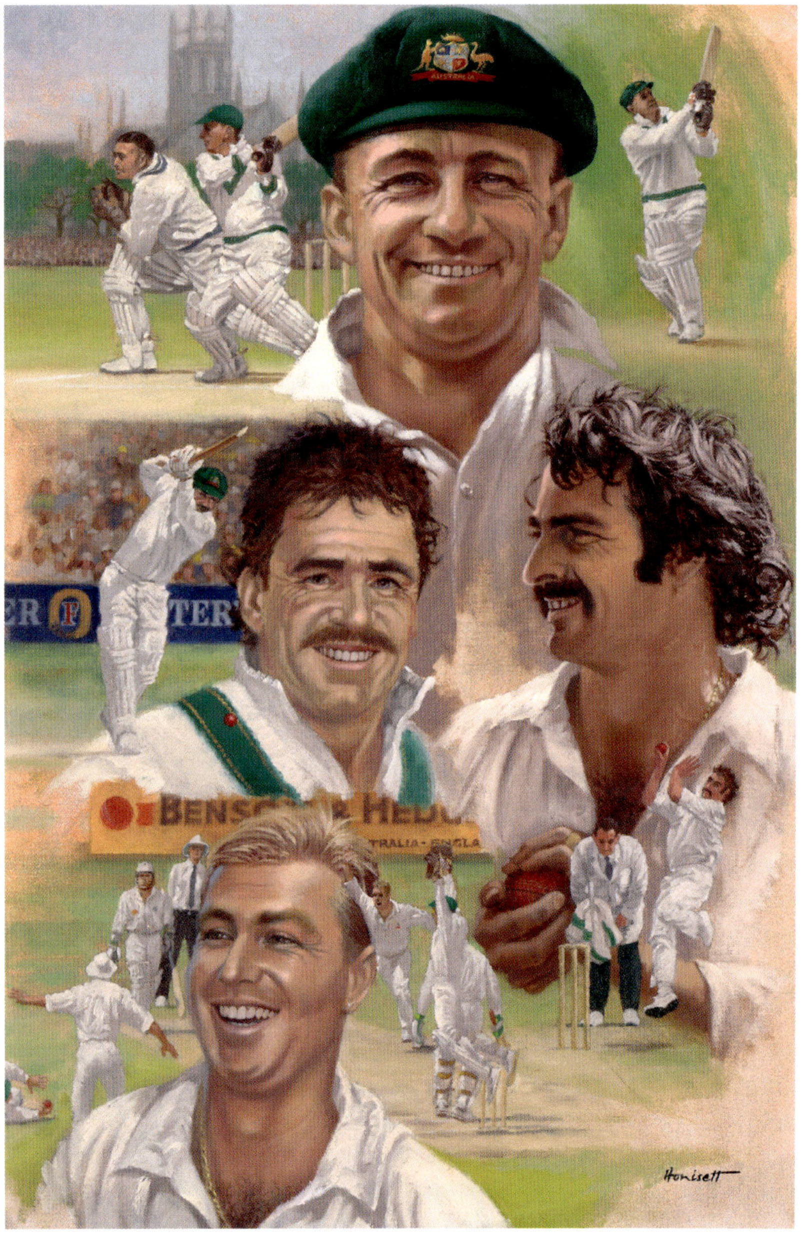

Left, Ray Honisett, ***Tennis Champions****, 2004*
Oil on canvas, 90 x 60 cm
Lachie Pitman Collection
From top, clockwise, Rod Laver, Margaret Court,
John Newcombe, Evonne Goolagong

Ray Honisett, ***Cricket Champions****, 1999*
Oil on canvas, 90 x 60 cm
Lachie Pitman Collection
From top, clockwise, Sir Donald Bradman, Dennis Lillee,
Shane Warne, Alan Border.

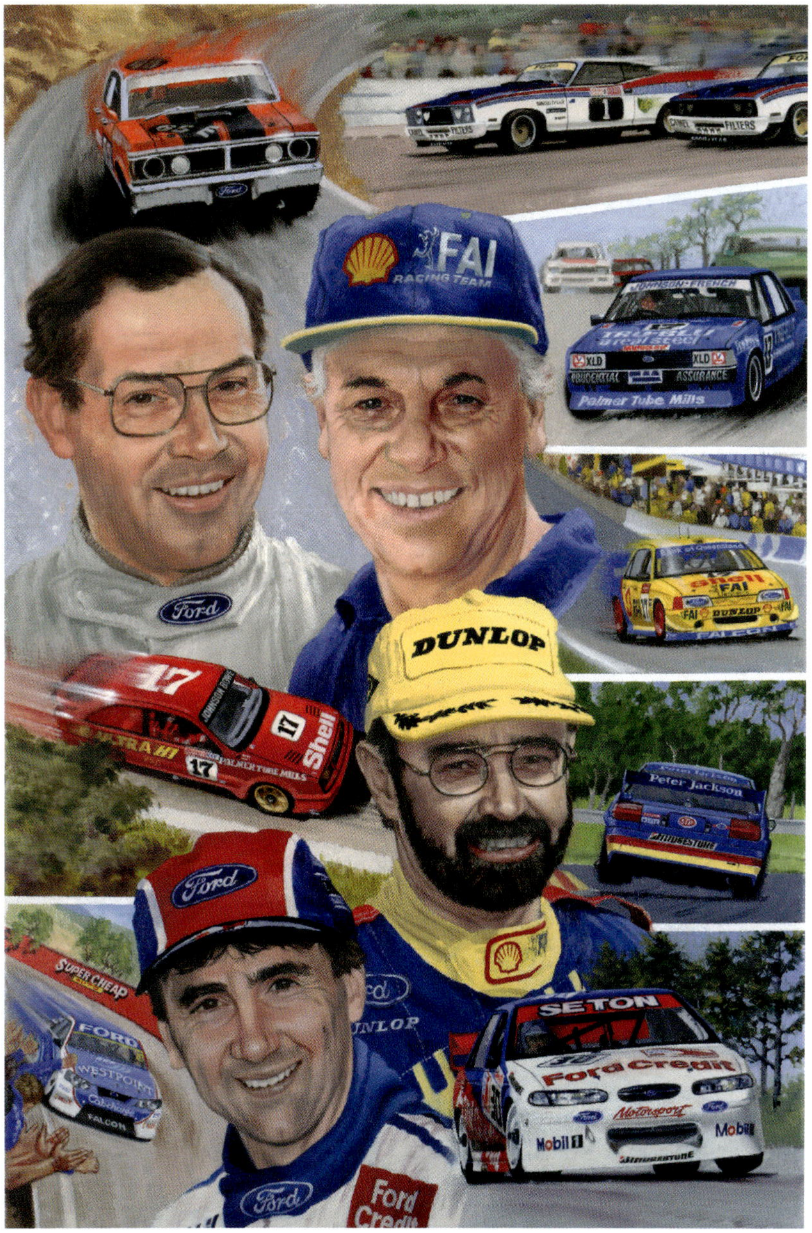

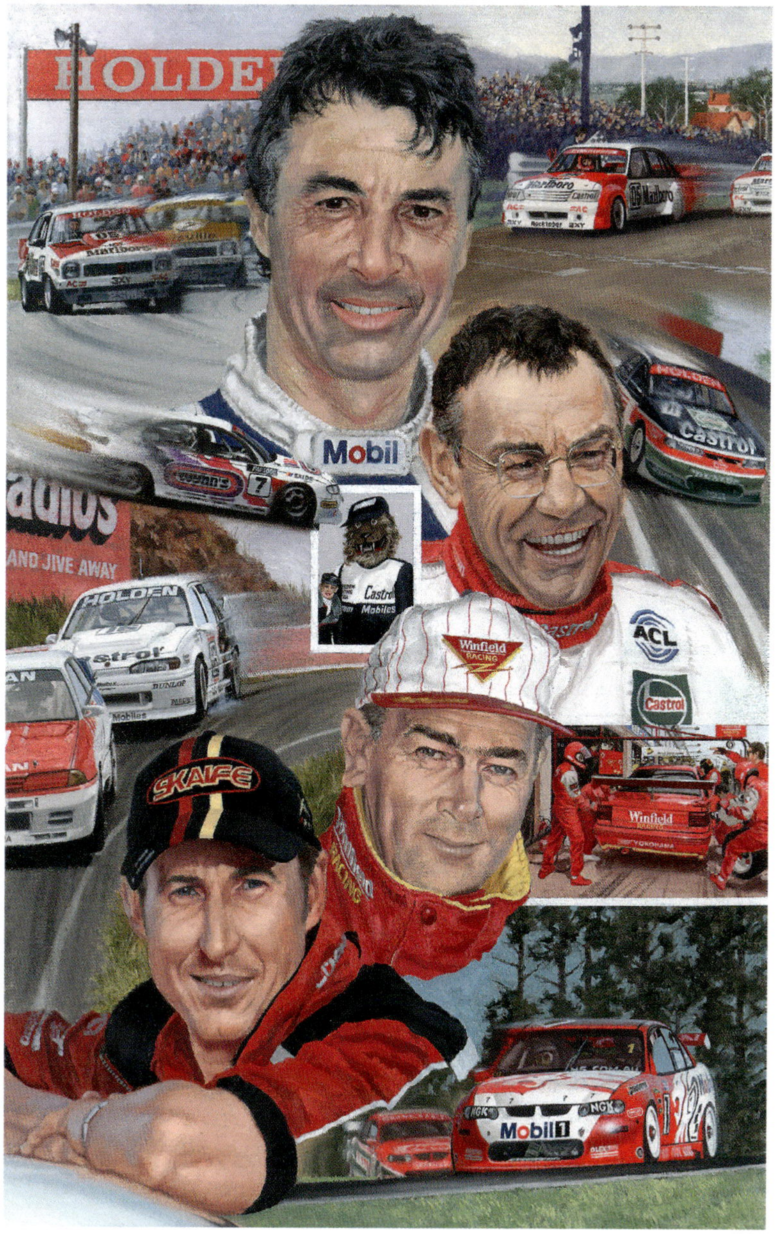

Ray Honisett, ***Ford Touring Car Champions****, 2004*
Oil on canvas, 90 x 60 cm
Lachie Pitman Collection
From top, clockwise, Alan Moffatt, Dick Johnson, John Bowe, Glenn Seton

Ray Honisett, ***Holden Touring Car Champions****, 2004*
Oil on canvas, 90 x 60 cm
Lachie Pitman Collection
From top, Peter Brock, Larry Perkins, Jim Richards. Mark Skaife

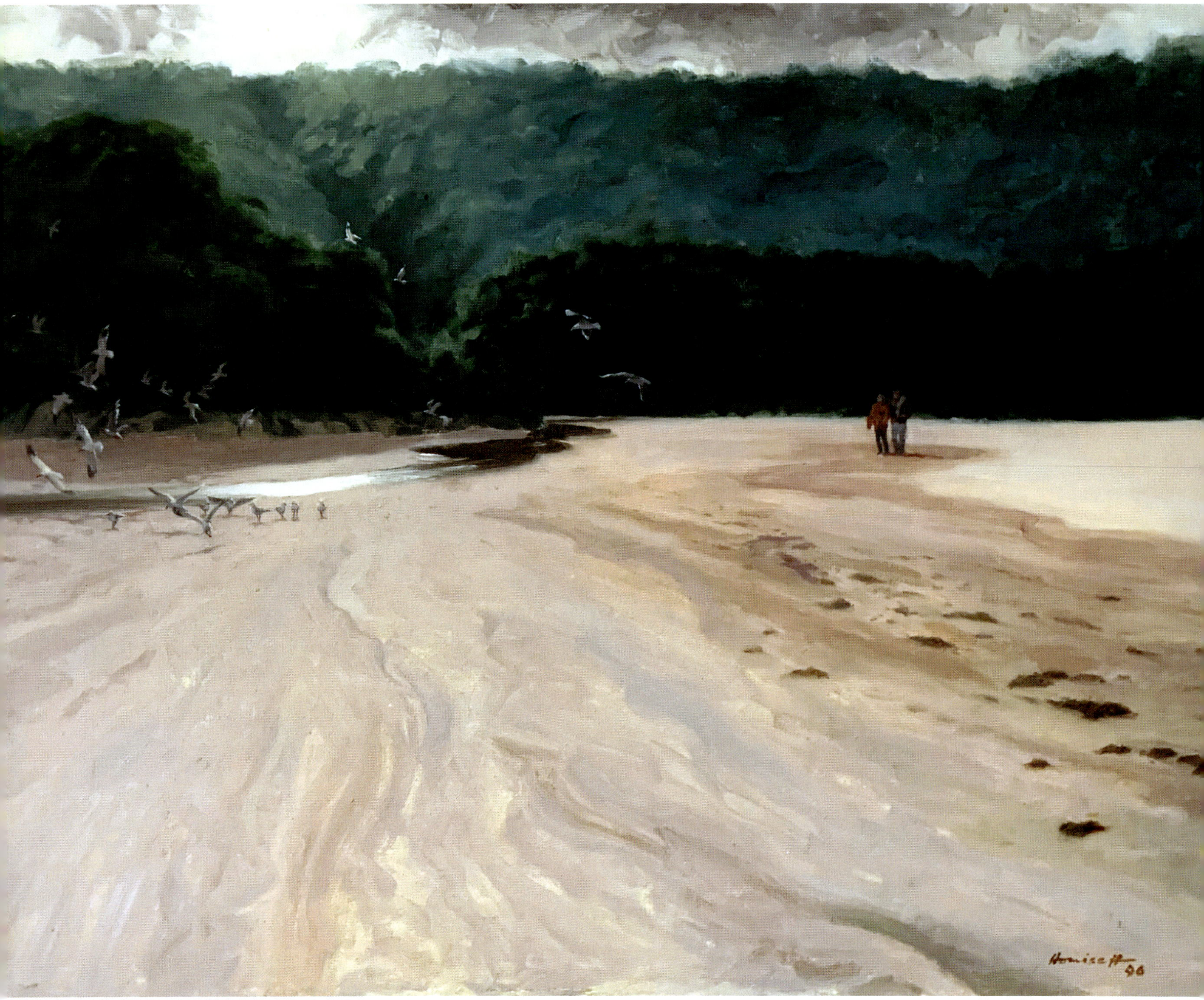

Ray Honisett, ***Tidal River, Wilson's Promontory****, 1996*
Oil on canvas
Private Collection, Melbourne

ON THE BEACH - AND BEYOND

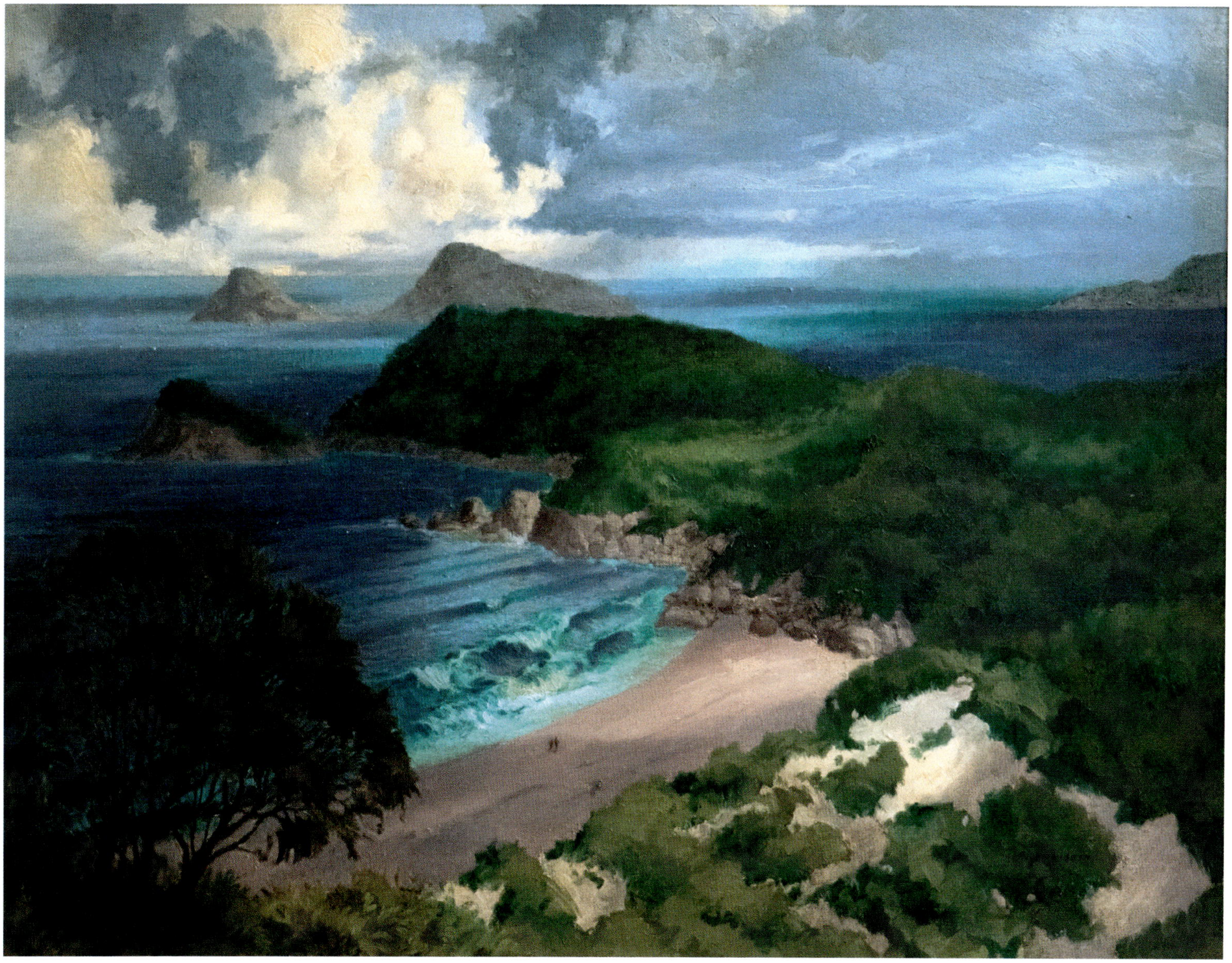

*Ray Honisett, **Squeaky Beach, Wilson's Promontory**, 1989*
Oil on canvas, 91 x 122 cm
Tsaklis Family Collection, Melbourne

After moving around a number of the outer eastern suburbs of Melbourne, Ray Honisett eventually settled into retirement at Rye on the Mornington Peninsula. For an artist who loved the sea it was the perfect place to paint, with easy access to both Port Philip Bay and the wilds of the back beaches facing Bass Strait. He remained in Rye for the rest of a long life, working in his home studio on all manner of projects and commissions. While landscape had always been an essential part of his aviation work, more and more he began to paint the landscape for its own sake, venturing east to the rugged shoreline of Wilson's Promontory and up into the snow country. These works found a ready market and he quietly carved out his own niche in the local art market.

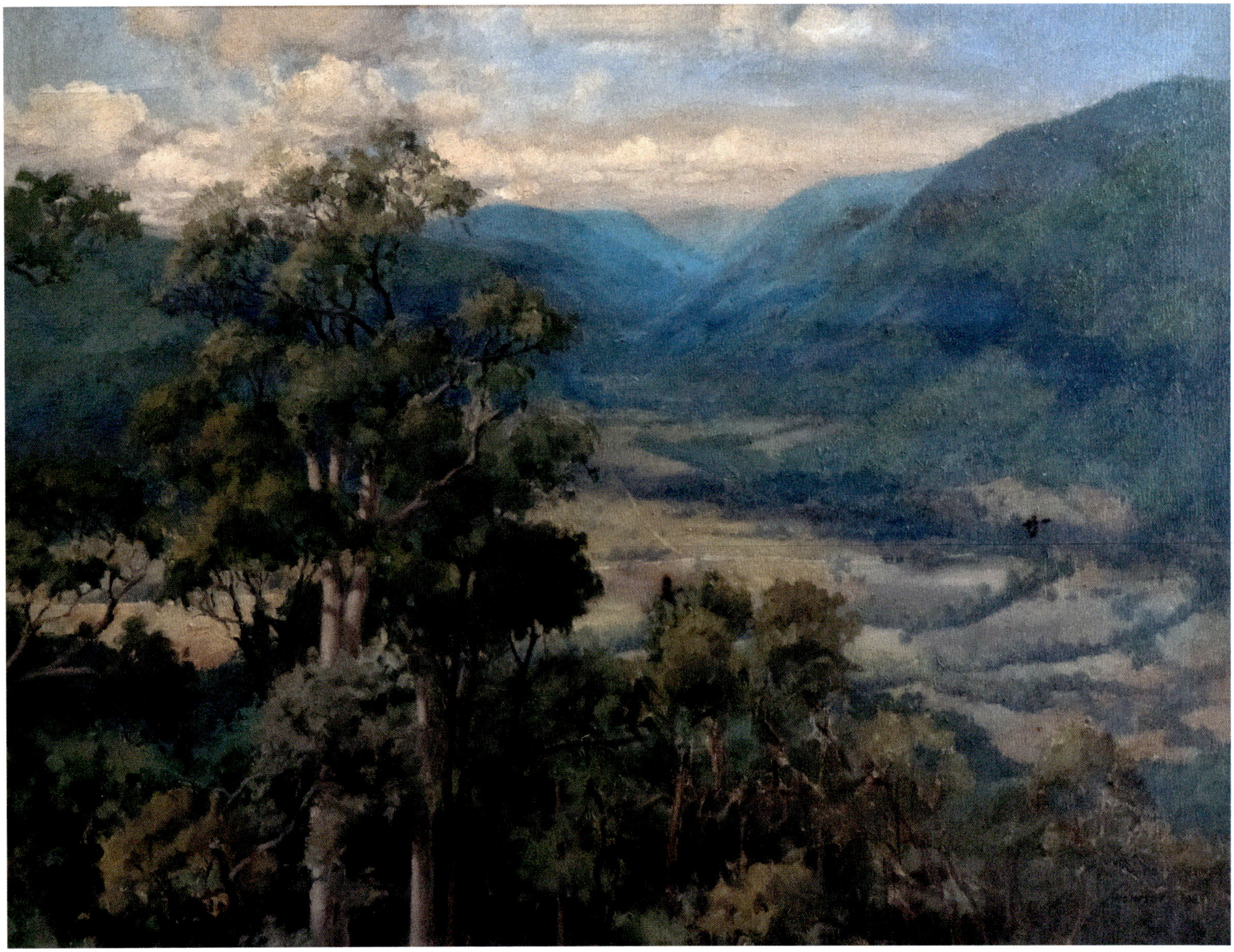

*Ray Honisett, **The Valley**, 1989*
Oil on canvas, 60 x 76 cm
Tsaklis Family Collection, Melbourne

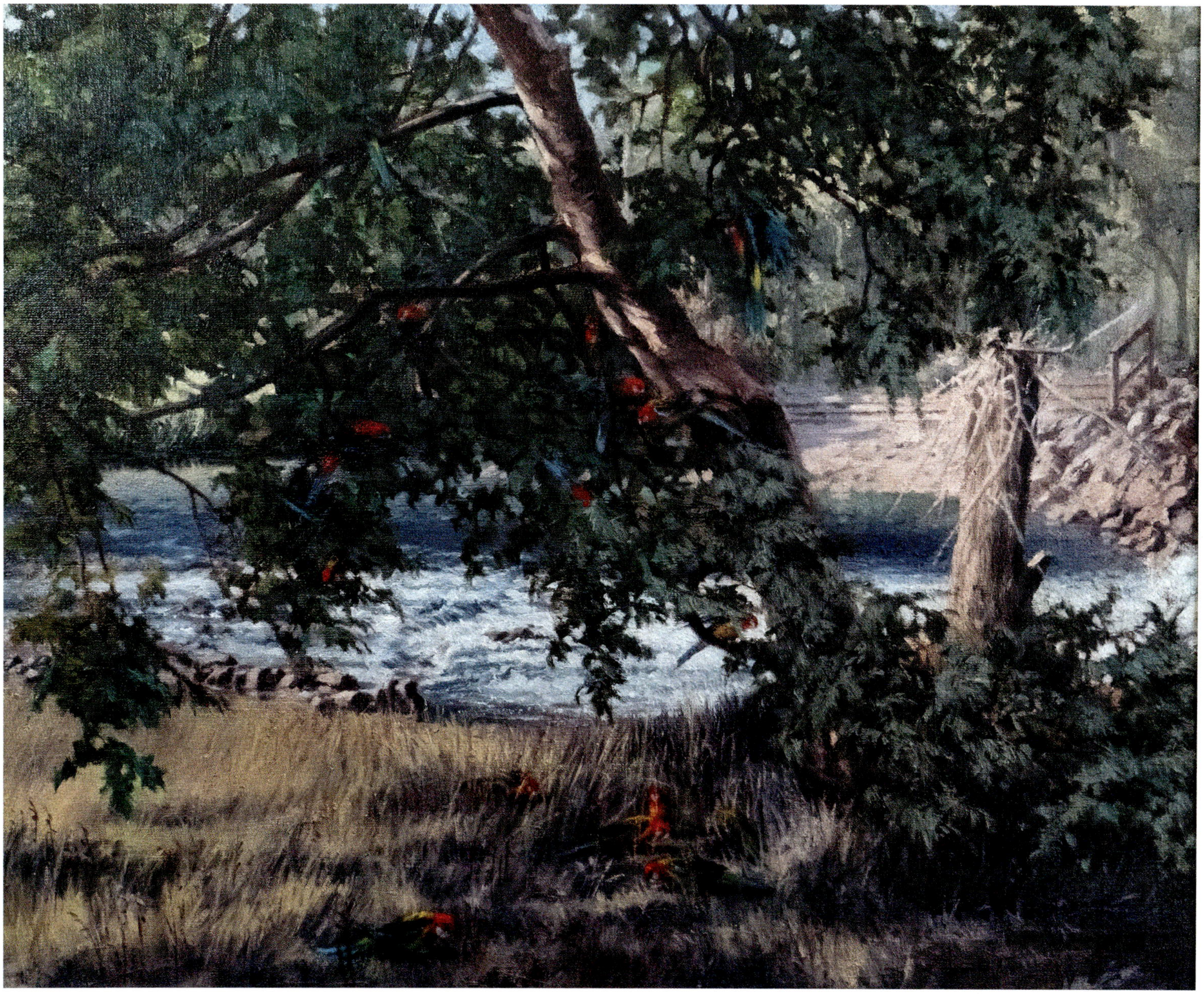

Ray Honisett, ***Rosellas****, 1989*
Oil on canvas, 60 x 76 cm
Tsaklis Family Collection, Melbourne

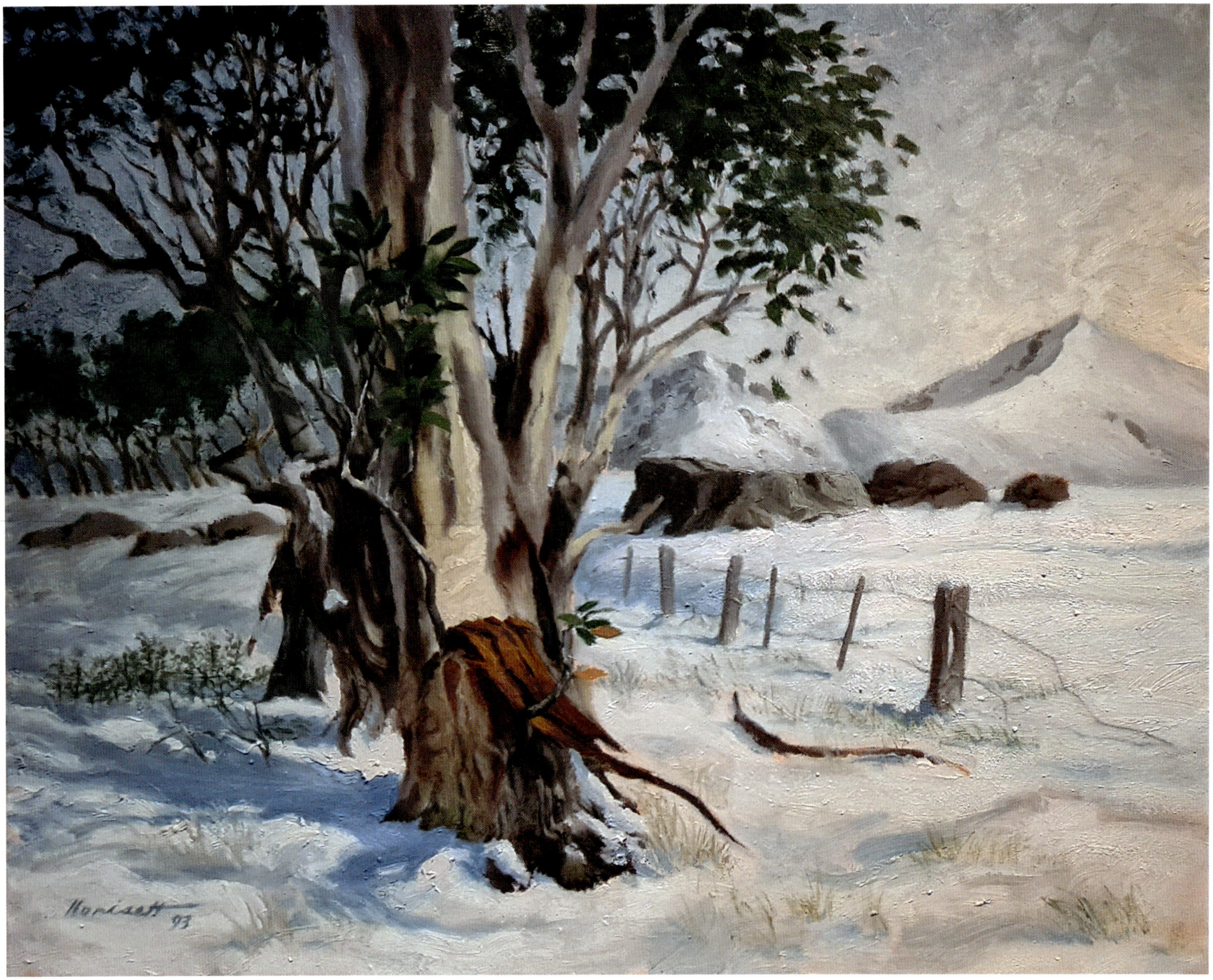

Ray Honisett, ***Mt Buffalo Snowscape****, 1993*
Oil on canvas, 61 x 76 cm
Tsaklis Family Collection, Melbourne

Opposite, Ray Honisett, ***The Inlet****, 1998*
Oil on canvas, 91 x 60 cm
Private collection, Melbourne

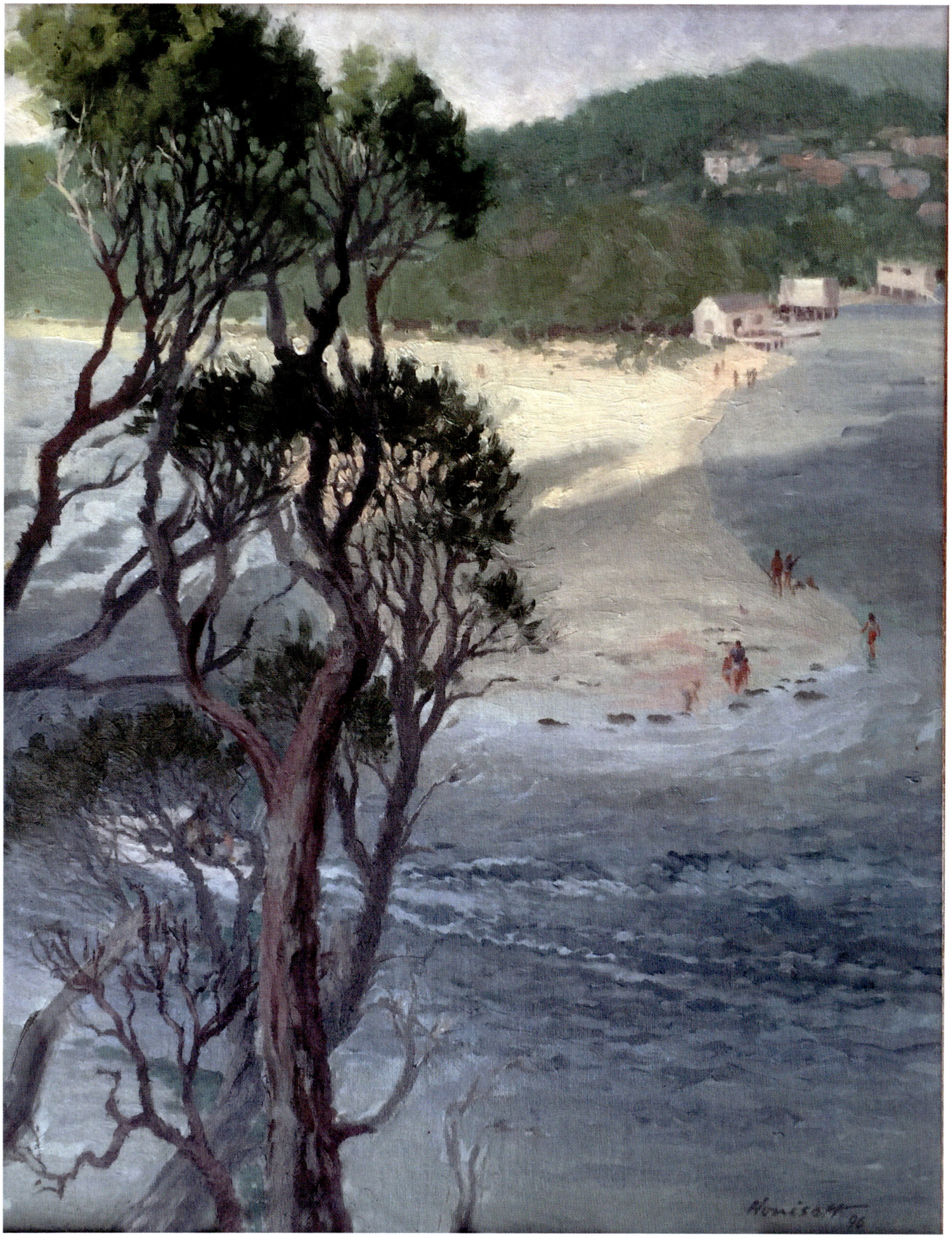

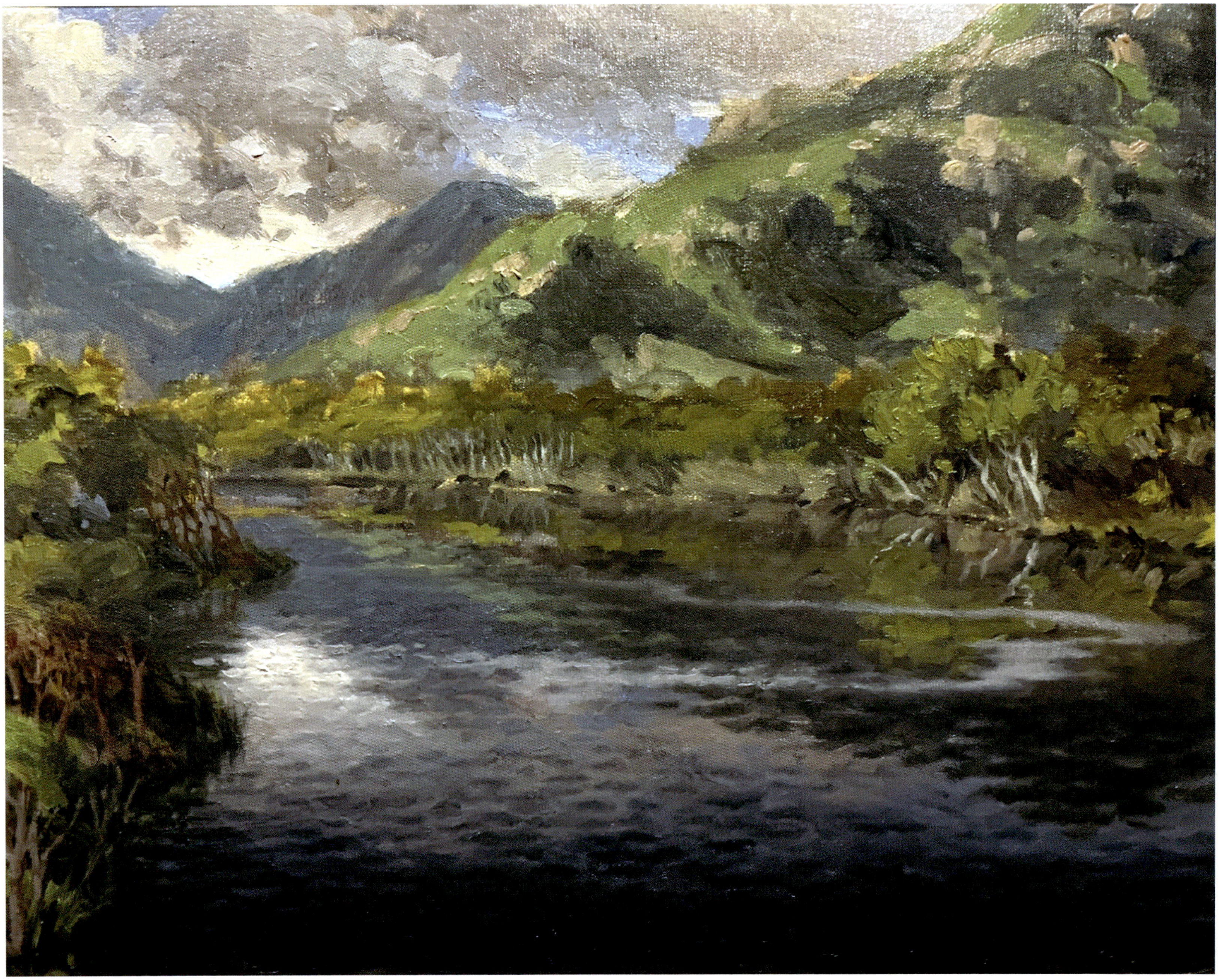

***Tidal River, Wilson's Promontory**, c.1996*
Oil on canvas
Private collection, Melbourne

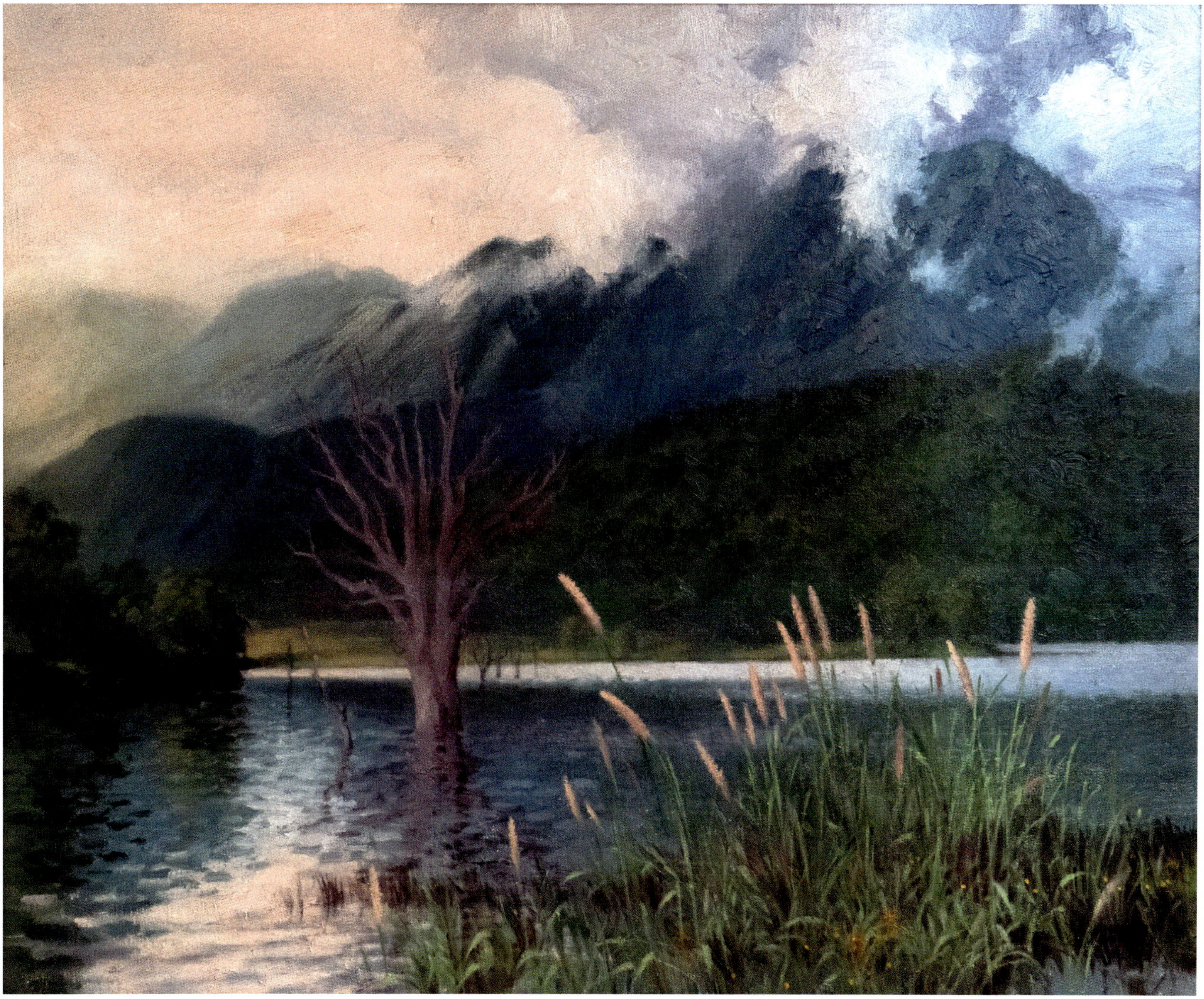

Ray Honisett, ***Reeds****, 1989*
Oil on canvas, 60 x 76 cm
Tsaklis Family Collection, Melbourne

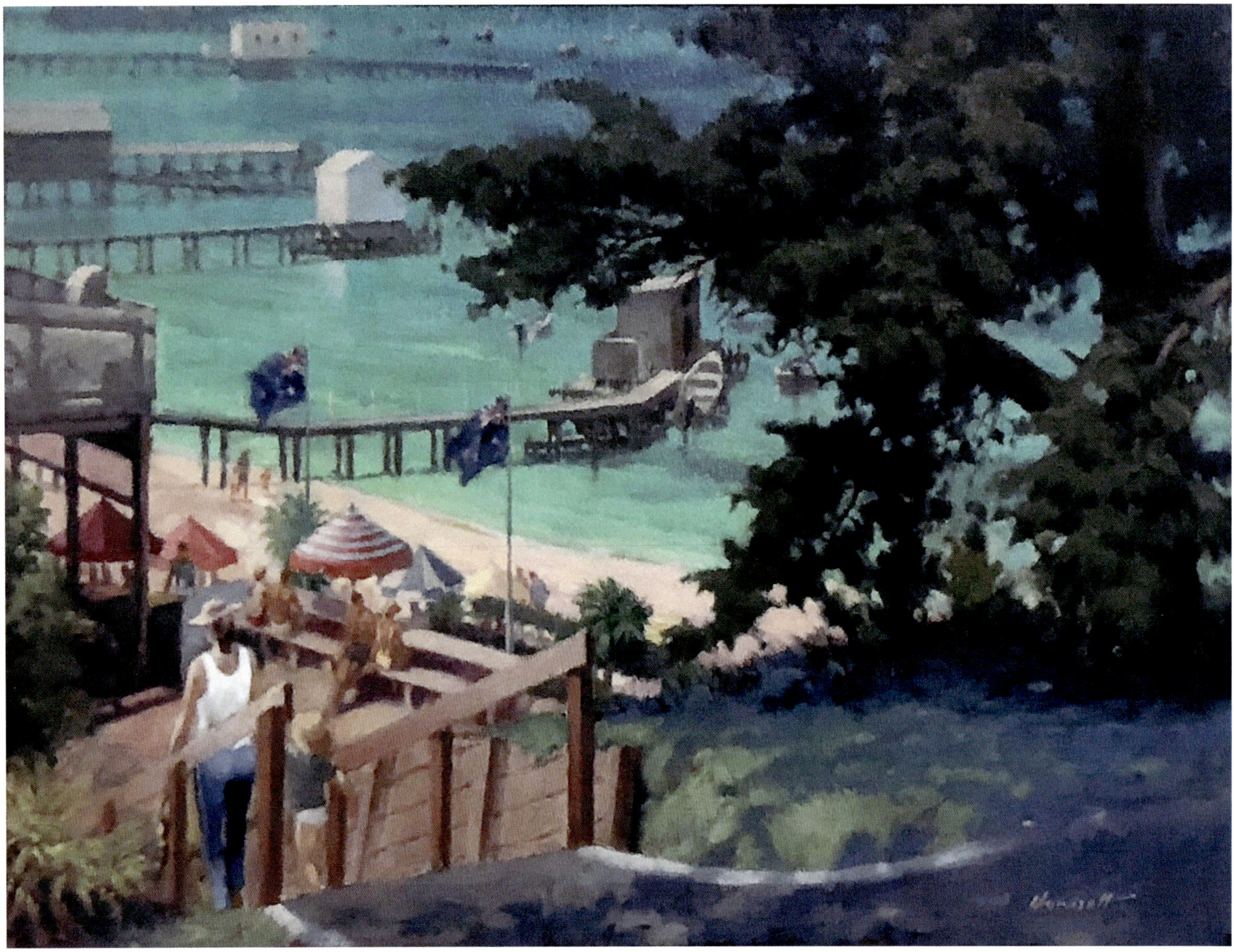

Ray Honisett, ***Beach at Sorrento***, *c.2010*
Oil on canvas, 44 x 60 cm
Collection, Caroline of Sorrento

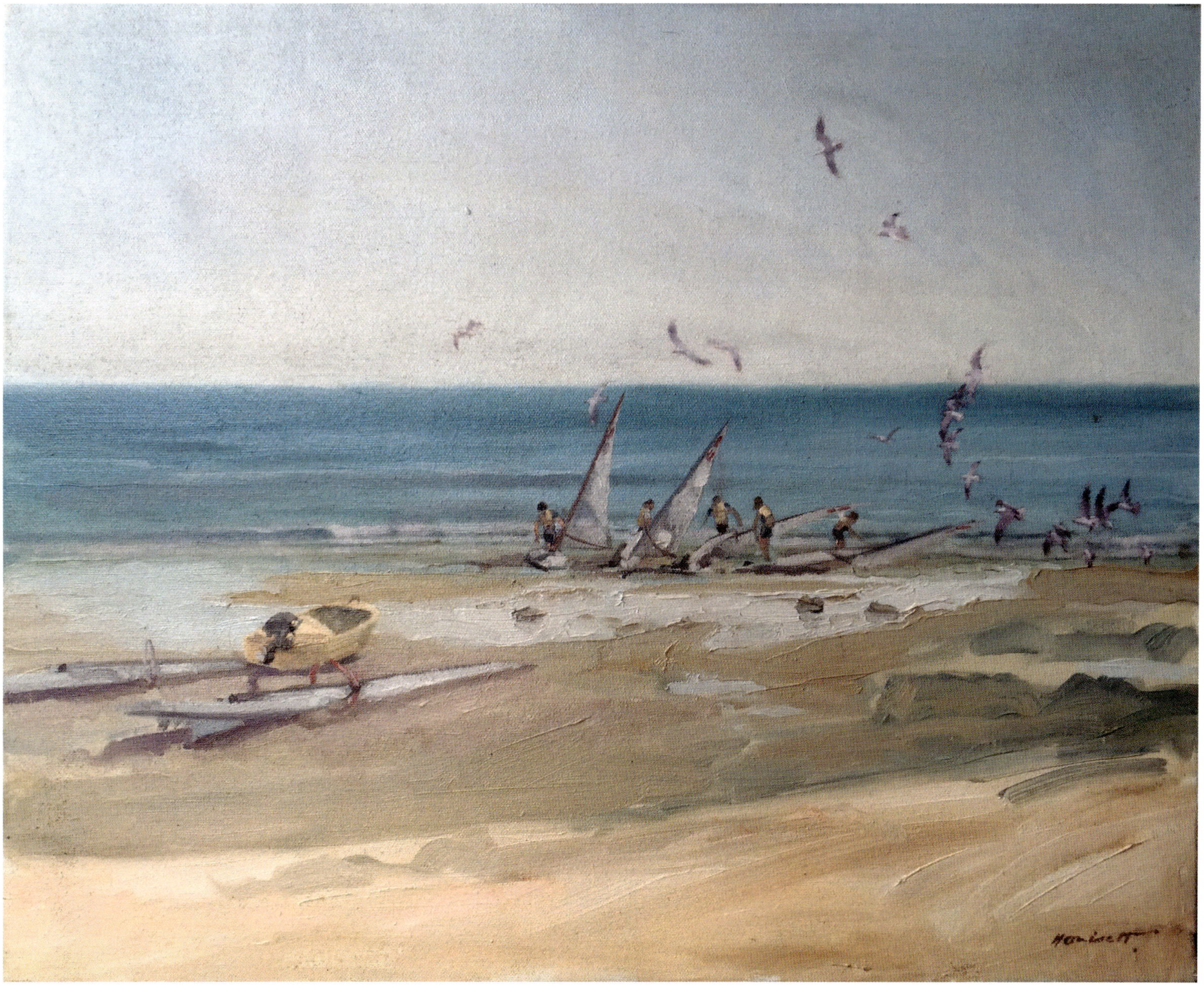

Ray Honisett, ***Wind surfers on the beach****, c.2000*
Oil on canvas
The Artist's Estate

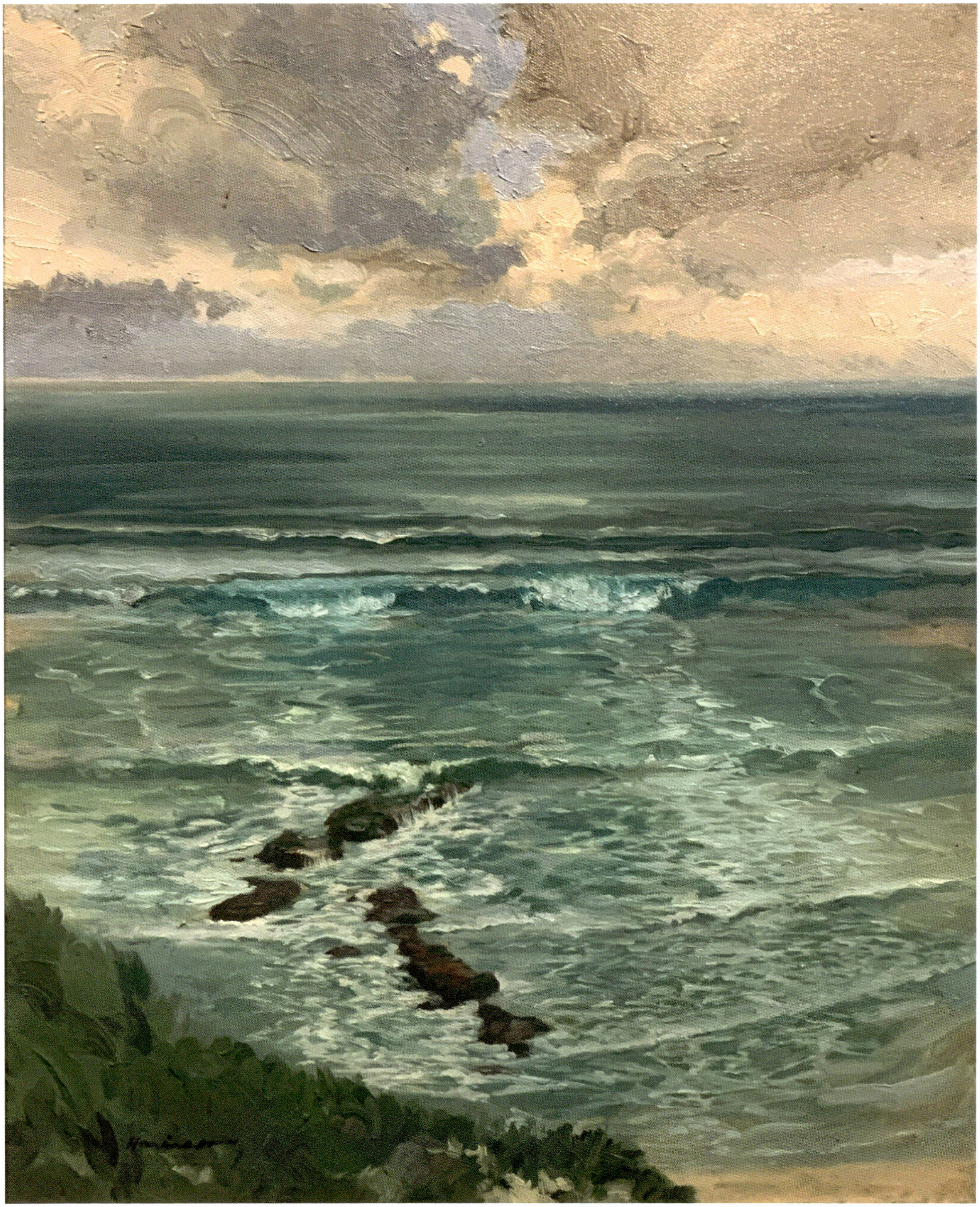

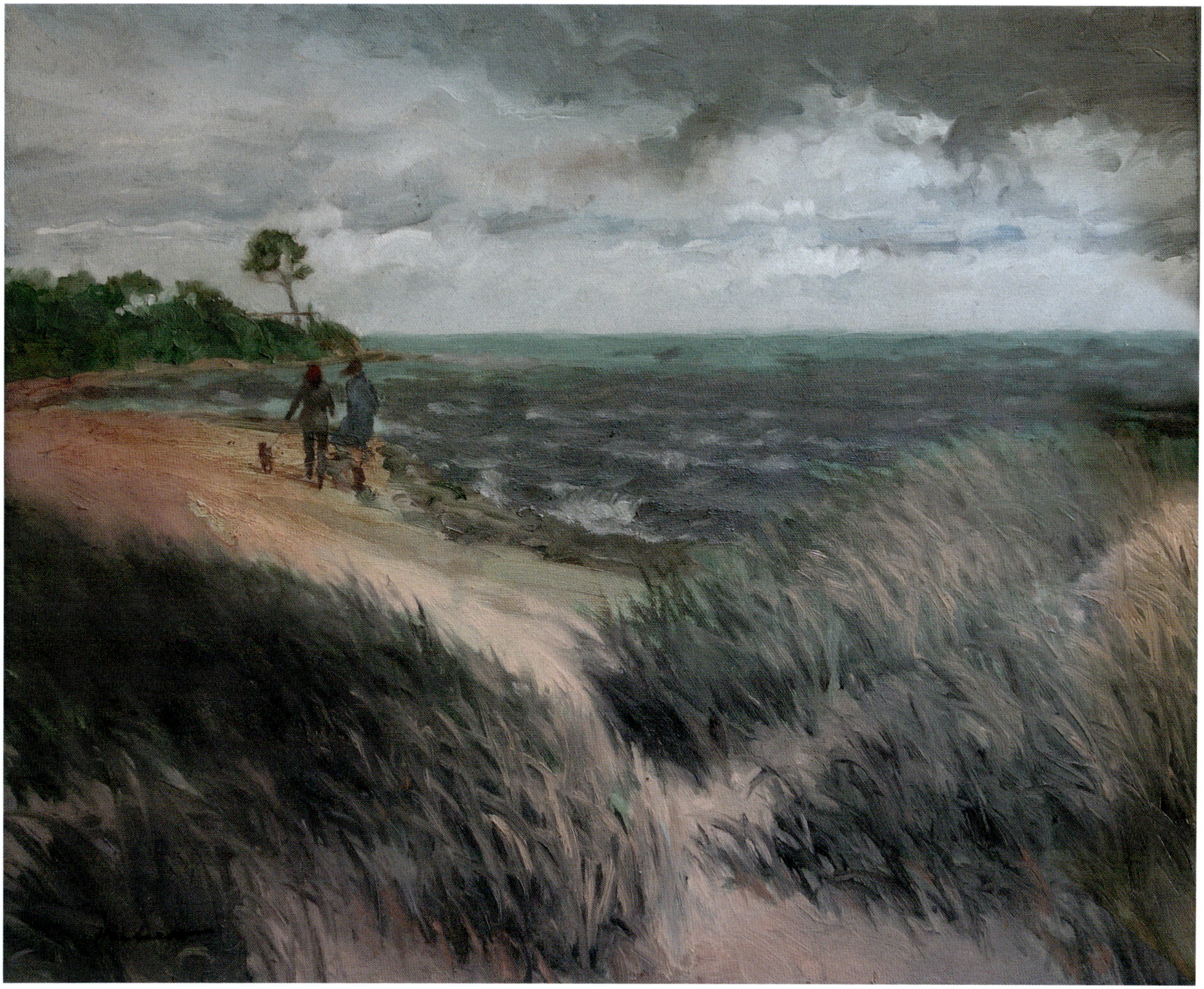

Ray Honisett, ***Windy day at Rye****, 1996*
Oil on canvas
Private collection

Opposite, *Ray Honisett,* ***Rye Back Beach****, c.1995*
Oil on canvas
Private collection

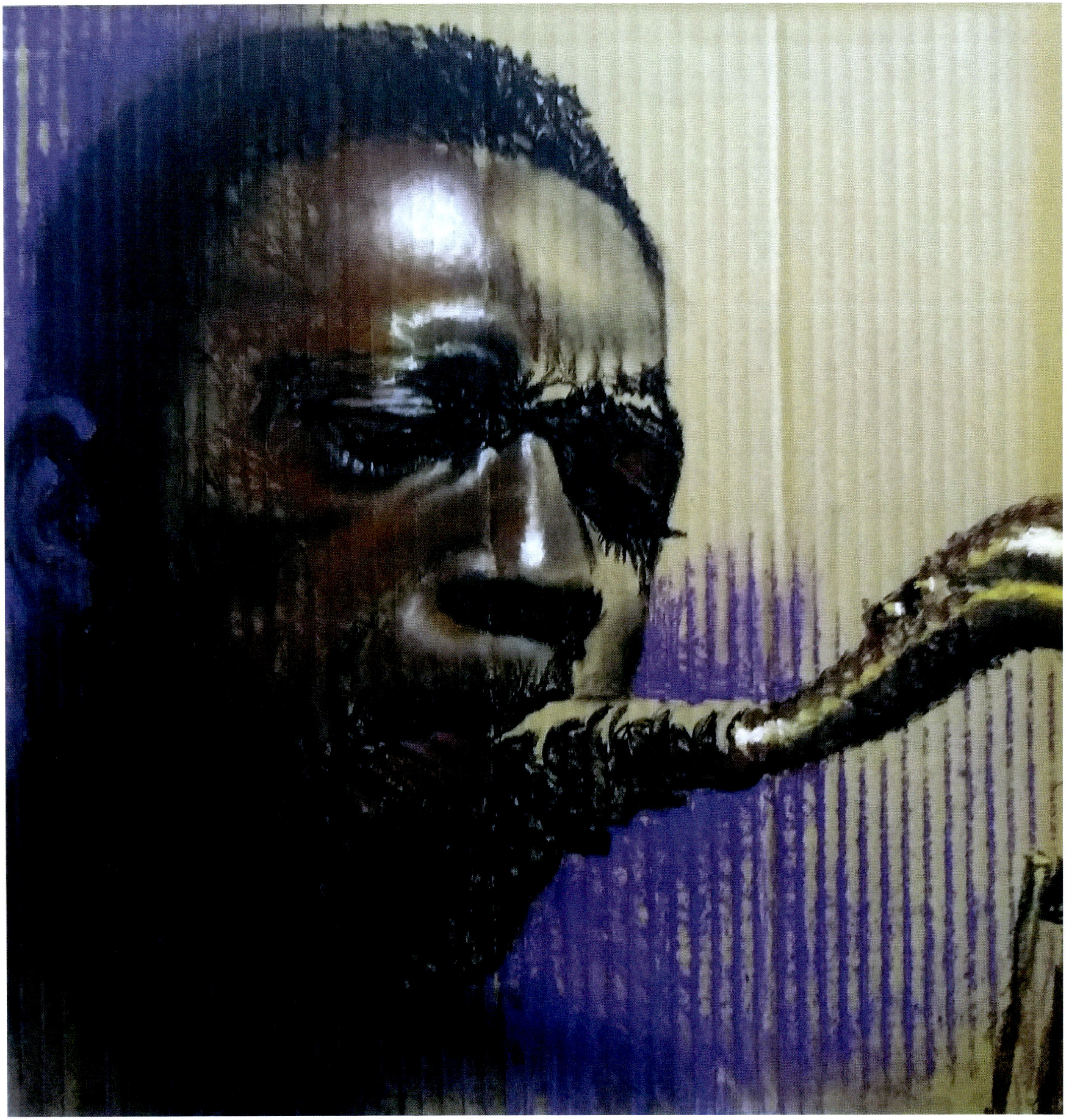

THE MUSICIANS

Like any artist working in the commercial domain, most of Ray Honisett's work was driven by the needs of clients and patrons, whether that was a national institution looking to commemorate some specific event or a private client fulfilling a long held dream for a certain picture on his wall. Just sometimes the opportunity would arise when an artist might respond to something personal, without a specific outcome in mind.

Ray's son Simon is a musician living in the United States, with an interest in jazz in its many forms and Ray made a drawing for his birthday. *"The first picture was of John Coltrane and I loved it and at some point we talked about him doing a bunch and maybe creating a market in the US. I had moved there in 1998 and started working at Powell's Books, the largest independent bookstore in the US. We organized an exhibition of his jazz portraits there in early 2000 and Ray worked up a bunch of portraits and came over to help promote it. I tried to find a way to market them, but it wasn't my area of expertise at all. I know that a lot of the jazz musicians he chose were done with an eye to what he thought people might like to purchase — he leaned toward female vocalists like Sarah Vaughan, Dinah Washington and Billie Holliday. He also did a number of likely sellers such as Duke Ellington, Count Basie, Miles Davis as well as Eric Dolphy and Charles Mingus, artists who I really loved and played for him. He was always very open to music, but never really sought it out, so I was always making tapes and cds for him. I asked him to illustrate portraits of some of my musical heroes through the years, including Buddy Holly, Elvis and Curtis Mayfield".*

Partly for the ease of postage to America, but also because it created an informal, gritty image, he chose to do many of the works in pastels on corrugated card. This was an effective, but vulnerable medium, requiring that the works be framed behind glass to prevent damage.

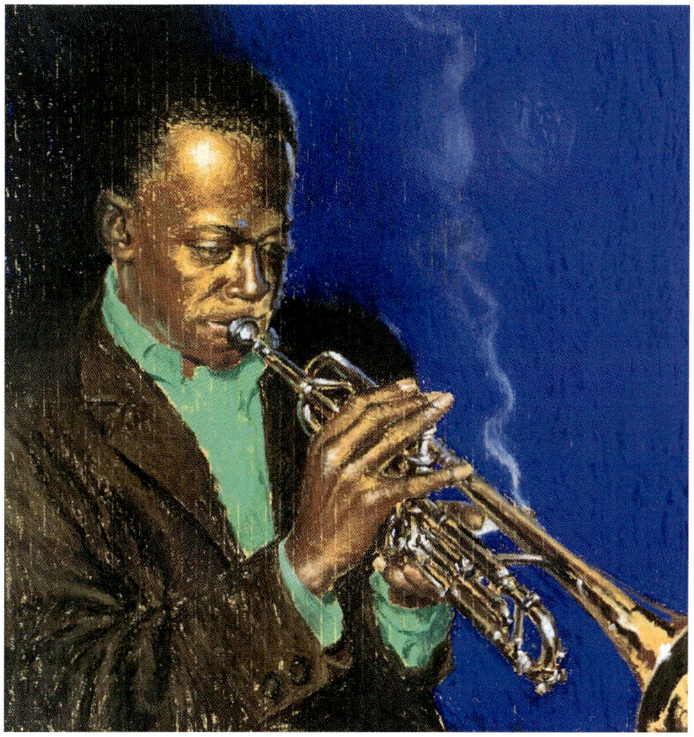

*Ray Honisett, **Miles Davis**, circa 2010*
Pastel on corrugated card
Private collection, Melbourne

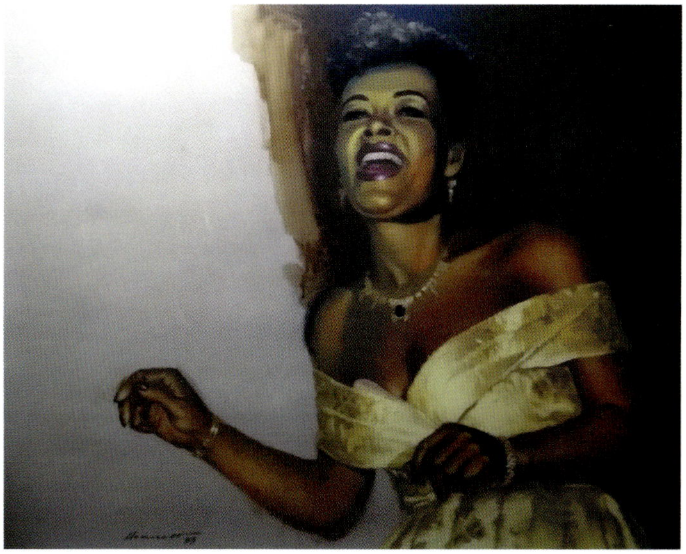

*Right, Ray Honisett, **Billie Holiday**, 1999*
Acrylic and pastel on paper, 63 x 78 cm
Private collection, USA

*Opposite, Ray Honisett, **John Coltrane**, circa 2010*
Pastel on corrugated cardboard
Private collection, USA

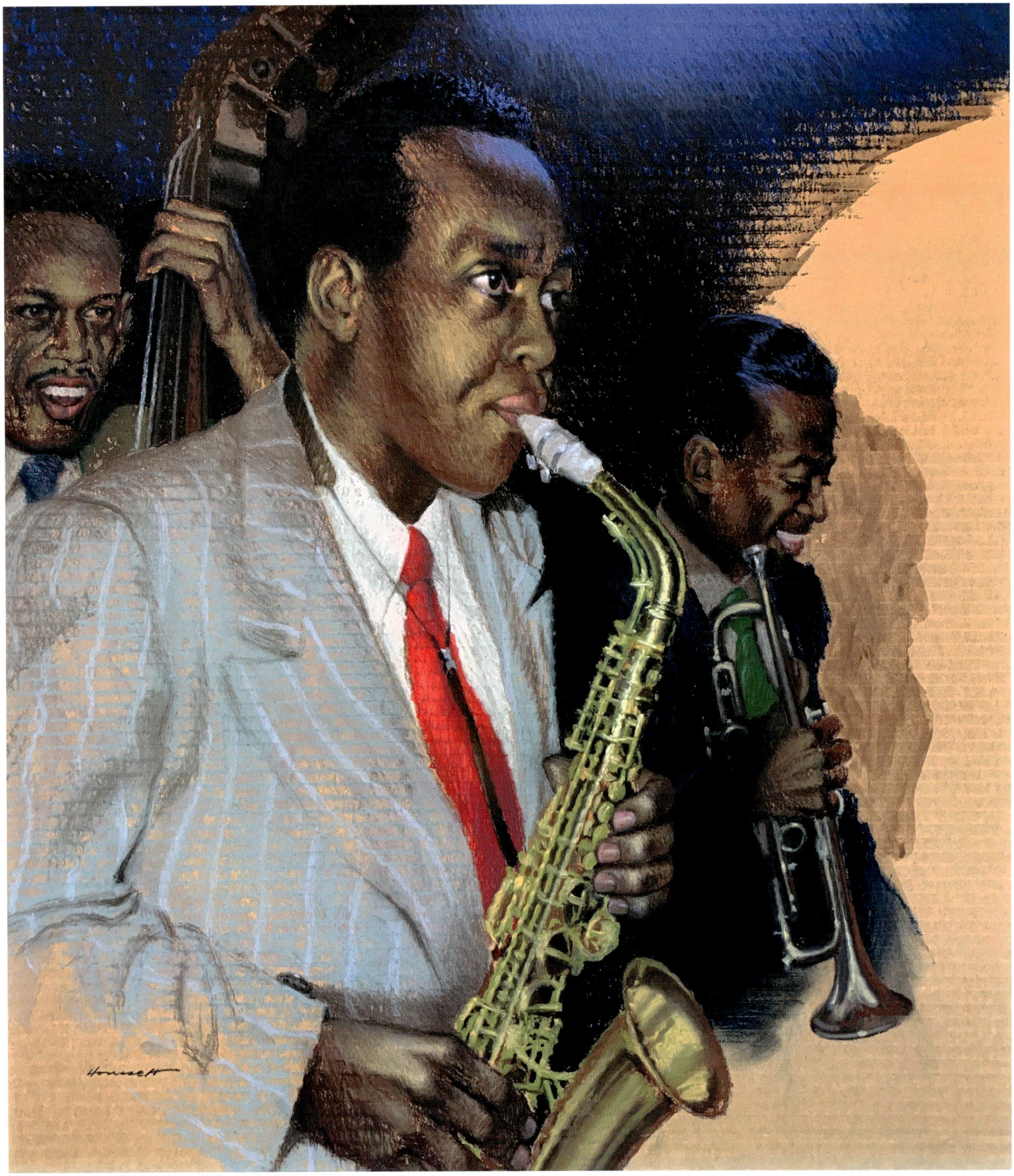

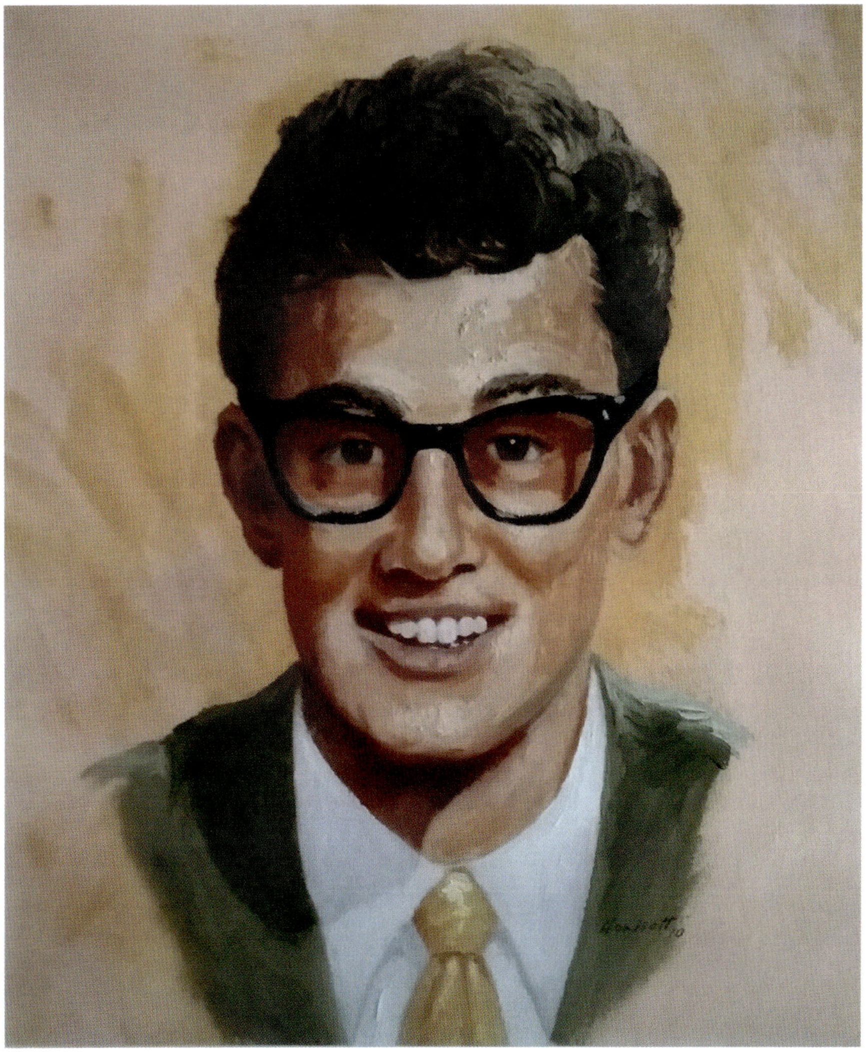

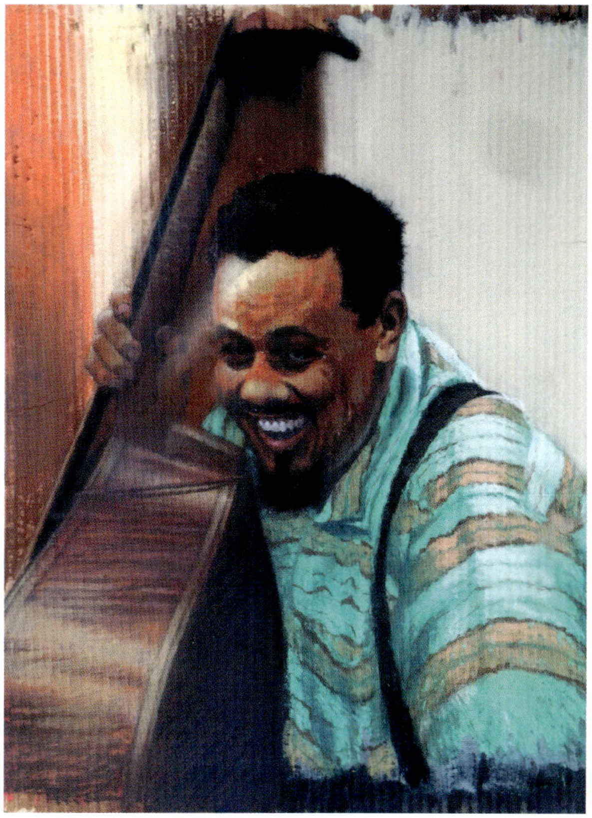

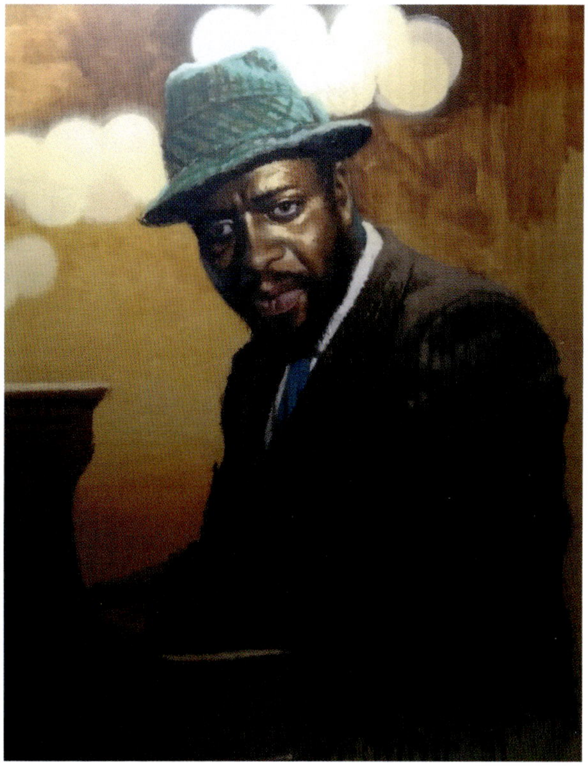

Above right , Ray Honisett, ***Charles Mingus****, c.2010*
Acrylic on corrugated cardboard
The Artist's Estate

Ray Honisett, ***Buddy Holly****, 2010*
Acrylic on card
Private Collection, USA

Right, Ray Honisett, ***Thelonious Monk****, c.2010*
Pastel on card, 60 x 45 cm,
Private Collection, USA

Opposite, Ray Honisett, ***Charlie Parker****, c. 2010*
Pastel on corrugated cardboard
Private collection, Melbourne

THE LATE WORKS

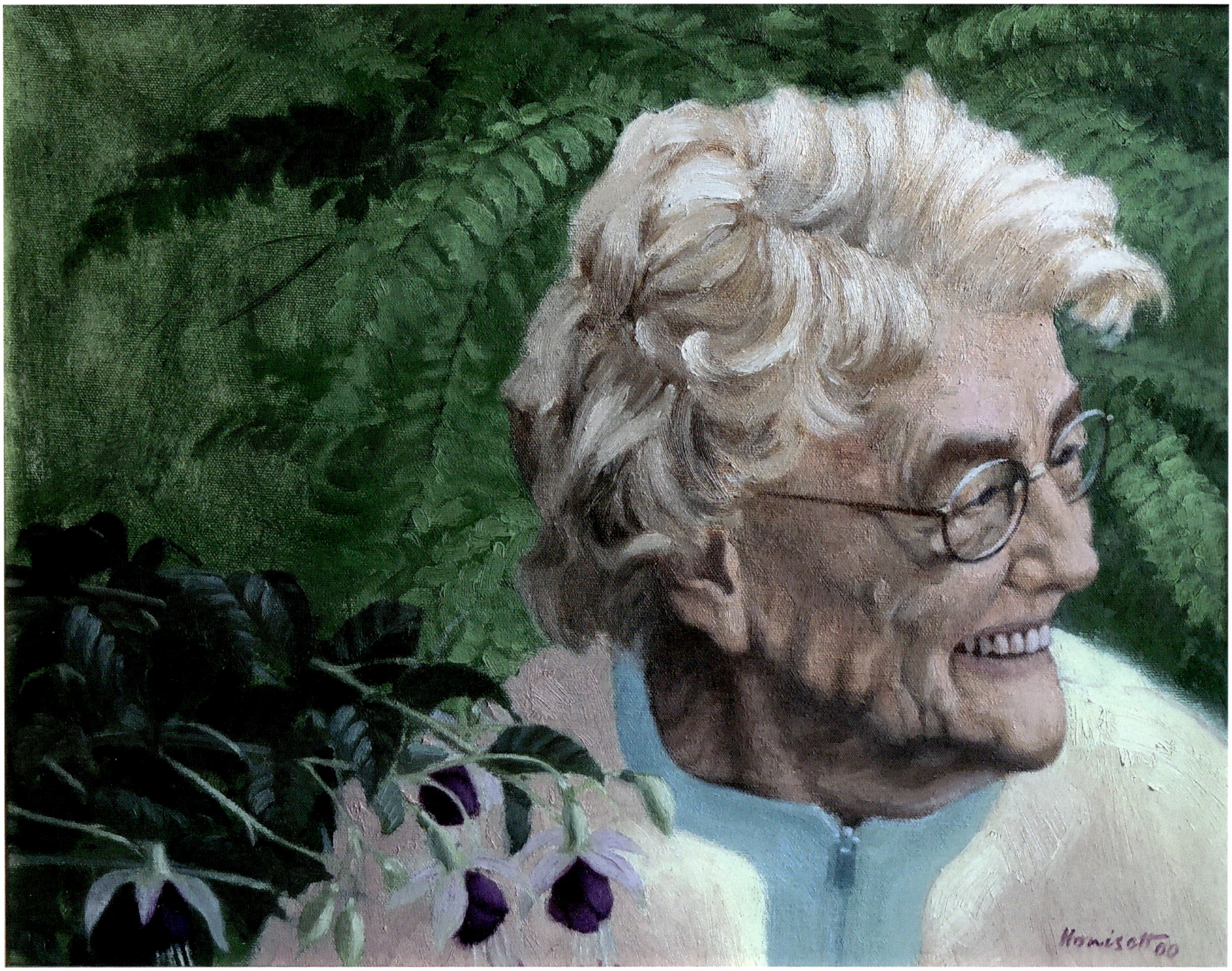

*Ray Honisett, **May McBain**, 2000*
Oil on canvas 45 x 60 cm

Another very personal work was a portrait of his daughter-in-law's grandmother, May McBain. The work is a warm and sympathetic response of one older person to another. It is neither sentimental nor patronising, but full of life and understanding — as one would imagine from an artist who had himself lived a long and at times challenging life.

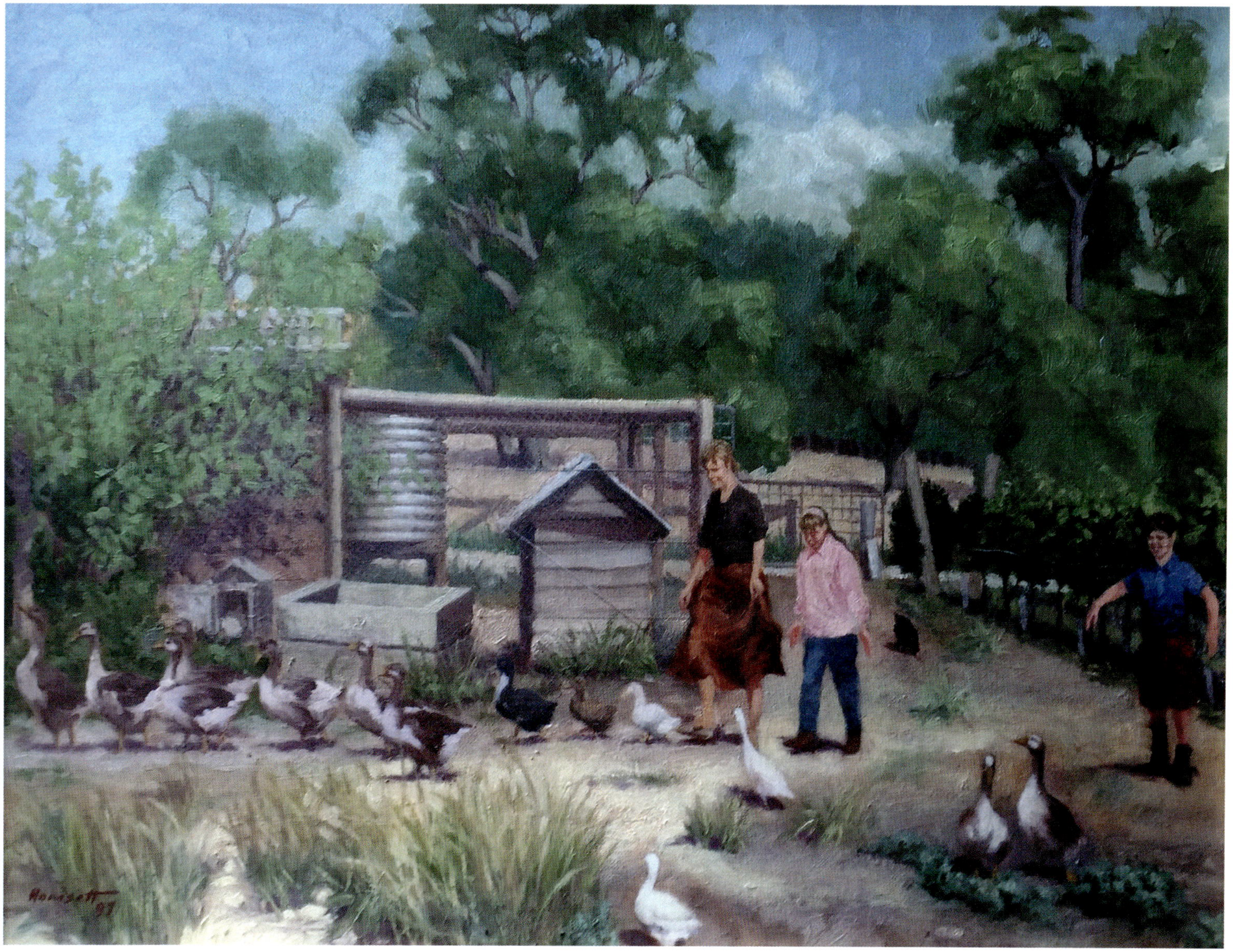

Ray Honisett, (The Vintner's Family) 1997
Oil on Canvas, 45 x 60 cm
Private collection, Melbourne

The Mornington Peninsula has become as famous for its wineries as it always has been for its beaches. Living at Rye brought Ray Honisett into contact with a local winemaker and he undertook a series of commissions depicting the people and places connected with the industry. This particular painting is not identified, but the vines to the right and the rural scene are typical of the winery paintings.

Right, the artist with a vineyard painting on the easel, Rye studio, circa 2000

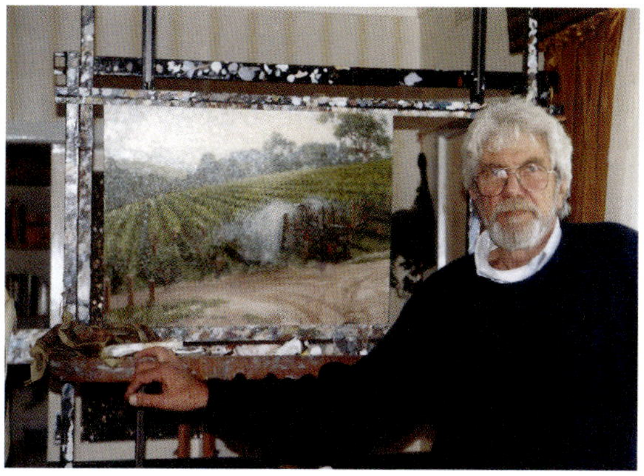

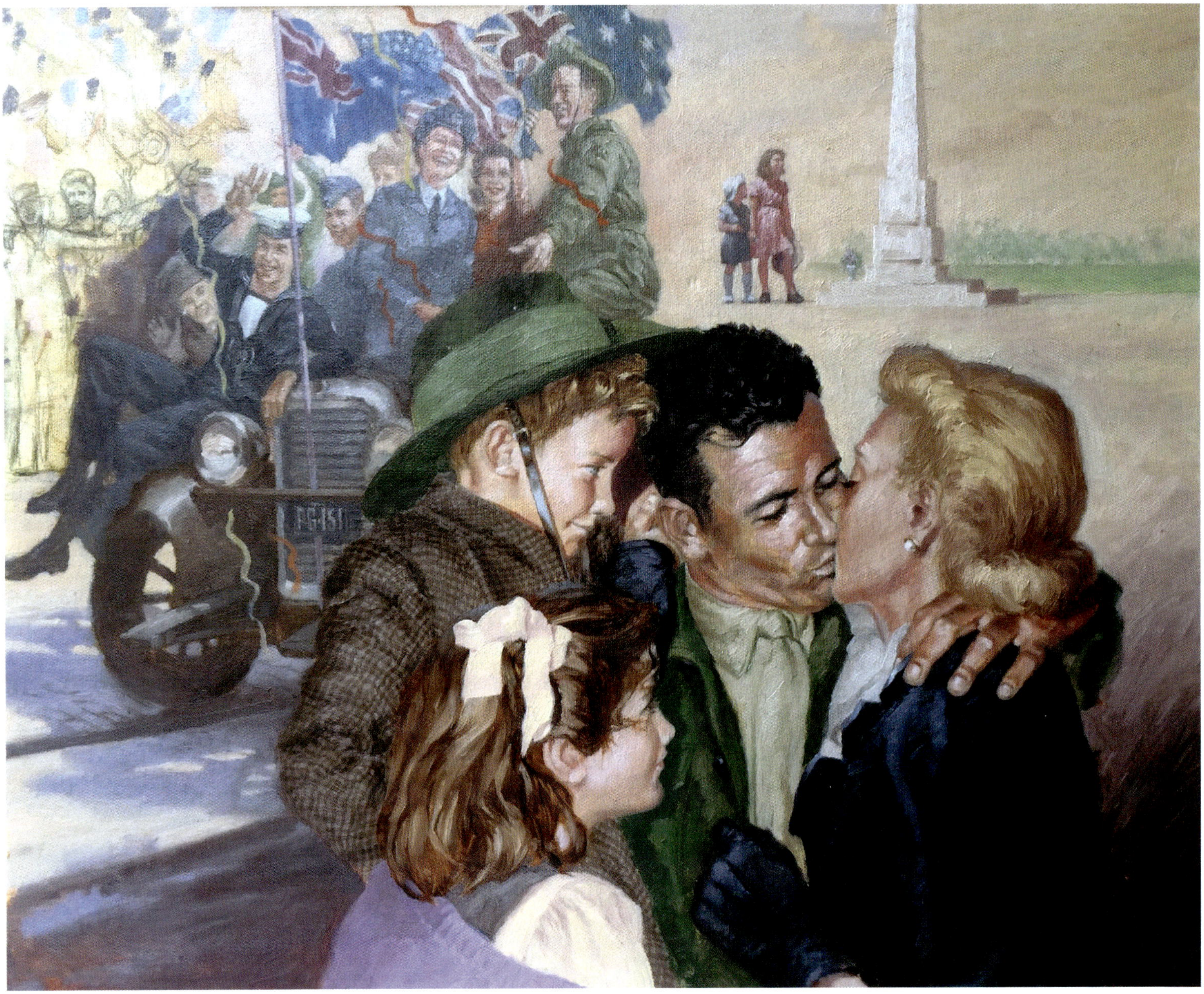

Ray Honisett, (The Soldier's Return) circa 2015
Oil on canvas
The Artist's Estate

A small mystery towards the end of the artist's life — an unfinished, or at least unsigned, work depicting a soldier's return at the end of World War Two. The work could have been created for a competition like the ANZAC art prize, or as a commission for a local RSL Club, containing as it does all the cues about return, celebration and remembrance. The multi-layered image reprises Ray's multiple portraits for the Australian War Memorial and the complex portrait groups of the sports series.

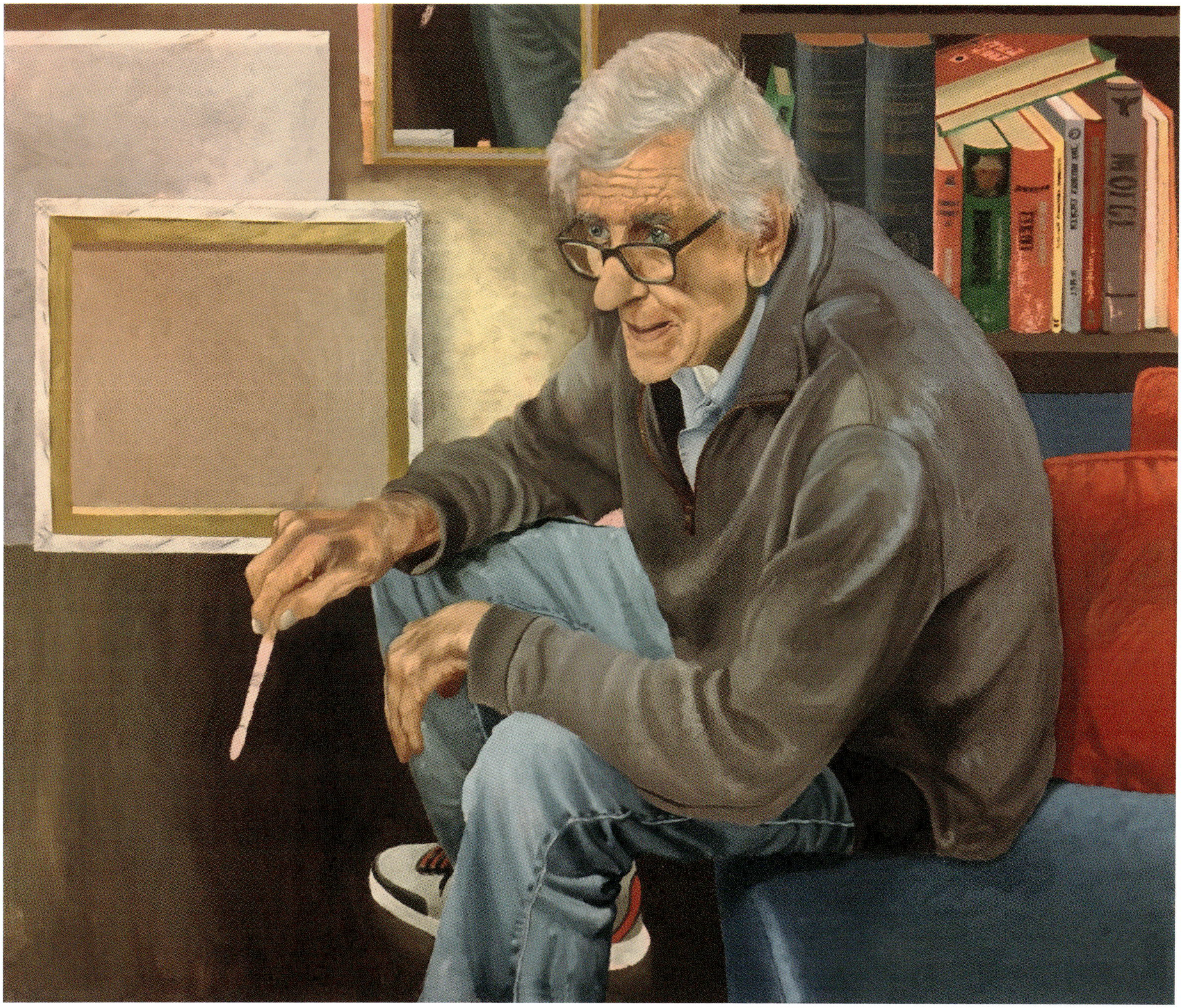

An artist until in the end — one of Ray's last paintings, a self-portrait of the thoughtful painter.

Ray Honisett, ***Self portrait****, circa 2018*
Oil on canvas
The Artist's Estate

ACKNOWLEDGEMENTS

Books of this kind always require the input and assistance of many people, especially when one is working at a considerable distance from the places where large public collections are held. This was doubly true in the extraordinary lock-down months of 2020. Firstly I would like to offer my sincere thanks to Ray's family, including children Simon, Jeremy, Samantha and Asha and also to Sherryl Mannerheim. This is the book I promised to Ray many years ago and I am sorry it took so long to see the light of day. Particular thanks are due to Alan Swann who has been tireless in chasing down material, especially in relation to the Australian Society of Aviation Artists. Ray's great supporter Lachie Pitman has supplied many important images, as well as personal memories of working with Ray on a number of projects. The staff at major institutions have been very supportive, especially in relation to negotiating reproduction access. Greta Wass and Ron Schroer at the Australian War Memorial, Kathy Hackett at the Powerhouse Museum, Emily Constantine at the RAAF Museum, Point Cook and the staff at the Australian National Maritime Museum have all been extremely helpful — and tolerant of my constant stream of emails and requests. Similarly the staff in the Australian War Memorial Research Centre were untiring in their efforts to have the necessary files cleared and available in time for my visit. The book is so much richer with the inclusion of works from those collections. Richard Beckon at Australia Post could not have been more helpful, giving so much of his time and valuable advice in relation to a crucial part of Ray's career. A special thank you to Diane Hatwell of the Dowerin District Museum, both for sourcing images and also for the fascinating ongoing conversations. A number of private collectors have been extremely generous with their time and the sourcing of images, including Mark Walland, Murray Pitman, Steven Tsaklis and the Tsaklis family.

As always, my sincerest thanks to Rebecca, who puts up with the endless requests to 'read this please … and have a look at this one' — and of course to Asha, whose constant presence on the lap ensures that I remain pinned to the work station.

Gavin Fry
2020

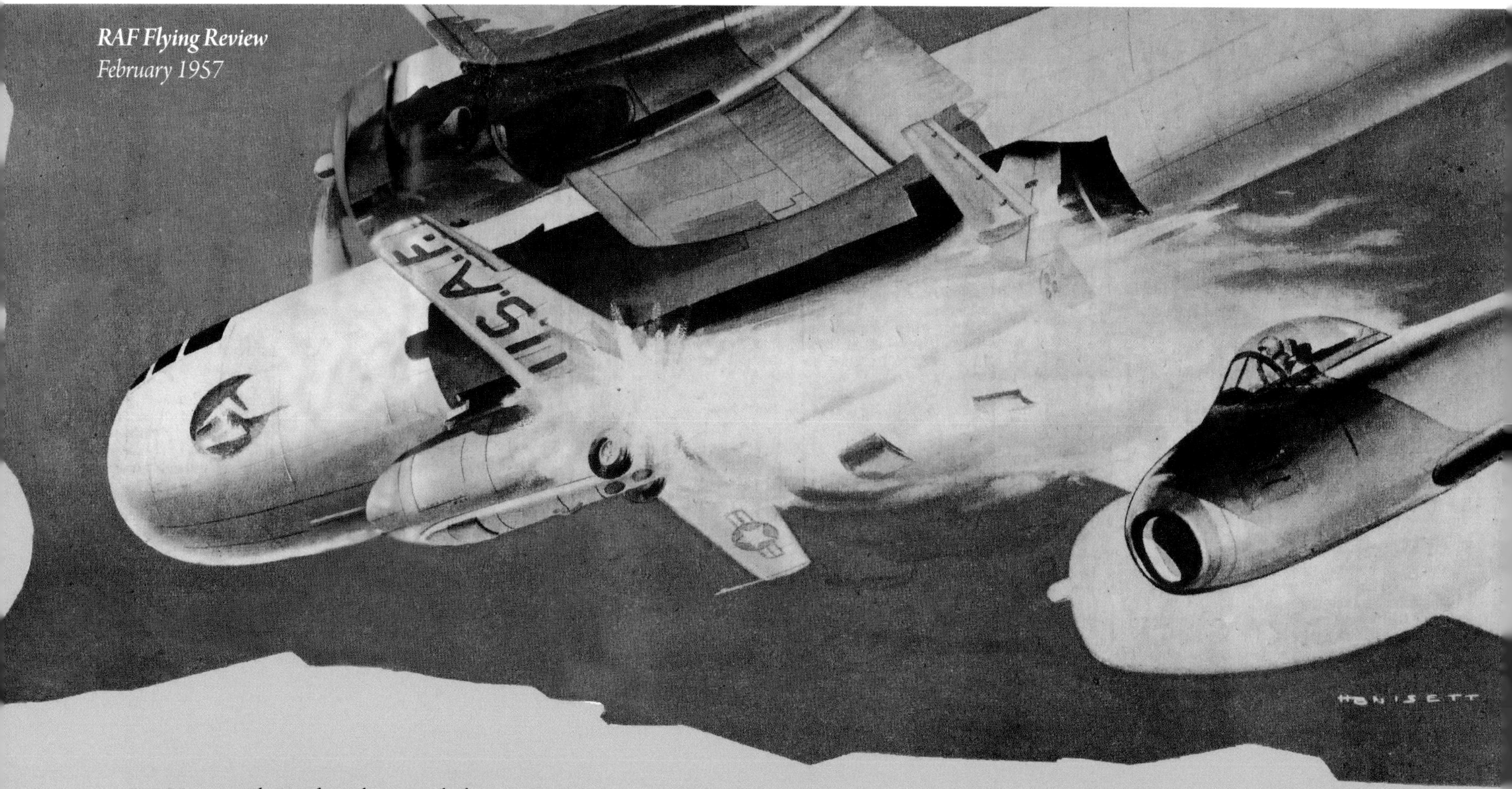

Ray Honisett, ***The Rocket Plane Explodes***
Royal Air Force Flying Review, February 1957

Following pages, ***Rider by the Yarra at Warrandyte***
Oil on canvas, 76 x 155 cm
Tsaklis Family Collection, Melbourne

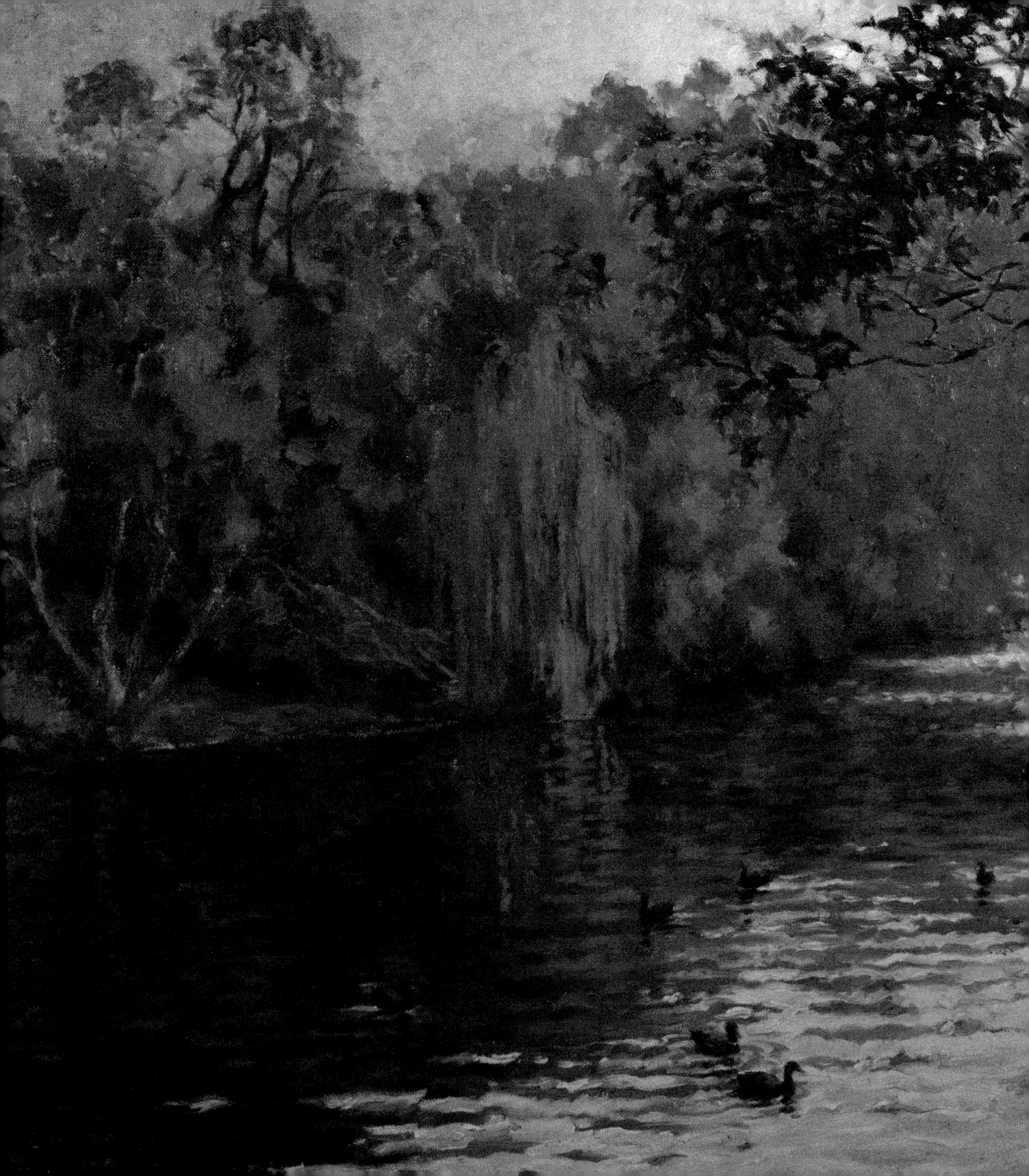

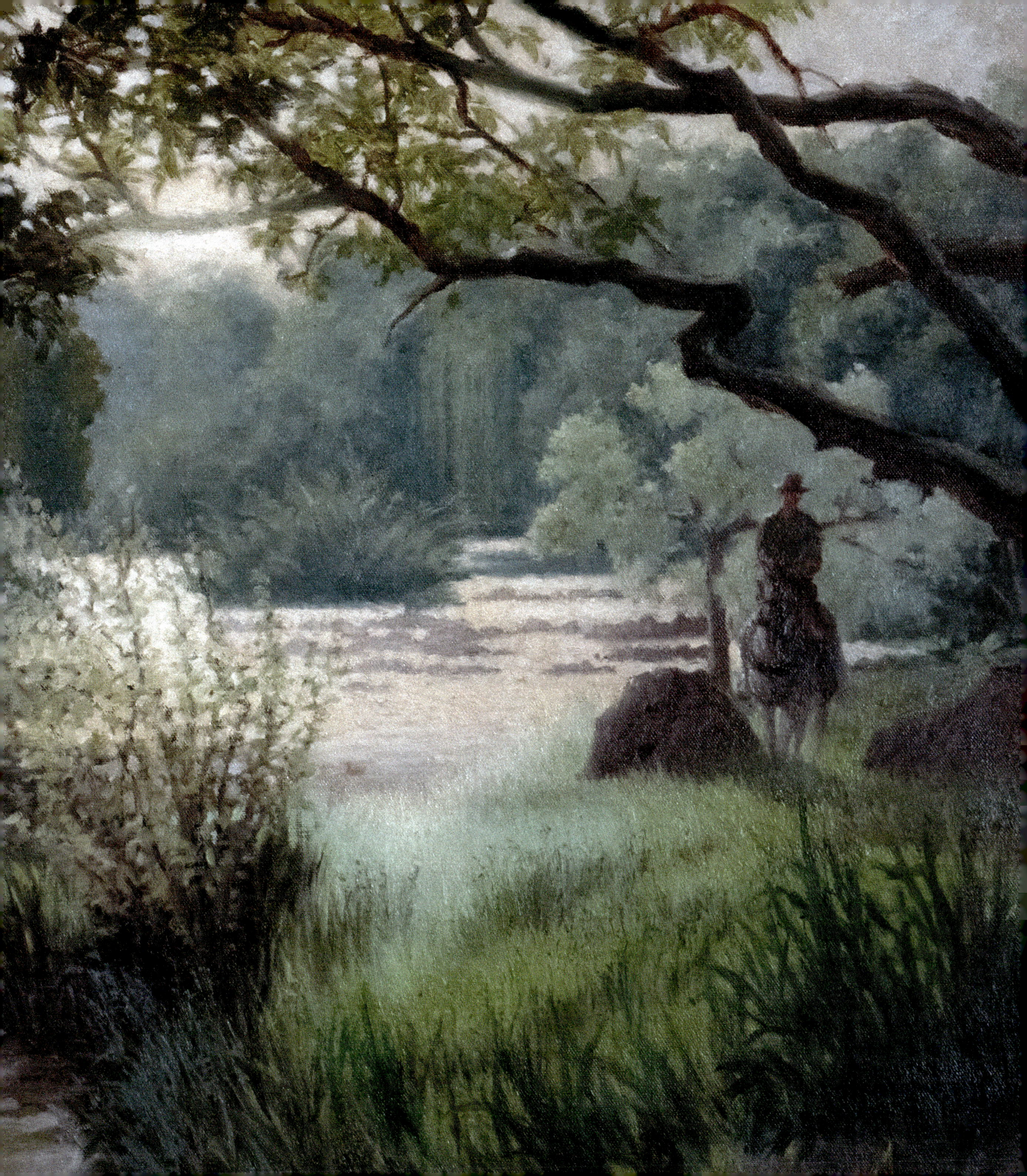